Latest Hydroforming Technology of Metallic Tubes and Sheets

Latest Hydroforming Technology of Metallic Tubes and Sheets

Editors

Yeong-Maw Hwang
Ken-ichi Manabe

MDPI • Basel • Beijing • Wuhan • Barcelona • Belgrade • Manchester • Tokyo • Cluj • Tianjin

Editors
Yeong-Maw Hwang
Department of Mechanical and Electro-mechanical Engineering
National Sun Yat-sen University
Kaohsiung
Taiwan

Ken-ichi Manabe
Department of Mechanical Systems Engineering
Tokyo Metropolitan University
Tokyo
Japan

Editorial Office
MDPI
St. Alban-Anlage 66
4052 Basel, Switzerland

This is a reprint of articles from the Special Issue published online in the open access journal *Metals* (ISSN 2075-4701) (available at: www.mdpi.com/journal/metals/special_issues/hydroforming_tubes_sheets).

For citation purposes, cite each article independently as indicated on the article page online and as indicated below:

LastName, A.A.; LastName, B.B.; LastName, C.C. Article Title. *Journal Name* **Year**, *Volume Number*, Page Range.

ISBN 978-3-0365-2354-5 (Hbk)
ISBN 978-3-0365-2353-8 (PDF)

Contents

About the Editors

Yeong-Maw Hwang

Yeong-Maw Hwang is currently a distinguished professor at National Sun Yat-sen University. He earned his Doctor's degree (1990) in industrial mechanical engineering from Tokyo University in Japan. He was a visiting professor at Ohio State University (1999–2000 and 2005–2006). He has been a professor in the Department of Mechanical and Electro-Mechanical Engineering (MEME), National Sun Yat-Sen University (NSYSU), Kaohsiung, Taiwan, since 1996. He has served as the department chair (2002 –2005 and 2017–2020) of MEME. His research interests have been in the areas of metal forming, design and analysis of micro-generators, machine design and mechanics analysis. He won the Best Paper Award (1992) and Outstanding Engineering Professor Award (2007) from Chinese Society of Mechanical Engineers in Taiwan. He earned the Fellow title from Japan Society for Technology of Plasticity (JSTP), Japan (2012) and Distinguished Professor of NSYSU (2012). He was elected as the president of Taiwan Society for Technology of Plasticity (TSTP) for two terms, equating to four years 2017–2020.

Ken-ichi Manabe

Ken-ichi MANABE has been an Emeritus Professor since 2017 and was awarded his Ph.D. in mechanical engineering in April 1985 form the Tokyo Metropolitan University (TMU), Japan. He had been a Professor at the TMU since 2002. Prof. Manabe has undertaken extensive research on the theory and modelling of tube/sheet metal forming processes and the intellectualization of their forming processes for over 45 years. Recently, his research interests extend toward tube microforming technology and deformation mechanics at the micro/meso scale. He has gained academic recognition from the Japan Society for Technology of Plasticity (JSTP), the Japan Society of Mechanical Engineers (JSME) and so on. He was a Conference chair (1993, 2011) and Co-chair (1995, 1997, 1999, 2017), and Honorary chair (2019) of the International Conference on Tube Hydroforming (TUBEHYDRO). He received the Best Paper Award in 1989, 2009 and 2012 from the JSTP. He received the JSTP Medal in 2010 and attained the grade of Fellow in 2009 from the JSTP. He was the President of the JSTP in 2015–2016.

 metals

MDPI

Editorial

Latest Hydroforming Technology of Metallic Tubes and Sheets

Yeong-Maw Hwang [1,*] and Ken-Ichi Manabe [2]

[1] Department of Mechanical and Electro-Mechanical Engineering, National Sun Yat-sen University, No. 70, Lien-Hai Rd., Kaohsiung 80424, Taiwan
[2] Department of Mechanical Systems Engineering, Graduate School of Systems Design, Tokyo Metropolitan University, 1-1 Minamiosawa, Hachioji, Tokyo 192-0397, Japan; manabe@tmu.ac.jp
* Correspondence: ymhwang@mail.nsysu.edu.tw; Tel.: +886-7-525-2000 (ext. 4233)

1. Introduction and Scope

Hydroforming processes of metal tubes and sheets are being widely applied in manufacturing because of the increasing demand for lightweight parts in sectors such as the automobile, aerospace, and ship-building industries. This technology is relatively new compared with rolling, forging, or stamping, so that there is limited knowledge available for the product or process designers. Compared to conventional manufacturing via stamping and welding in particular, tube hydroforming offers several advantages, such as (1) a decrease in workpiece cost, tool cost, and product weight, (2) an improvement of structural stability and an increase of the strength and stiffness of the formed parts, (3) a more uniform thickness distribution, (4) fewer secondary operations, etc. However, this technology suffers some disadvantages, such as slow cycle time, expensive equipment, and the lack of an effective database for tooling and process design.

Compound forming, which involves hydroforming and other forming processes such as crushing or preforming, is implemented to achieve a lower clamping force and forming pressure, as well as to ensure a uniformly distributed thickness of the formed product. Other tube hydroforming related effects like hydro-piercing, hydro-joining, hydro-flanging and hydro-inlaying are also important topics.

The aim of this Special Issue is to present the latest achievements in various tube and sheet hydroforming processes together with other tube processing technology and innovation. Through this Special Issue, a comprehensive understanding of the present status and future trends of tube/sheet hydroforming technology are to be expected. Thus, all researchers in this field were invited to contribute their research works to this special issue.

This special issue consists of some extended papers presented at The 9th International Conference on Tube Hydroforming (TUBEHYDRO 2019) held in Kaohsiung, Taiwan in 2019, and some papers newly submitted from authors with and without attending the conference.

Citation: Hwang, Y.-M.; Manabe, K.-I. Latest Hydroforming Technology of Metallic Tubes and Sheets. *Metals* **2021**, *11*, 1360. https://doi.org/10.3390/met11091360

Received: 2 August 2021
Accepted: 25 August 2021
Published: 30 August 2021

2. Contributions to the Special Issue

The contributions are generally divided into three basic groups according to the workpiece geometry and forming methods. The first group is tube hydroforming (THF) [1–9], in which a hydraulic media and dies are used to deform a tube workpiece. The second group is tube forming [10–14], in which only dies are used to deform a tube workpiece. The third group is sheet hydroforming [15,16], in which dies and/or hydraulic media are used to deform a sheet workpiece.

In the first group, many different forming technologies and methodologies related to tube hydroforming were used to overcome the forming difficulty to successfully obtain the desired product dimensions and material properties. For example, Yasui et al. [1] developed a warm hydroforming system and used this system to examine experimentally and numerically the influence of internal pressure and axial compressive displacement on the formability of small-diameter ZM21 magnesium alloy tubes in warm tube hydroforming. Supriadi et al. [2] proposed a vision-based fuzzy control algorithm and an image processing technology to manufacture bellows with a semi-dieless forming process. Morishima and Manabe [3] used finite element analysis combined with a fuzzy control model to carry out optimization of symmetrical temperature distributions and process loading paths for the warm T-shape forming of magnesium alloy AZ31B tube. Hwang and Tsai [4] developed a tube hydroforming process using a novel movable die design for decreasing the internal pressure to manufacture irregular bellows with small corner radii and sharp angles. Hwang et al. [5] proposed a new hydro-flanging process combining hydro-piercing and hydro-flanging to investigate the effects of punch shape and loading path in the hydro-flanging processes of aluminum alloy tubes. Bach et al. [6] used the functionality of the tools and the heating strategy for a curved component as well as a measurement technology to investigate the heat distribution in a component during hot metal gas forming (HMGF). Yoshimura et al. [7] proposed a local one-sided rubber bulging method of metal tubes to evaluate various strain paths at an aimed portion and measured the forming limit strains of metal tubes at the place of the occurrence of necking under biaxial deformation. Han and Feng [8] investigated the circumferential material flow using overlapping blanks with axial constraints in tube hydroforming of a variable-diameter part. Alexandrov et al. [9] proposed a simple analytical solution for describing the expansion of a two-layer tube under plane-strain conditions with an arbitrary pressure-independent yield criterion and a hardening law. The results can be applied to the preliminary design of hydro-expansion processes.

The second group consists of manuscripts related to tube forming with dies without hydraulic media inside the tube. Kajikawa et al. [10] proposed a tube expansion drawing method to effectively produce a thin-wall and large-diameter tube. Hirama et al. [11] proposed a new ball spin forming equipment, which can form a reduced diameter section on the halfway point of a tube. The effects of forming process parameters on the surface integrity and deformation characteristics of the product were investigated. Arai and Gondo [12] proposed a method of forming a tube into an oblique/curved shape by synchronous multipass spinning, in which the roller moves back and forth along the workpiece in the axial direction to gradually deform a blank tube into a target shape. Nakajima et al. [13] investigated the deformation properties and suppression characteristics of an ultra-thin-walled rectangular tube in rotary draw bending with a laminated mandrel. Tomizawa et al. [14] investigated the crash characteristics of partially quenched curved products by three-dimensional hot bending and direct quench. The results can be used as fundamental research of the design for improving energy absorption.

The third group includes only two articles, one is related to sheet hydroforming and the other is related to sheet friction tests. Hong et al. [15] developed a pneumatic experimental apparatus to evaluate strain rate sensitive forming limits of 7075 aluminum alloy sheets under biaxial stretching modes at elevated temperature. Hwang and Chen [16] investigated the frictional behaviors of sheets at variant speeds using a self-developed sheet friction reversible test machine. The effects of various parameters, including sliding speeds, contact angles, sheet materials, and lubrication conditions on friction coefficients at the sheet–die interface were discussed.

2.1. Tube Hydroforming Processes

Magnesium and its alloys have been widely applied in the automotive, aircraft and telecommunication industries for their excellent characteristics, such as light weight and high strength. In addition, magnesium alloy has been expected to be employed as a material in medical devices, owing to its outstanding biocompatibility. Yasui et al. [1] developed a warm hydroforming system and used this system to examine the effects of internal pressure and axial compressive displacement on the formability of small-diameter ZM21 magnesium alloy tubes in warm tube hydroforming. The deformation behavior of ZM21 tubes, with a 2.0 mm outer diameter and 0.2 mm wall thickness, was evaluated in taper-cavity and cylinder-cavity dies. The simulation code used was the dynamic explicit finite element code, LS-DYNA 3D. The experiments were conducted at 250 °C. The deformation characteristics, forming defects, and forming limit of ZM21 tubes were investigated. Their deformation behavior in the taper-cavity die was affected by the axial compressive direction. Additionally, the occurrence of tube buckling could be inferred by changes of the axial compression force, which were measured by the load cell during the processing. In addition, grain with twin boundaries and refined grain were observed at the bended areas of tapered tubes. The hydroformed samples could have a high strength. Moreover, wrinkles, which are caused under a lower internal pressure condition, were employed to avoid tube fractures during the axial feeding. The tube with wrinkles was expanded by a straightening process after the axial feed. It was found that the process of warm THF of the tubes in the cylinder-cavity die was successful.

Metal bellows consist of convoluted metal tube that provides high flexibility in various directions. They have been widely applied in the flexible joining of piping systems for water, oil, and gas provisions. Metal bellows are usually produced through a hydroforming or a gas-forming process. Supriadi et al. [2] proposed earlier a novel semi-dieless bellows forming process with a local heating technique and axial compression. However, with this technique it is extremely difficult to maintain the output quality due to its sensitivity to the processing conditions. The product quality mainly depends on not only the temperature distribution in the radial and axial direction but also the compression ratio during the semi-dieless bellows process. A finite element model clarified that a variety of temperatures produced by unstable heating or cooling will promote an unstable bellows formation. An adjustment to the compression speed is adequate to compensate for the effect of the variety of temperatures in the bellows formation. Therefore, it is necessary to apply a real-time process for this process to obtain accurate and precise bellows. In this paper, they proposed a vision-based fuzzy control to control bellows formation. Since semi-dieless bellows forming is an unsteady and complex deformation process, the application of image processing technology is suitable for sensing the process because of the possible wide analysis area afforded by applying multi-sectional measuring. A vision sensing algorithm was developed to monitor the bellows height from the captured images. An adaptive fuzzy was verified to control bellows formation from 5 mm stainless steel tube in a bellows profile up to 7 mm bellows height and a processing speed up to 0.66 mm/s. Appropriate compression speed paths guide the bellows formation following deformation references. The results show that the bellows shape accuracy between the target and experiment increase becomes 99.5% under the given processing ranges.

The warm tube hydroforming (WTHF) process of lightweight materials such as magnesium alloy contributes to a remarkable weight reduction. The success of the WTHF process strongly depends on the loading path with internal pressure and axial feeding and other process variables including temperature distribution. Optimization of these process parameters in this special forming technique is an important issue to be resolved. Morishima and Manabe [3] used finite element analysis combined with a fuzzy control model to carry out the optimization of symmetrical temperature distributions and process loading paths for the warm T-shape forming of magnesium alloy AZ31B tube. The results show that a satisfactory good agreement of the wall thickness distribution of the samples formed under the optimum loading path condition can be obtained between the finite element analyses and the experimental results. Based on the validity of the finite element model, the proposed optimization method was applied to other material (AZ61) and forming shape (cross-shape), while the applicability was also discussed.

Nowadays, tube hydroforming technologies have been widely applied in automotive and aerospace industries for manufacturing stronger and lighter products. However, higher pressures and complicated loading paths are required in the manufacturing of complex shape products. If the loading path or the relationship between the internal pressure and axial feeding is not controlled appropriately, various defects such as bursting and wrinkling, etc., would probably occur. Metal bellows have been widely applied in various industries, such as chemical plants, power systems, heat exchangers, automotive vehicle parts, etc., for absorbing the irregular expansions of the pipes and damping vibration of the circumference and mechanical movements. Conventional regular metal bellows with multiple convolutions are usually manufactured by hydraulic bulging and die folding. However, for irregular bellows, in which the outer diameter of the bellows is not much larger than its inner diameter, the conventional manufacturing method combining hydraulic bulging and die folding cannot be applied to this kind of irregular bellows. This kind of irregular bellows can only be made by hydraulically bulging the tube into the desired shape under an irregular closed die set. Hwang and Tsai [4] developed a tube hydroforming process using a novel movable die design for decreasing the internal pressure to manufacture irregular bellows with small corner radii and sharp angles. A finite element simulation software "DEFORM 3D" was used to analyze the plastic deformation of the tube within the die cavity using the proposed movable die design. Forming windows for sound products using different feeding types were also investigated. Finally, tube hydroforming experiments of irregular bellows were conducted and experimental thickness distributions of the products compared with the simulation results to validate the analytical modeling with the proposed movable die concept.

Recently, tube hydroforming processes sometimes incorporate piercing, flanging, or joining processes to become hydro-piercing, hydroflanging, or hydro-joining, which are more efficient compared with a single process and can reduce the total weight of the final product. Hwang et al. [5] proposed a new hydro-flanging process combining hydro-piercing and hydro-flanging to investigate the effects of punch shape and loading path in hydro-flanging processes of aluminum alloy tubes. Three kinds of punch head shapes were designed to explore the thickness distribution of the flanged tube and the fluid leakage effects between the punch head and the flanged tube in the hydro-flanging process. A finite element code DEFORM 3D was used to simulate the tube material deformation behavior and to investigate the formability of the hydro-flanging processes of aluminum alloy tubes. The effects of various forming parameters, such as punch shapes, internal pressure, die hole diameter, etc., on the hydro-flanged tube thickness distributions were discussed. Hydro-flanging experiments were also carried out. The die hole radius was designed to make the maximum internal forming pressure needed smaller than 70 MPa, so that a general hydraulic power unit could be used to implement the proposed hole flanging experiments. The flanged thickness distributions were compared with simulation results to verify the validity of the proposed models and the designed punch head shapes.

Climate targets set by the EU, including the reduction of CO_2, are leading to the increased use of lightweight materials for mass production such as press hardening steels. Besides sheet metal forming for high-strength components, tubular or profile forming (hot metal gas forming—HMGF) allows for designs that are more complex in combination with a lower weight. Bach et al. [6] used the functionality of the tools and the heating strategy for the curved component as well as the measurement technology to investigate the heat distribution in the component during hot metal gas forming. This paper particularly examined the application of conductive heating of the component for the combined press hardening process. The previous finite-element-method (FEM)-supported design of an industry-oriented, curved component geometry allowed the development of forming tools and process peripherals with a high degree of reliability. This work comprised a description regarding the functionality of the tools and the heating strategy for the curved component as well as the measurement technology used to investigate the heat distribution in the component during the conduction process. Subsequently, forming tests were carried out, material characterization was performed by hardness measurements in relevant areas of the component, and the FEM simulation was validated by comparing the tube thickness distributions to the experimental values.

During tube forming, tube materials are subjected to complex and severe deformation and, thus, some forming defects such as cracking and buckling often occur. To avoid such forming defects, the formability of the tube materials should be evaluated appropriately.

Yoshimura et al. [7] proposed a local one-sided rubber bulging method of metal tubes to evaluate various strain paths at an aimed portion and measured the forming limit strains of metal tubes at the place of the occurrence of necking under biaxial deformation. Using this method, since rubber was used to give pressure from the inner side of the tube, no sealing mechanisms were necessary unlike during hydraulic pressure bulging. An opening was prepared in front of the die to locally bulge a tube at only the evaluation portion. To change the restriction conditions of the bulged region for biaxial deformation at the opening, a round or square cutout, or a slit was introduced. The test was conducted using a universal compression test machine and simple dies rather than a dedicated machine. On considering the experimental results, it was confirmed that the strain path was varied by changing the position and size of the slits and cutouts. Using either a cutout or a slit, the strain path in the side of the metal tubes can be either equi-biaxial tension or simple tension, respectively.

Tube hydroforming of overlapping blanks is a forming process where overlapping tubular blanks instead of regular tubes are used to enhance the forming limits and improve the thickness distributions. A distinguishing characteristic of hydroforming of overlapping blanks is that the tube material can flow along the circumferential direction easily. Han and Feng [8] investigated circumferential material flow using overlapping blanks with axial constraints in tube hydroforming of a variable-diameter part. AISI 304 stainless steel blanks were selected for numerical simulation and experimental research. The circumferential material flow distribution was obtained from the profile at the edge of the overlap. The peak value was located at the middle of the cross-section. In addition, the circumferential material flow could also be reflected in the variation of the overlap angle. The variation of the overlap angle kept increasing as the initial overlap angle increased. There was an optimal initial overlap angle to minimize the thinning ratio.

Alexandrov et al. [9] proposed a simple analytical solution for describing the expansion of a two-layer tube under plane-strain conditions. Each layer's constitutive equations consist of an arbitrary pressure-independent yield criterion, its associated plastic flow rule, and an arbitrary hardening law. The elastic portion of strain was neglected. The method of solution was based on two transformations of space variables. First, a Lagrangian coordinate was introduced instead of the Eulerian radial coordinate. Then, the Lagrangian coordinate was replaced with the equivalent strain. The solution reduced to ordinary integrals that, in general, should be evaluated numerically. However, for two hardening laws of practical importance, these integrals were expressed in terms of special functions. Three geometric parameters for the initial configuration, a constitutive parameter, and two arbitrary functions classified the boundary value problems. The illustrative example demonstrated the effect of the outer layer's thickness on the pressure applied to the inner radius of the tube.

2.2. Tube Forming Processes

Kajikawa et al. [10] proposed a tube expansion drawing method to effectively produce a thin-wall and large-diameter tube. In the proposed method, the tube end is flared by pushing a plug into the tube, and the tube is then expanded by drawing the plug in the axial direction while the flared end is chucked. The forming characteristics and effectiveness of the proposed method were investigated through a series of finite element analyses and experiments. The finite element simulation results show that the expansion drawing effectively reduced the tube thickness with a smaller axial load when compared with the conventional method. According to the experimental results, the thin-walled tube was produced successfully by the expansion drawing. The maximum thickness reduction ratios for a carbon steel (STKM13C) and an aluminum alloy (AA1070) were 0.15 and 0.29 when the maximum expansion ratios were 0.23 and 0.31, respectively. The above results suggest that the proposed expansion drawing method is effective for producing thin-walled tubes.

Tubes with variable diameters in the axial direction are in demand but it is costly to manufacture them. For instance, changing the tube diameter is achieved by connecting different diameter tubes using joints. It takes time and effort to connect, and in some cases, the connection often causes low airtightness. Therefore, the demand for different diameter continuous tubes (DDC tubes) and the process for DDC tubes without joining processes is high. Hirama et al. [11] proposed a new ball spin forming equipment, which can form a reduced diameter section on the halfway point of a tube. The effects of forming process parameters on the surface integrity and deformation characteristics of the product were investigated. The proposed method can reduce the diameter in the middle portion of the tube, and the maximum diameter reduction ratio was over 10% in one pass. When the feed pitch of the ball die was more than 2.0 mm/rev, spiral marks remained on the surface of the tube. Torsional deformation, axial elongation, and an increase in thickness appeared in the tube during the forming process. All of them were affected by the feed pitch and feed direction of the ball die, while they were not affected by the rotation speed of the tube. When the tube was pressed perpendicularly to the axis without axial feed, a diameter reduction ratio of 21.1% was achieved without defects using a ball diameter of 15.9 mm. The polygonization of the tube was suppressed by reducing the pushing pitch. The ball spin forming has a big advantage in flexible diameter reduction processing on the halfway point of the tube for producing different diameter tubes.

Arai and Gondo [12] proposed a method of forming a tube into an oblique/curved shape by synchronous multipass spinning, in which the roller moves back and forth along the workpiece in the axial direction to gradually deform a blank tube into a target shape. The target oblique/curved shape is expressed as a series of inclined circular cross sections. The contact position of the roller and the workpiece is calculated from the inclination angle, center coordinates, and diameter of the cross sections, considering the geometrical shape of the roller. The blank shape and the target shape are interpolated along normalized tool paths to generate the numerical control command of the roller. Aluminum tubes were formed experimentally into curved shapes with various radii of curvature, and the forming accuracy, thickness distributions, and strain distributions were examined.

Nakajima et al. [13] investigated the deformation properties and suppression characteristics of an ultra-thin-walled rectangular tube in rotary draw bending with a laminated mandrel. Aluminum alloy rectangular tubes with a height of 20 mm, width of 10 mm, and wall thickness of 0.5 mm were used. The deformation properties after rotary draw bending were investigated. The results show that deformation in the height direction of the tube was suppressed on applying the laminated mandrel, whereas a pear-shaped deformation peculiar to the ultra-thin wall tube occurred. Because axial tensions and lateral constraints were applied and the widthwise clearance of the mandrel was adjusted appropriately, the pear-shaped deformation was suppressed and a more accurate cross-section was obtained.

Recently, improvement of hybrid and electric vehicle technologies, equipped with batteries, continues to contribute to solving energy and environmental problems. Lighter weight and crash safety are required in these vehicle bodies. In order to meet these requirements, three-dimensional hot bending and direct quench (3DQ) technology, which enables hollow tubular automotive parts to be formed with a tensile strength of 1470 MPa or over, has been developed. Tomizawa et al. [14] investigated the crash characteristics of partially quenched curved products by three-dimensional hot bending and direct quench. The main results are as follows: (1) for partially quenched straight products in the axial crash test, buckling that occurs at the non-quenched portion could be controlled; (2) for the nonquenched conventional and overall-quenched curved products, buckling occurred at the bent portion at the initial stage in axial crash tests, and its energy absorption was low; (3) by partially optimizing quench conditions, buckling occurrence could be controlled; and (4) in this study, the largest energy absorption was obtained from the partially quenched curved product, which was 84.6% larger than the energy absorption of the conventional nonquenched bent product in the crash test.

2.3. Sheet Forming Processes

Hong et al. [15] developed a pneumatic experimental apparatus to evaluate the strain rate sensitive forming limits of 7075 aluminum alloy sheets under biaxial stretching modes at elevated temperature. For optimization of the die shape design, the ratio of minor to major die radius (k) and profile radius (R) were parametrically studied. The final shape of the die was determined by whether the history of the targeted deformation mode was well maintained and if the fracture was induced at the pole (specimen center), to prevent unexpected failure at other locations. As a result, a circular die with k = 1.0 and an elliptic die with k = 0.25 were selected for the balanced biaxial mode and near plane strain mode, respectively. Lastly, the pressure inducing fracture at the targeted strain rate was studied as the process design. An analytical model previously developed to maintain the optimized strain rate was modified for this designed model. The results of the integrated design were compared with the experimental results. The shape and thickness distributions of numerical simulation results show good agreement with those of the experiments.

Friction at the interface between sheet and dies is an important factor influencing the formability of strip or sheet forming. Hwang and Chen [16] investigated the frictional behaviors of sheets at variant speeds using a self-developed sheet friction reversible test machine. This friction test machine, stretching a strip around a cylindrical friction wheel, was used to investigate the effects of various parameters, including sliding speeds, contact angles, strip materials, and lubrication conditions on friction coefficients at the sheet–die interface. The friction coefficients at the sheet–die interface were calculated from the drawing forces at the sheet on both ends and the contact angle between the sheet and die. A series of friction tests using carbon steel, aluminum alloy, and brass sheets as the test piece were conducted. From the friction test results, it became known that the friction coefficients could be reduced greatly with lubricants on the friction wheel surface while the friction coefficients were influenced by the sheet roughness, contact area, relative speeds between the sheet and die, etc. The friction coefficients obtained under various friction conditions can be applied to servo deep drawing or servo draw-bending processes with variant speeds and directions.

3. Conclusions

This Special Issue and Book "Latest Hydroforming Technology of Metallic Tubes and Sheets" includes 16 papers, which cover the state of the art of forming technologies in the relevant topics in the field. The technologies and methodologies presented in these papers will be very helpful for scientists, engineers, and technicians in product development or forming technology innovation related to tube hydroforming processes.

Funding: This research received no external funding.

Acknowledgments: The Guest Editors would like to thank all who contributed to the development of this Special Issue. Thanks to the authors who submitted manuscripts to share results of their research activity, and to the reviewers who agreed to read them and gave constructive suggestions to improve the final quality of the papers. The Guest Editors would also like to send the warmest gratitude to the *Metals* editorial team and, particularly, to Sunny He for their assistance and support during the preparation of this special issue.

Conflicts of Interest: The authors declare no conflict of interest.

References

1. Yasui, H.; Miyagawa, T.; Yoshihara, S.; Furushima, T.; Yamada, R.; Ito, Y. Influence of Internal Pressure and Axial Compressive Displacement on the Formability of Small-Diameter ZM21 Magnesium Alloy Tubes in Warm Tube Hydroforming. *Metals* **2020**, *10*, 674. [CrossRef]
2. Supriadi, S.; Furushima, T.; Manabe, K. A Vision-Based Fuzzy Control to Adjust Compression Speed for a Semi-Dieless Bellows-Forming. *Metals* **2020**, *10*, 720. [CrossRef]
3. Morishima, T.; Manabe, K. Warm Hydroforming Process under Non-Uniform Temperature Field for Magnesium Alloy Tubes. *Metals* **2021**, *11*, 901. [CrossRef]

4. Hwang, Y.M.; Tsai, Y.J. Movable Die and Loading Path Design in Tube Hydroforming of Irregular Bellows. *Metals* **2020**, *10*, 1518. [CrossRef]
5. Hwang, Y.M.; Pham, H.N.; Tsui, H.S.R. Investigation of Punch Shape and Loading Path Design in Hydro-Flanging Processes of Aluminum Alloy Tubes. *Metals* **2021**, *11*, 636. [CrossRef]
6. Bach, M.; Degenkolb, L.; Reuther, F.; Psyk, V.; Demuth, R.; Werner, M. Conductive Heating during Press Hardening by Hot Metal Gas Forming for Curved Complex Part Geometries. *Metals* **2020**, *10*, 1104. [CrossRef]
7. Yoshimura, H.; Nakahara, K.; Otsu, M. Local One-Sided Rubber Bulging Test to Measure Various Strain Paths of Metal Tube. *Metals* **2021**, *11*, 751. [CrossRef]
8. Han, C.; Feng, H. Circumferential Material Flow in the Hydroforming of Overlapping Blanks. *Metals* **2020**, *10*, 864. [CrossRef]
9. Alexandrov, S.; Lyamina, E.; Lang, L. Description of the Expansion of a Two-Layer Tube: An Analytic Plane-Strain Solution for Arbitrary Pressure-Independent Yield Criterion and Hardening Law. *Metals* **2021**, *11*, 793. [CrossRef]
10. Kajikawa, S.; Kawaguchi, H.; Kuboki, T.; Akasaka, I.; Terashita, Y.; Akiyama, M. Tube Drawing Process with Diameter Expansion for Effectively Reducing Thickness. *Metals* **2020**, *10*, 1642. [CrossRef]
11. Hirama, S.; Ikeda, T.; Gondo, S.; Kajikawa, S.; Kuboki, T. Ball Spin Forming for Flexible and Partial Diameter Reduction in Tubes. *Metals* **2020**, *10*, 1627. [CrossRef]
12. Arai, H.; Gondo, S. Oblique/Curved Tube Necking Formed by Synchronous Multipass Spinning. *Metals* **2020**, *10*, 733. [CrossRef]
13. Nakajima, K.; Utsumi, N.; Saito, Y.; Yoshida, M. Deformation Property and Suppression of Ultra-Thin-Walled Rectangular Tube in Rotary Draw Bending. *Metals* **2020**, *10*, 1074. [CrossRef]
14. Tomizawa, A.; Hartanto, S.S.; Uematsu, K.; Shimada, N. Crash Characteristics of Partially Quenched Curved Products by Three-Dimensional Hot Bending and Direct Quench. *Metals* **2020**, *10*, 1322. [CrossRef]
15. Hong, J.H.; Yoo, D.; Kwon, Y.N.; Kim, D. Pneumatic Experimental Design for Strain Rate Sensitive Forming Limit Evaluation of 7075 Aluminum Alloy Sheets under Biaxial Stretching Modes at Elevated Temperature. *Metals* **2020**, *10*, 1639. [CrossRef]
16. Hwang, Y.M.; Chen, C.C. Investigation of Effects of Strip Metals and Relative Sliding Speeds on Friction Coefficients by Reversible Strip Friction Tests. *Metals* **2020**, *10*, 1369. [CrossRef]

MDPI

Article

Warm Hydroforming Process under Non-Uniform Temperature Field for Magnesium Alloy Tubes

Toshiji Morishima and Ken-Ichi Manabe *

Department of Mechanical Systems Engineering, Tokyo Metropolitan University, 1-1 Minamiosawa, Hachioji, Tokyo 192-0397, Japan; toshiji.morishima@gmail.com
* Correspondence: manabe@tmu.ac.jp; Tel.: +81-42-675-3059

Abstract: The warm tube hydroforming (WTHF) process of lightweight materials such as magnesium alloy contributes to a remarkable weight reduction. The success of the WTHF process strongly depends on the loading path with internal pressure and axial feeding and other process variables including temperature distribution. Optimization of these process parameters in this special forming technique is a great issue to be resolved. In this study, the optimization of the symmetrical temperature distribution and process loading path for the warm T-shape forming of magnesium alloy AZ31B tube was carried out by finite element (FE) analysis using a fuzzy model. As a result, a satisfactory good agreement of the wall thickness distribution of the samples formed under the optimum loading path condition can be obtained between the FE analysis result and the experimental result. Based on the validity validation of FE analysis model, the optimization method was applied to other materials and forming shapes, and applicability was discussed.

Keywords: magnesium alloy tube; warm hydroforming; non-uniform temperature field; protrusion type forming; wall thickness distribution; coupled thermal-structural analysis; optimization

Citation: Morishima, T.; Manabe, K.-I. Warm Hydroforming Process under Non-Uniform Temperature Field for Magnesium Alloy Tubes. *Metals* **2021**, *11*, 901. https://doi.org/10.3390/met11060901

Academic Editor: Badis Haddag

Received: 5 April 2021
Accepted: 21 May 2021
Published: 31 May 2021

1. Introduction

In recent years, the world has begun to move toward carbon neutrality as one of the solutions to prevent global warming. In response to the trend, along with the acceleration toward electric vehicles, the weight of the entire vehicle including the vehicle body is being further reduced. Tube hydroforming (THF), which uses a tubular material with a hollow cross section for high rigidity, is considered to be promising for reducing the weight of vehicles, in addition to material replacement with lightweight materials and high-strength materials, and is actually applied in the automotive industry.

Aluminum, magnesium, and fiber-reinforced polymers have been taken up as typical lightweight materials. Among these, magnesium is the lightest of all practical metals and is attracting attention as one of the important metals for realizing a low-carbon society due to its unique physical and mechanical properties. However, magnesium alloys have been treated as one of the hard-to-form metals at room temperature. So far, several metal forming processes for magnesium alloys using a heat-assisted approach have been introduced in order to enhance the formability and reduce the environmental load.

Most of the research on forming processes for magnesium alloy tubes has been conducted using heat-assisted processing since the mid-2000s, and started with hot spin forming. Yoshihara et al. [1] examined the dome-shape forming characteristics of the tube end while locally heating the formed part with a gas burner for hot spin forming of an AZ31 tube, and Murata et al. [2] investigated the compression spinning process of an AZ31 tube using the heated roller. For the high temperature bulge forming, Okamoto et al. [3] succeeded in warm T-shape forming of an AZ31B tube at 400 °C using nitrogen gas. Manabe et al. [4,5] performed T-shape forming (WTHF) at 250 °C using silicone oil as a pressure medium for AZ31 tube and compared the forming characteristics by experiment and FE analysis. Hwang and Wang [6] have succeeded in a Y-shape forming experiment of AZ61 tube at 250 °C, and it was shown that Y molding can be performed even at 150 °C. He et al. [7] investigated and evaluated the formability in a basic formability test of the AZ31B tube at high temperature. For the drawing process of magnesium alloy tube, Furushima et al. developed a high-frequency induction heating device system to locally heat and cool the AZ31 tube, and it was examined to improve the processing time of dieless drawing and the drawability [8,9]. An in vitro corrosion test [10] of the dieless drawn tube of AZ31B and the dieless drawing process by a laser heating device [11] were also performed.

Other heat-assisted processing technologies applied to hard-to-form metals other than magnesium alloy tubes include warm and hot hydroforming [12,13] for aluminum alloy tubes. Hot gas bulge forming was performed on aluminum alloy tubes [14,15] and high-strength steel tubes [16,17] using a direct resistance heating technique. In addition, profile forming [18], utilizing a bulge deformation under axial compression by local heating using a high-frequency induction device for a double-layer metal tube was performed. As a new heat-assisted forming technology, a novel dieless bellows forming that continuously creates a convolution at the heating portion with a difference in axial feeding speed on both tube sides while locally heating with a high-frequency induction device was developed for the stainless steel SUS304 tubes [19,20]. Additionally, the dieless bending process of tubes [21] is a typical heat-assisted processing technology that applies the local heating and cooling technique, and it has been introduced into practical use since the latter half of the 1960s and is the most widely used processing technology in the industry. Recently, a three-dimensional hot bending and direct quench (3DQ) technology [22] was developed as the latest technology that can be applied to high-strength steel tubes of 1470 MPa class or higher, based on basically a push bending process with a local heating and cooling system using a high-frequency induction device.

The various forming processes described above are classified in the heat-assisted forming technology from the viewpoints of "steady deformation", "unsteady deformation", "continuous/sequential processes", "intermittent process", and "dieless process" without die for forming objects. The above-mentioned dieless drawing, dieless bellows forming, and dieless bending process belong to sequential/continuous deformation process at the local heating portion, respectively.

The heat-assisted forming technology is classified according to whether or not the heating temperature is uniform over the entire forming portion. If the temperature field is uniform, the main difference in the forming characteristics is only the variation in the physical characteristics of the material.

Meanwhile, in the case of the non-uniform temperature field, there are two types of non-uniform temperature fields. (1) The first is the case where the temperature distribution is formed on the coordinates fixed on the metal blank, and (2) the second is the case where the temperature distribution is formed on the coordinates fixed on the space such as tooling. In sheet metal forming, the former (1) is used in stretch forming and bending, and in the latter (2), the local heating/cooling deep drawing method has been performed since the 1950s.

However, so far, there are very few studies on WTHF for non-uniform temperature fields. The above-mentioned study on the non-uniform temperature field in (1) was reported by Liu et al. [23]. This is an investigation on a warm axisymmetric bulge forming process with die in the non-uniform temperature field. The effectiveness of the non-uniform temperature distribution was confirmed for the AZ31 tube.

On the other hand, the branch protrusion type THF is an asymmetric deformation process and shows more complicated deformation behavior than the axisymmetric bulge forming. It is a similar deformation mode as the deep drawing process in sheet metal forming. Thus, it can be said that the branch protrusion type THF in this non-uniform temperature field has the same forming principle as the local heating/cooling deep drawing method.

The main deformation domain reduces drawing resistance by heating and softening, while the other load-carrying domain is cooled to prevent failure defects including fracture, improving the forming limit and the quality and accuracy of the product. More importantly, the blank material flows through the heating and cooling domains.

Yoshihara et al. [24] adopted the local heating/cooling deep drawing method to improve the deep drawability of AZ31 sheet and achieved a large drawing limit of $LDR > 5$.

In general, for the warm T-shape THF in the uniform temperature field [4,5], the wall thickness of the blank tube is locally thickened at the place of contact with an axial feeding punch and the bottom on the opposite of protruded branch part. To make the wall thickness distribution of the formed product uniform at the same time, instead of processing under a uniform temperature field, the forming temperature should be kept low at the part where deformation should be suppressed, wall thickness should be suppressed, and the material flow at the other part should be promoted.

Manabe et al. [25,26] investigated the effect of enhancing the hydroforming limit and improving the wall thickness distribution under a non-uniform temperature field by locally heating/cooling warm T-shape forming of an AZ31 magnesium alloy. However, even for the non-uniform temperature field in the previous report [25,26], the above technological issues were not solved enough yet and remain.

The objective of this study is to elucidate the optimum temperature distribution for further improving the formability and wall thickness distribution uniform of magnesium alloy tubes with low hydroformability at room temperature. Therefore, focusing on the non-uniform temperature field in WTHF process, the effect of the temperature distribution in the tooling on the wall thickness distribution of the formed sample is investigated in detail, and the optimum non-uniform temperature field and the optimum loading path for different magnesium materials (AZ31B and AZ61) and forming shapes (T-shape and cross-shape) are examined.

2. Experiments

2.1. Materials

The tubes used in this study were made of AZ31B and AZ61 magnesium alloy with an outer diameter of 42.7 mm and a wall thickness of 1.0 mm, and the test piece for the hydroforming test was 200 mm in length. Table 1 shows the chemical composition of these magnesium alloy tubes. A uniaxial tensile test was carried out using an arc-shaped test piece made by cutting out the tubular material in the longitudinal direction. The tensile test piece used was half the size of JIS 13B. In the tensile test, the test temperature was 20 °C, 100 °C, 150 °C, 200 °C, 250 °C, and 300 °C under 6 conditions, and the tensile speed was 1.5 mm/min, 15 mm/min, and 150 mm/min under 3 conditions (initial strain rate was 0.001 s^{-1}, 0.01 s^{-1}, and 0.1 s^{-1}, respectively). Figure 1 shows the typical stress-strain curves at different temperatures and tensile speeds for AZ31B and AZ61. From these results, the elongation of both materials increases as the temperature rises, and the tensile strength decreases. In addition, although the strain rate dependence is slightly observed as the temperature rises from room temperature, no significant strain rate dependence is observed in the test range, and the temperature dependence is large. Comparing AZ31B and AZ61, AZ61 has a slightly higher strength (deformation resistance) up to about 150 °C, and there is no big difference in elongation.

Table 1. Chemical composition of AZ31B and AZ61 magnesium alloy tubes (mass %).

Material	Al	Zn	Mn	Si	Fe	Cu	Ni	Ca	Mg
AZ31B	3.28	0.75	0.32	0.010	0.002	0.001	0.001	0.001	Bal.
AZ61	5.5–6.5	0.5–1.5	0.15–0.4	≦0.1	≦0.005	≦0.05	≦0.005	-	Bal.

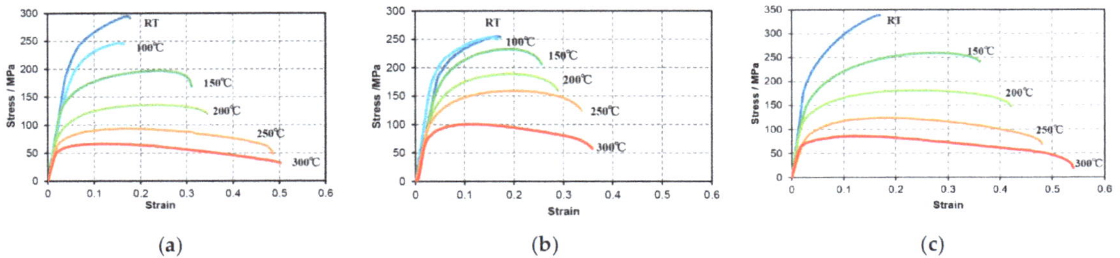

Figure 1. Stress-strain curves of AZ31 and AZ61 at different temperatures and initial strain rates. (**a**) AZ31B initial strain rate: 0.001 s^{-1}; (**b**) AZ31B, 0.1 s^{-1}; (**c**) AZ61, 0.01 s^{-1}.

2.2. Experimental Setup and Procedure

Figures 2 and 3 show the schematic illustration of the T-shape forming tooling used in this study and the configuration of the hydraulic circuit and the closed loop control system for the T-shape WTHF process, respectively [2–4]. To conduct this experimental work, a new T-shape warm forming machine with a local heating/cooling apparatus for the die was developed. This machine is an improved version of the previously developed equipment [3,4], and it enables controlling the temperature distribution of the die as well. In this testing machine, each cylinder and pressure booster is operated by a command from a computer, and closed-loop control can be performed by sending the measured values of the axial punch displacement and internal pressure from moment to moment to the computer.

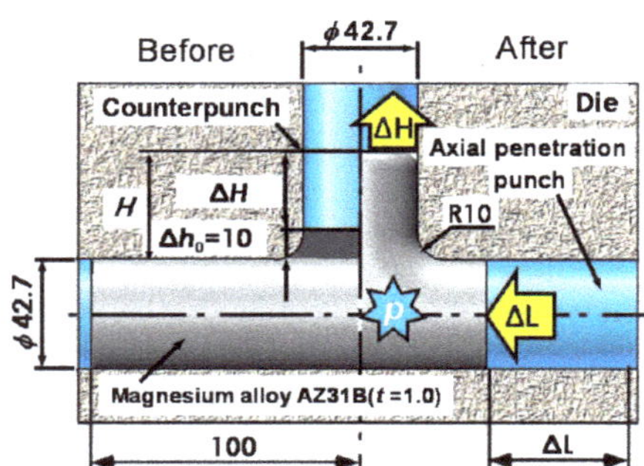

Figure 2. Outline of the tooling for T-shape forming (Unit: mm).

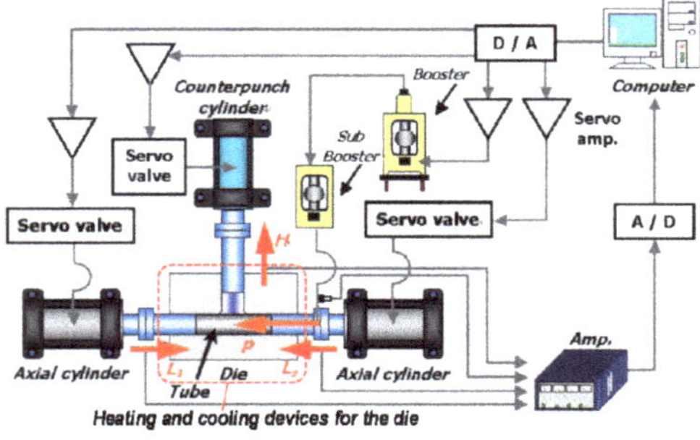

Figure 3. Configuration of a hydraulic circuit and its closed loop control system in WTHF.

Figure 4 shows the appearance of the heating and cooling device. For heating and cooling the die, six through holes are opened for inserting cartridge heaters into the upper and lower dies, and heating for uniform temperature field (Figure 4a) can be performed with a total of 12 cartridge heaters with maximum total output of 11 kW. The forming die was fully covered with the insulator boards.

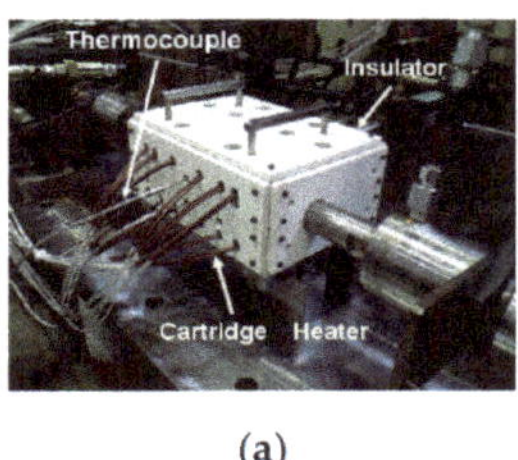

(a)

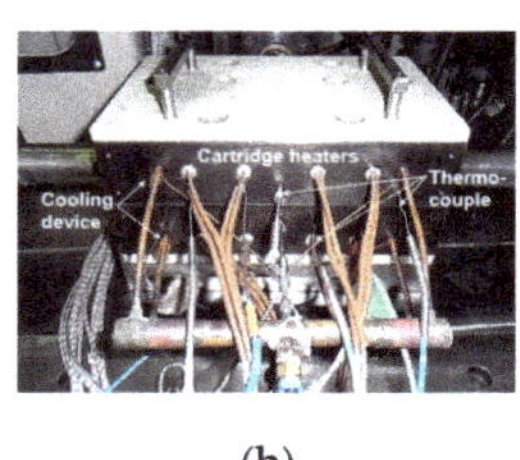

(b)

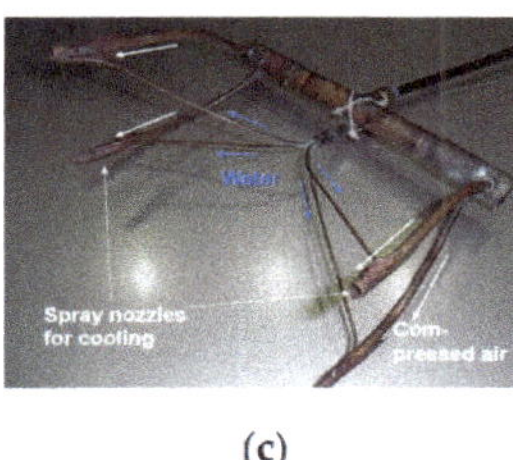

(c)

Figure 4. Heating and cooling system for the warm T-shape die. (**a**) For uniform warm temperature field, the entire circumference of the die is surrounded by heat insulator material; (**b**) for non-uniform temperature field; (**c**) cooling device with spray nozzle.

As the cooling method, we adopted a method of passing compressed air through the four through holes at the upper and lower ends of the heater and a method of cooling by using the heat of vaporization of water mist (Figure 4b). In the case of the water mist cooling, compressed air of about 70 kPa was provided through a larger pipe from a compressor, and a small amount of water was flowed from a tip of a smaller pipe to spray it into a mist for cooling the die (Figure 4c). In addition, there were six other holes on the side of the die for temperature measurement (Figure 4b). A sheath-type K thermocouple was inserted into each hole, and they measured a variation in the temperature distribution during the forming process.

To create a non-uniform temperature field, after heating to 270 °C with all 12 heaters, only the total of eight heaters in the middle continued to be heated, and in the meantime, the cooling devices were attached to the four holes at both ends of the die. After about 30 min, the temperature field became stable. Since the die temperature uniformly shifted down about around 15 °C during the total forming time of about 5 min, the forming experiment started when the temperature at the center reached at around 260 °C to realize the set temperature distribution in the middle of forming process.

The temperature of upper and lower dies was controlled independently using temperature controllers. To confirm uniform temperature distribution over the tubular blank, thermocouples with 0.1 mm diameter are fixed into a groove made on the tube.

Figure 5 shows the typical temperature distributions of the die for T-shape forming realized by using a local heating and cooling system.

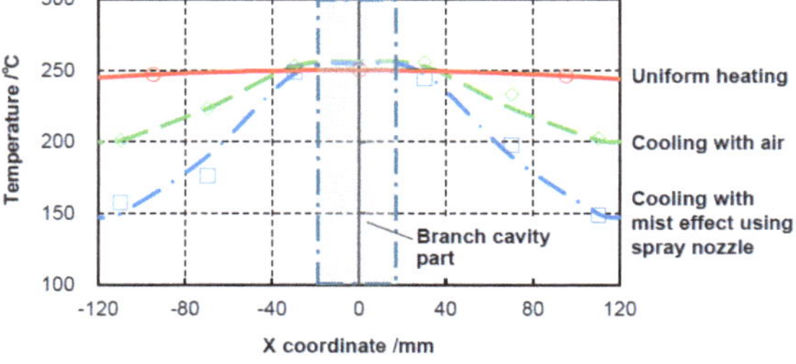

Figure 5. Temperature distributions used in WTHF experiment.

The pressure medium was methylphenyl silicone liquid KF-54 (Shin-Etsu Chemical Co., Ltd., Tokyo, Japan), and a dry fluorine lubricant (spray type) was used for the lubrication between the tube and the die.

3. Finite Element Modeling

Finite element (FE) analysis is performed by a thermal-structural coupled dynamic explicit FE commercial code, LS-DYNA ver. 971. (Ansys Inc., Canonsburg, PA, USA). Using a quarter-symmetrical model for T-shape and a one-eighth model for the cross-shape, the tubular blank was divided into two in the wall thickness direction, 60 in the circumferential direction, and 100 in the axial direction. Figure 6 shows the FE model for warm T-shape hydroforming. The material properties obtained from the tensile test in Section 2.1 were employed in the FE analysis. Other input values are the density $\rho = 1.78\ \mathrm{g/cm^3}$, Poisson's ratio $\nu = 0.35$, linear expansion coefficient $\alpha = 26.5 \times 10^{-6}/\mathrm{K}$, and specific heat $c = 1.05$ kJ/kgK.

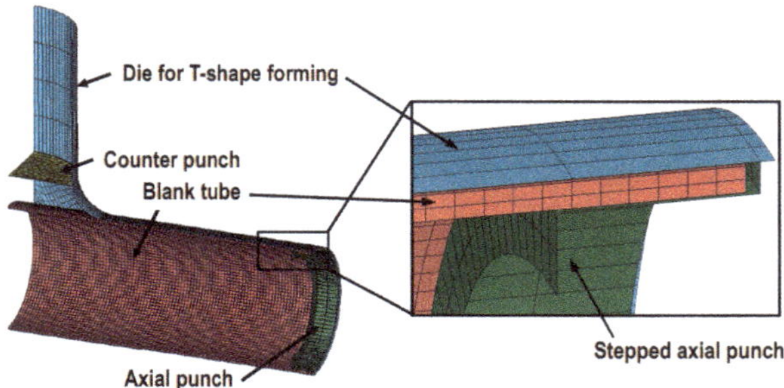

Figure 6. FE model for warm T-shape hydroforming.

In the material model considering the temperature dependence, the equivalent stress-equivalent plastic strain relationship was defined by inputting the deformation resistance curve for each temperature, and at the intermediate temperature, the deformation resistance was determined by interpolating each curve. In general, magnesium alloy has a strain rate dependence in the high temperature range. The processing time of this forming experiment, however, was 5 min, and the forming process was performed slowly. In the previously reported FE analysis [2], it was possible to sufficiently predict the detail forming characteristics in the same T-shape forming experiment with a material model that does not consider the strain-rate dependence. Moreover, from the material test results shown in Section 2.2 above, no large strain-rate dependence was observed when the forming temperature conditions and processing time were taken into consideration. Furthermore, under the non-uniform temperature field of 250 °C or less in this study, the region of the formed part where the strain-rate dependence appears changes with time, the temperature conditions there were also different, and the strain-rate dependence was considered to become small. Therefore, we used a material model that did not consider the strain-rate dependence in this study.

The coefficient of friction between the tool and the blank was static friction coefficient $\mu_s = 0.02$, the kinetic friction coefficient was $\mu_d = 0.01$, and the heat transfer coefficient was 1400 W/m^{2}K.

4. Results and Discussion

4.1. Process Window in Proportional Loading Path under Uniform Temperature Field and Validation of the FE Modeling

In order to clarify the process window in warm T-shape forming of AZ31B magnesium alloy and the FE modeling validation, WTHF experiments and FE analysis were performed in a proportional loading path under a uniform temperature field. Figure 7 shows the process window for T-shape forming for AZ31B under 250 °C obtained by the experiment and FE analysis. The onset point of "bursting" in the FE analysis was determined as when the wall thickness reduction began to increase rapidly without using any fracture criterion, or when the analysis was stopped. On the other hand, the occurrence of buckling was visually evaluated by visually displaying the deformed shape of the blank tube using the graphic processing of the post processor. It is confirmed that the bursting is predominant under higher internal pressure loading, the buckling is dominant when the axial penetration is predominant, and the successful region between both becomes narrow as the process proceeds.

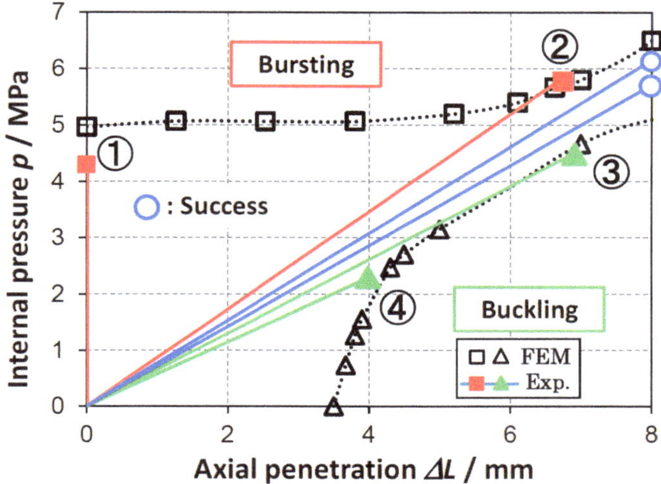

Figure 7. Process window and forming limits of T-shape forming for AZ31B tube in a proportional loading under 250 °C. The ①, ②, ③, and ④ in the figure indicate different loading paths (experiment).

Figure 8 shows the comparisons of final bulged heights and their shapes at the forming limit between experimental results and FE simulation results in Figure 7. The failure modes in Figure 8a,b are both bursting, as shown in Figure 7, and their formed shapes and bulged heights are in good agreement. Figure 8c,d show different buckling modes, both of which are in good agreement with the experimental and FE analysis results in a uniform temperature field at 250 °C. The results confirmed the validity of this FE model and demonstrated the importance of loading path design using FE analysis.

At the same time, it was confirmed that the loading path has a strong influence on the forming limit. When investigating the effects of temperature fields, the need to determine the appropriate loading paths to avoid bursting and buckling can be understood.

4.2. Effects of Temperature Distribution on Wall Thickness Distribution in Proportional Loading Path under Non-Uniform Temperature Field

The effectiveness of non-uniform temperature distribution for the uniform wall thickness of formed products in the warm T-shape forming process for the AZ31B magnesium alloy tube was reported by a comparison of the FE analysis result by a material model obeying a linear hardening law [4]. In this section, aiming for a more accurate analysis, the effect of the temperature distribution on the wall thickness distribution of formed products is examined by FEM analysis based on the strain hardening characteristics of the material obtained in the tensile test, as described in the previous chapter.

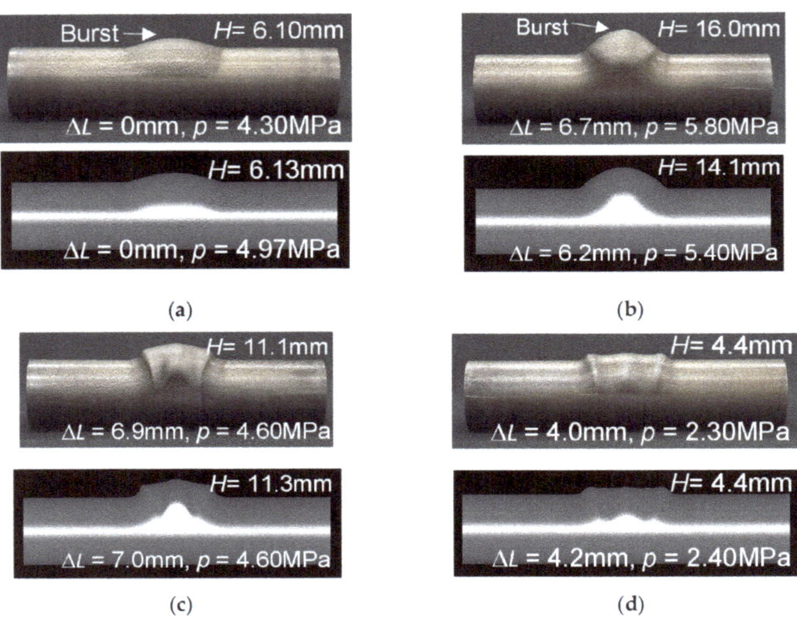

Figure 8. Comparisons of final bulged shape between experimental and simulated results at T-shape forming limit for AZ31B tube in a proportional loading under 250 °C. (**a**) Path ① in Figure 7; (**b**) Path ②; (**c**) Path ③; (**d**) Path ④.

Figure 9 shows the outline of temperature distributions for the die applied for that purpose, one is a uniform temperature field of 250 °C and another is a non-uniform temperature distribution of 250 °C to 150 °C in Figure 5. Figure 10 shows the comparison of experiments and FE analysis results on the wall thickness distribution along the bulged branch side and the bottom side in the T-shape forming under the two different temperature fields shown in Figure 9. Firstly, Figure 10a shows the wall thickness distributions along the bulging branch side of the T-shaped product. The FE analysis results in a uniform temperature field (red solid line) of 250 °C are quantitatively very consistent with the experimental results (red dotted line) except for the seal portions at both ends of the blank tube. As shown in the FE model of Figure 6 above, the shape of the axial punch used in the experiment is stepped to insert into the tube end and seal the internal pressure applied. Therefore, at the ends of the tube where the punch inserts, the deformation shape is constrained. In the non-uniform temperature field (blue line), since this part is a low temperature part at 150 °C, the deformation resistance is larger and the deformation is suppressed and smaller, and it is considered that this region is deformed in the same way as the FE analysis result. On the other hand, in a uniform warm temperature field (red line), both tube-end portions are also easily deformed, so it is considered that a local irregular large deformation occurred near the stepped portion where the punch is inserted. On the other hand, on the bottom side shown in Figure 10b, the experiment and FE analysis results show qualitatively good agreement in a uniform temperature field of 250 °C (red line). In the non-uniform field, both show a good quantitative agreement except for the central part. The T-shape forming process is an asymmetric deformation that shows a complicated deformation behavior and is a difficult deformation target as compared with an axisymmetric bulge deformation. Due to the asymmetry, the tube material also can flow significantly in the circumferential direction even at the bottom, as shown in Figure 10b during the forming process, and the other parts also flow in the circumferential direction accordingly. The difference between the analysis of the wall thickness distribution in Figure 10b and the experiment is considered to be due to this difference in material flow in the circumferential direction. There are various possible reasons for this difference, such as processing conditions and material-side factors, but the major problem is that the stress-strain relationship in the large strain range has not been experimentally obtained for the material properties. Additionally, in the result of this experiment in Figure 10b, the wall thickness strain in the central part reached about 1.0 in true strain. However, its relationship over a large strain range cannot currently be determined by material testing, and it is presumed that extrapolation to this large strain range is also one of the major reasons.

Figure 9. Temperature distribution conditions in warm T-shape forming process for AZ31B. (**a**) Uniform temperature field at 250 °C; (**b**) Non-uniform temperature field.

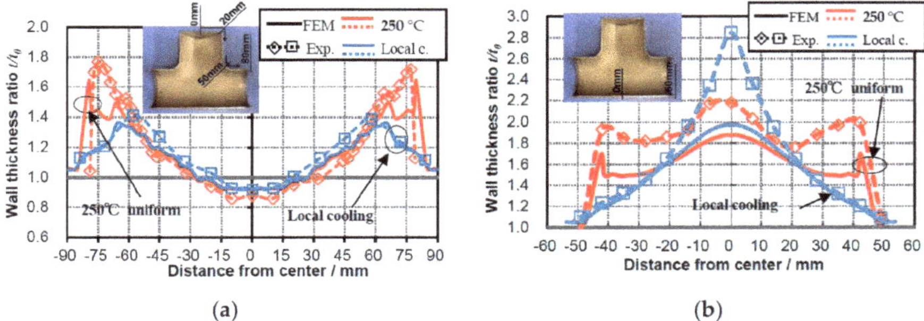

Figure 10. Comparison of wall thickness distribution between uniform temperature and non-uniform temperature fields. (**a**) Bulged side; (**b**) bottom side. Solid line: FE analysis result; dashed line: experimental result; red color: 250 °C; blue color: non-uniform temperature.

Consequently, the FE analysis is in good agreement with the experimental results. From this, it can be said that the analysis accuracy is improved by approximating the material model to multiple points according to the actual work hardening characteristics. The validity of this FE model for the evaluation of the wall thickness distribution is confirmed.

4.3. Determination and Its Validation of Optimum Loading Path for Improving Formability and Bulged Shape in Warm T-shape Forming Process under Uniform Temperature Field

In THF, the loading path has a great influence on hydro-formability, forming shape, wall thickness distribution, and so on. In the actual forming process, so far the optimum loading path of the internal pressure and axial feeding is empirically determined. In our previous study [27], the optimum loading path using a fuzzy model was predicted and confirmed. Here, firstly the optimum loading path for improving bulged shape and bulged height is investigated, and the obtained loading path is used to clarify the optimum temperature conditions for the non-uniform temperature field.

The final internal pressure p_f in the T-shape forming process is given from the following empirical formula [28].

$$P_f = k\frac{2t_0\sigma_y}{D_0} \quad (1)$$

Here, t_0 is the wall thickness, σ_y is the yield stress, D_0 is the outer diameter of the bulging part, k is the experimental constant, and 4.7 is used for T-shape forming. By substituting the dimensions and material properties of the magnesium alloy tube at 250 °C into the above equation, p_f = 10 MPa was determined.

The axial penetration and the counter punch displacements during the forming process were determined every moment by the same fuzzy inference model in the previous report [27]. Basically, the following typical evaluation functions were employed for evaluating the wavy buckling and the contact length with the counter punch with increasing the internal pressure during the process.

Figure 11 shows the evaluation functions used in the fuzzy inference model for T-shape forming. Figure 11a shows an evaluation function Φ defined as an index for evaluating the risk of wavy buckling in order to suppress the occurrence of the buckling deformation near the die inlet shoulder part during free bulge deformation at the initial forming stage. R_{max} on the right side of the equation of the evaluation function Φ in the figure is the maximum bulged height near the die inlet shoulder part, and R_{min} is the minimum bulged height near there.

Figure 11. Evaluation functions for T-shape forming. (**a**) Evaluation function Φ for the risk of wavy buckling; (**b**) Evaluation function Ψ for the contact length with counter punch. R_{max}: maximum bulged height (radius) near the die inlet shoulder part; R_{min}: minimum bulged height (radius) near there; C_l: contact length with a counter punch; C_{ref}: assumed reference contact length. A red rectangle indicates a maximum bulged part region near the die inlet.

Since the decrease in wall thickness of the bulged part can be suppressed by aggressively feeding the axial punch from the contact of the material with the counter punch to the latter stage of the forming process, the evaluation function Ψ of the contact length with the counter punch is defined as an index for evaluating the risk of the forming failure.

C_l on the right side of the equation of the evaluation function Ψ in Figure 11b is the contact length with counter punch, and C_{ref} is the assumed reference contact length in which the contact length with counter punch increases in proportion to the bulge height.

Input and output variables for fuzzy rules were fuzzified so that the input variables Φ, Φ' (differential value of Φ), and Ψ, Ψ' (differential value of Ψ) are "Small" and "Large", and the output variables were ruled for the axial punch displacement increment ΔAPS and the counter punch displacement increment ΔCPS.

As an example of the membership functions that determines the output variables, Figure 12 shows the membership functions for the evaluation function Φ that represents the risk of wavy buckling at the initial stage of forming process. The input and output membership functions (ΔAPS and ΔCPS) at the main forming stage are similarly ruled.

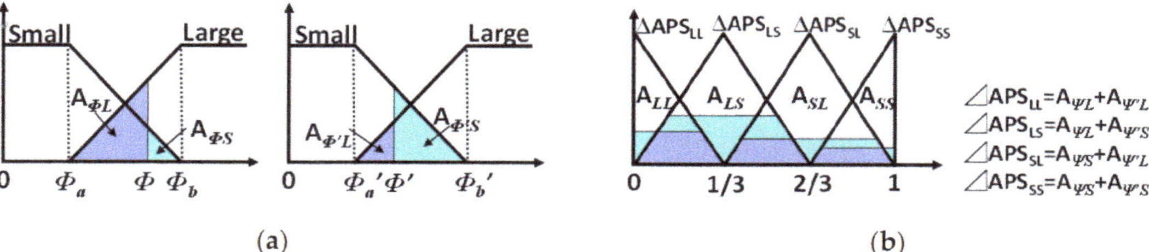

Figure 12. Membership functions for the early stage. (**a**) Input membership function for Φ; (**b**) output membership function for axial penetration increment ΔAPS.

Fuzzy inference was performed based on the membership functions for each forming stage, and the output value for each forming stage by defuzzification was determined by the centroid method.

The two output variables, ΔAPS and ΔCPS, are sequentially determined during the process to maintain an appropriate contact state.

Figure 13 shows a comparison between the loading path at 250 °C assumed from the authors' past experience and the loading path obtained by the fuzzy inference model. From the comparison of the two, the p of the fuzzy inference result in the loading path is larger than the p of the assumed path after the point A. As a result, in the ΔH-ΔL curve for the counter punch, the fuzzy inference model early reaches the maximum setting bulge height of ΔH = 30 mm (point B). Therefore, in the assumed loading path, an axial feeding displacement of ΔL = about 50 mm was required to obtain the final shape, but in the loading path obtained by the fuzzy inference model, the final shape can be obtained by feeding the axial punch to ΔL = about 45 mm (point C). It can be said that efficient forming process can be performed by deriving the loading path using a fuzzy model system.

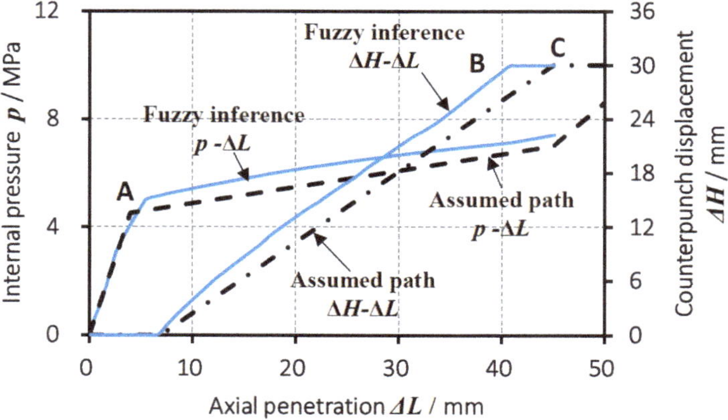

Figure 13. Comparison of loading path between the manual set from experience and the fuzzy inference model (250 °C), p: internal pressure, ΔH: counterpunch displacement, ΔL: axial penetration.

A T-shape forming experiment was conducted to verify the validity. Figure 14 shows a comparison of the final shape and dimensions with the FE analysis results. Quantitatively good agreements are found on the shape, bulge height, and axial length, demonstrating the effectiveness of the loading path determination by this system.

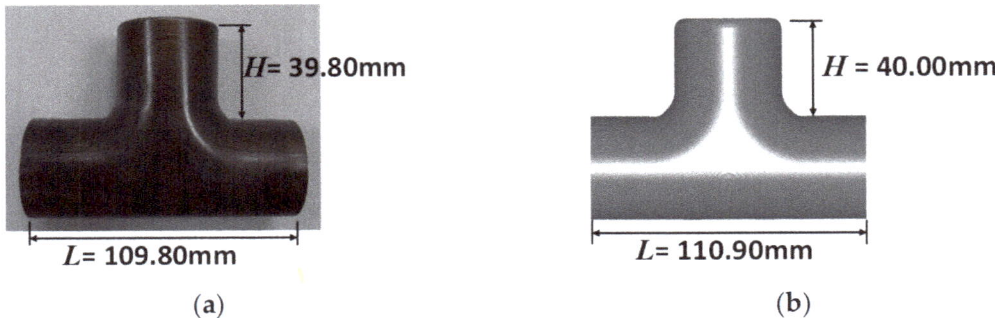

Figure 14. Comparison between the final shape and dimensions of T-shape formed products. (**a**) Experiment; (**b**) FE analysis.

4.4. Optimum Temperature Distribution for Improving Wall Thicknes Distribution in Warm T-Shape Forming Process Using the Optimum Loading Path under Non-Uniform Temperature Field

Since the loading path affecting hydro-formability was optimized in the previous section, the optimum temperature distribution under the no-uniform temperature field will be examined in that loading path condition.

4.4.1. Effect of Temperature of Counter Punch on Thinning Behavior of the Bulged Part under Non-Uniform Temperature Field

In the bulging part including a bulge head, wall thinning occurs due to the internal pressure applied, but it is thought that this can be suppressed by cooling the counter punch and increasing the deformation resistance. Figure 15 shows the effect of the temperature T_{CP} at the center of the counter punch on the wall thinning rate of the bulging part. It can be seen that the wall thinning of the bulging part can be suppressed by reducing T_{CP}. However, since buckling/wrinkling in Figure 15b occurred near the die side and die shoulder part at a temperature of 160 °C or less, it is considered that there is an appropriate temperature range that suppresses wall thinning without causing forming defects, and T_{CP} in T-shape forming of AZ31B. It is suggested that the optimum temperature of the counter punch is 170 °C.

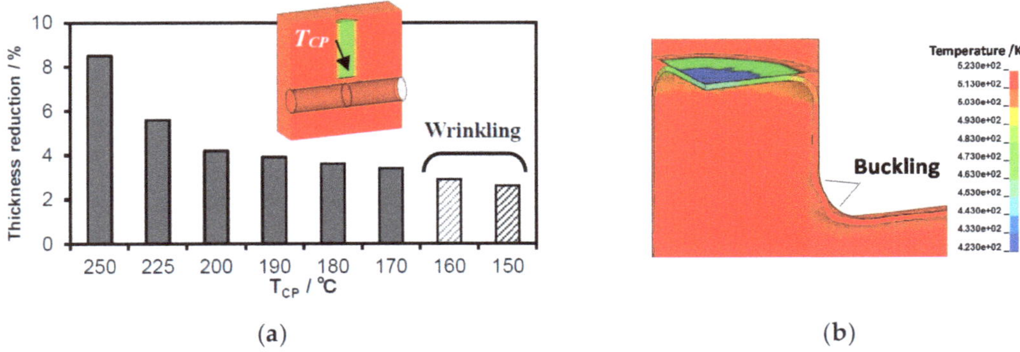

Figure 15. Effect of counter punch temperature *TCP* on wall thickness reduction at the center of counter punch. (**a**) Effect of TCP; (**b**) buckling occurred near the die shoulder.

4.4.2. Effect of Temperature of Die Bottom on Wall Thickness Distribution under Non-Uniform Temperature Field

In order to make the wall thickness distribution on the bulging side and its bottom side uniform at the same time, the authors will consider the case of further cooling to the bottom center. For the purpose, the temperature on the straight tube ends side is set to 150 °C, and also the central part on the bottom side is locally cooled to provide a temperature distribution to the central part of the die in the circumferential direction. In this case, in order to quantitatively evaluate the uniformity of the wall thickness distribution and determine the optimum temperature conditions, the root mean square of the bottom wall thickness distribution in each T_{bottom}, t_{RMS}, was calculated by the following formula,

$$t_{RMS} = \sqrt{1/N \sum_{i=1}^{N} {t_i}^2} \tag{2}$$

Figure 16 shows the effect of the bottom center temperature T_{bottom} on the uniformity of the wall thickness distribution. In this figure, a comparison based on the t_{RMS} is performed. It can be seen that the uniformity of the wall thickness distribution is improved by decreasing the T_{bottom}. On the other hand, the buckling shown in Figure 16b was confirmed in the bulging part when the temperature of T_{bottom} was 120 °C or less, so it is possible that there is an appropriate temperature condition for T_{bottom} to make the wall thickness distribution uniform without causing forming defects. It is suggested that the optimum temperature of T_{bottom} in T-shape forming of AZ31B is 130 °C. By the way, the difference in the t_{RMS} between the "250 uniform" and the T_{bottom} "250" in Figure 16 is caused by the temperature distribution of the die. When T_{bottom} = 250 °C, there is still a temperature distribution of the die in the axial direction of the tube, and the part where the temperature on the tube end side remains low. Due to the influence of the non-uniform temperature field, there is a difference in t_{RMS} from the "250 uniform" for the uniform temperature field.

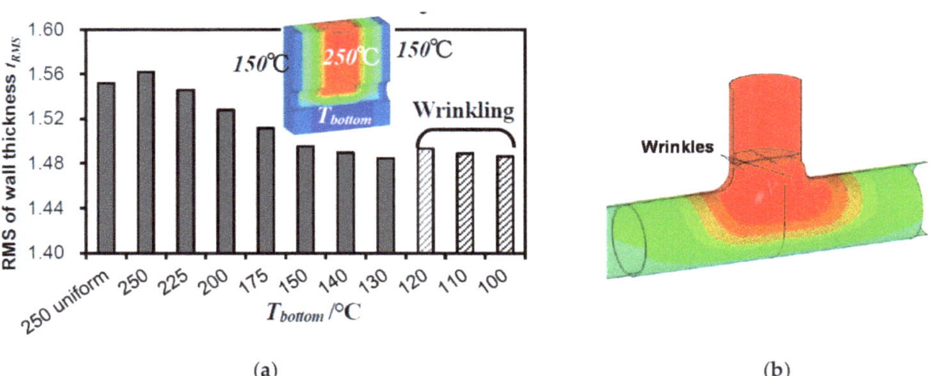

(a)

(b)

Figure 16. Effect of T_{bottom} on RMS of wall thickness distribution along the bottom part. (**a**) The t_{RMS}; (**b**) wrinkles occurred at bulging head.

From the above results, by creating a complex non-uniform temperature field lowering not only the die temperature in the direction of the tube axis but also to 130 °C to the opposite bottom of the bulge branch, it is found that the wall thickness distribution of the entire T-shaped product is further improved compared to under the uniform temperature field. For the T-shape forming of AZ31B, the optimized temperature conditions for the uniformity of the wall thickness of the formed product may be as shown in Figure 17.

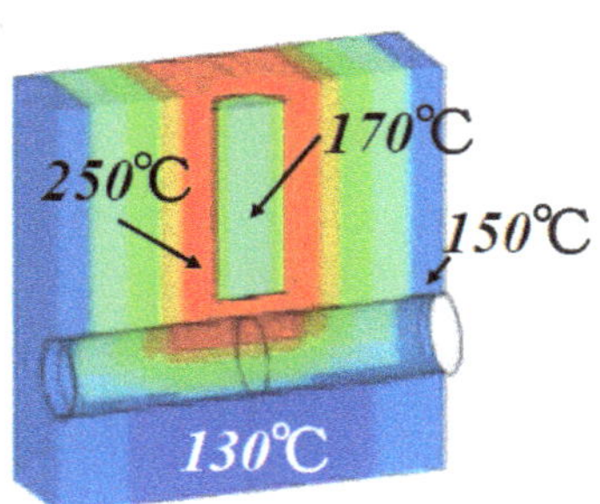

Figure 17. Optimum temperature conditions in T-shape forming for AZ31B magnesium alloy.

4.5. Wall Distribuntion in the Optimum Loading Path under Optimum Temperature Distribution

Figure 18 shows a comparison of the wall thickness distribution of the obtained formed product between the results of the conventional manual loading path and the optimum loading path. From these figures, it can be seen that a formed product having a more uniform wall thickness distribution is obtained on both the bulging side and the bottom side. It is shown that it is possible to determine appropriate processing conditions for making the formed product uniform by performing the optimum process path in the hydroforming of the T-shaped sample of the AZ31B magnesium alloy tube.

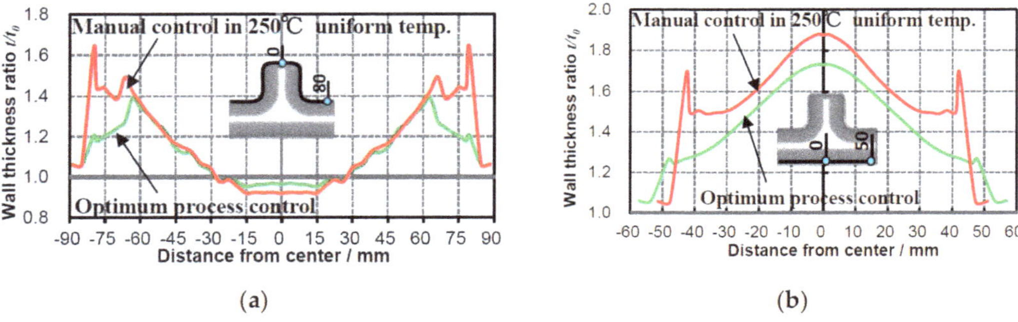

Figure 18. Comparison of wall thickness distribution between manual control and optimum process control (FE analysis). (**a**) Bulge side; (**b**) bottom side.

4.6. Application of the Optimization Methods to AZ61 Alloy Tube and Cross-Shape Forming

Up to the previous section, the authors optimized the temperature distribution and loading path in a non-uniform temperature field and demonstrated numerically the possibility of uniform wall thickness distribution and improvement of formed shape and accuracy. Thereby, in this section, the authors will investigate and compare the case where this method is applied to the AZ61 tube of higher strength magnesium alloy and the case where it is applied to cross-shape forming as a forming shape other than T-shape forming. For cross-shape forming, the forming shape is different from T-shape forming; a new fuzzy model for cross-shape forming was created, and the various fuzzy parameters were modified accordingly.

4.6.1. Optimum Loading Path for Different Materials and Formed Shapes

Figure 19 shows the comparison of optimum loading path for T-shape forming and cross-shape forming of AZ31B and AZ61 at 250 °C. Figure 19a shows the effect of the different strength between AZ31B and AZ61 on the optimum loading path in T-shape forming. At the initial stage of the p-ΔL curve, the axial punch is feeding at the same time, while the p is applied in order to suppress the thinning of the bulge apex during free bulge deformation, and the p-ΔL curve increases almost linearly. On the other hand, in the ΔH-ΔL curve for the counter punch, the punch keeps stopping at the initial position of Δho = 10 mm until the blank tube contacts the counter punch. After that, the counter punch starts to move behind the feeding of the axial punch. The counter punch moves almost proportionally as the bulge deformation progresses and stops when it reaches the set maximum bulge height (point B), and in that state, the axial punch feeding is further progressing to the point C in order to attain more uniform wall thickness and improvement of the die filling rate and the dimension accuracy of the product.

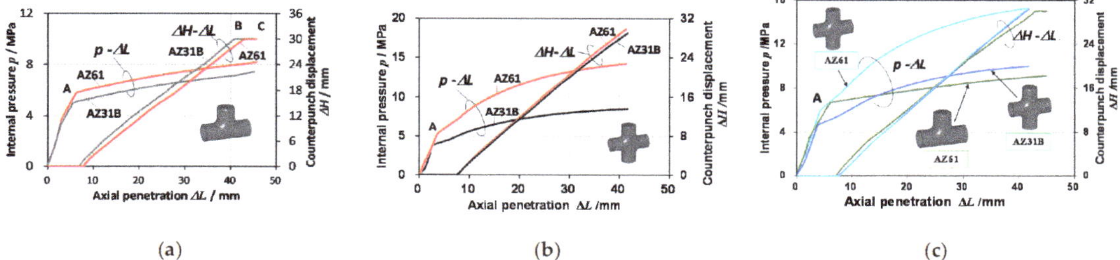

Figure 19. Comparison of optimum loading path for T-shape forming and cross-shape forming of AZ31B and AZ61 at 250 °C. (**a**) T-shape forming; (**b**) cross-shape forming; (**c**) comparison between T-shape forming and cross-shape forming. p: internal pressure, ΔH: counterpunch displacement, ΔL: axial penetration.

Regarding the effect of the material on the p-ΔL curve, the strength of AZ61 is higher than that of AZ31B, so the p at point A increases corresponding to the increase in strength. The required p is high while being maintained even after the point A. On the other hand, in the ΔH-ΔL curve, the effect of the difference in material strength is that the position of the axial punch at which the counter punch starts to move is slightly delayed in AZ61, and then it is almost the same as in AZ31B.

Figure 19b shows the effect of the difference between AZ31B and AZ61 on the optimum loading path in cross-shape forming. The p-ΔL curve of the cross-shape forming is significantly different in shape from the T-shape forming in Figure 19a. For T-shape forming, it is represented by two straight lines at point A, but for cross-shape forming, the latter half forming region is dominant except for the initial stage of forming, and the p increases in an upwardly convex quadratic curve. The effects of AZ31B and AZ61 on this quadratic p-ΔL curve are similar to those in Figure 19a at the early stage of the process. In the dominant latter half, the effect of the strength difference expands as the forming progresses. It can be seen that the high-strength AZ61 requires about 1.5 times higher internal pressure. On the other hand, in cross-shape forming, the ΔH-ΔL curve shows almost no difference between AZ31B and AZ61.

By the way, the effects of T-shape forming and cross-forming on the p-ΔL curves, which have different forming shapes, can be clearly seen from the result of AZ61 in Figure 19c, which expresses the vertical axis on the same scale. At the early stage of forming process, there is a slight difference in the p-ΔL curve between the two, but there is a slight difference in the p value at point A. It can be seen that the increase in the p is large for the cross-shape forming, and it requires nearly 1.8 as much as in the latter half of T-shape forming. Meanwhile, the ΔH-ΔL curve does not differ depending on the forming shape.

Until now, the loading path has been determined by experience and trial and error, but it is expected that this method will be applied to the optimization of processing conditions and the process control.

4.6.2. Optimum Temperature Distribution for Different Forming Shapes

By applying the optimization method for T-shape forming of AZ31B described above, optimization of the non-uniform temperature field was performed for AZ61 tube and cross-shape forming by the same FE analysis. In T-shape forming for AZ61 tube, buckling occurred at the bulged head area when the temperature of T_{bottom} was 140 °C or lower. The difference in deformation behavior between AZ31B and AZ61 is considered to be due to the deformation resistance of both. The difference in deformation resistance between the two is not large in the high temperature range of 200 °C or higher, but the difference is large in the temperature range of 150 °C or lower. Therefore, in the WTHF of AZ61, it is considered that it becomes difficult to form the bulging part by reducing the temperature T_{bottom} at the center of the bottom. From these results, it can be said that the optimum temperature of T_{bottom}, which makes the bottom wall thickness distribution uniform in T-shape forming of AZ61 magnesium alloy, is 150 °C.

In cross-shape forming, the wall thinning of the bulging part can be suppressed by lowering the temperature T_{CP} at the center of the counter punch, but wrinkles occur near the die shoulder by reducing the temperature. The temperature condition at which wrinkles began to occur was $T_{CP} < 200$ °C for both AZ31B and AZ61. By setting $T_{CP} = 200$ °C, it is possible to suppress the wall thinning of the bulging part by about 50% compared to the case of uniform temperature field, and the effect of suppressing the wall thinning by cooling the counter punch was confirmed in cross-shape forming for AZ61 as well as for AZ31B.

Figure 20 shows the optimum temperature distribution for T-shape forming of AZ61 and for cross-shape forming of both magnesium alloy tubes. For T-shape forming of AZ61 in Figure 20a, the optimum temperature distribution is different from that of AZ31B in Figure 18. The bottom temperature for AZ61 is 150 °C and 20 °C higher than that of AZ31B. On the other hand, for cross-shape forming shown in Figure 20b, in the cross-shape forming of AZ31B and AZ61, the optimum temperature distribution was the same. It was shown that the die was 250 °C and the counter punch was 200 °C, and for the temperature distribution cooling, only the bulge head was sufficient.

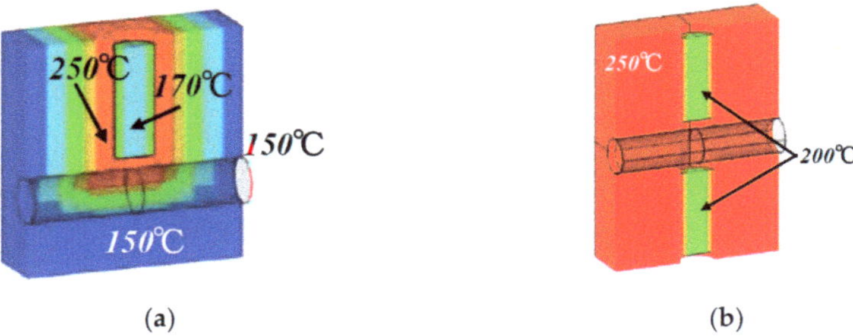

Figure 20. Optimum temperature conditions to make the wall thickness distribution more uniform. (**a**) T-shape forming for AZ61; (**b**) cross-shape forming for AZ31B and AZ61.

5. Conclusions

In this study, local heating/cooling in the warm T-shape and cross-shape forming processes was carried out for the AZ31B and AZ61 magnesium alloy tube, which is a hard-to-form material at room temperature, to improve the formability, wall thickness distribution, and formed shape accuracy. For this purpose, a new local heating and cooling apparatus was developed and adopted to achieve a high wall-thickness quality of the hydroformed T-shape and cross-shape products. Under experimental constraints with a forming time of 5 min at 250 °C or below, optimization of temperature distribution and loading path during process using fuzzy inference model previously developed by the authors was attempted. As a result, firstly, it was verified that the local heating/cooling approach to creating a non-uniform temperature field is more effective in the hydroforming process for the hard-to-form materials and enhancing the formability and making wall thickness uniform than that under a uniform temperature field. In addition, optimum symmetric and asymmetric temperature distributions for the T-shape and cross-shape forming of tubes were shown. Its effectiveness is considered to be confirmed under other conditions from the verification results obtained by the experiment and FE simulation for the T-shape forming of AZ31B.

It was suggested that the loading path and temperature distribution have important effects on the WTHF process and that these optimizations are necessary to achieve optimum process design and process control.

The influence of the loading path and the temperature condition on each other is not large, and both the optimizations can be realized independently.

Author Contributions: Conceptualization, K.-I.M.; formal analysis, T.M.; investigation, T.M.; validation, T.M.; writing—original draft preparation, T.M.; writing—review and editing, K.-I.M.; supervision, K.-I.M. All authors have read and agreed to the published version of the manuscript.

Funding: This research received no external funding.

Data Availability Statement: Not applicable.

Acknowledgments: The authors sincerely express our thanks to Tsutomu Murai and Humiyuki Nakagawa of Sankyo Tateyama Aluminum Ltd. For providing the test materials, Ryouden Kasei Ltd. for providing the thermal insulation boards, and Tetsuya Yagami and Masamitsu Suetake for their invaluable advice.

Conflicts of Interest: The authors declare no conflict of interest.

References

1. Yoshihara, S.; Mac Donald, B.; Hasegawa, T.; Kawahara, M.; Yamamoto, H. Design improvement of spin forming of magnesium alloy tubes using finite element. *J. Mater. Process. Technol.* **2004**, *153–154*, 816–820. [CrossRef]
2. Murata, M.; Kuboki, T.; Murai, T. Compression spinning of circular magnesium tube using heated roller. *J. Mater. Process. Technol.* **2005**, *162*, 540–545. [CrossRef]
3. Okamoto, A.; Naoi, H.; Kuwahara, Y. Study on Hot Bulge Forming for Tees of Magnesium Alloy Pipe Joints. In Proceedings of the TUBEHYDRO 2007 (Tube Hydroforming Technology), Harbin, China, 4–5 June 2007; pp. 121–128.
4. Manabe, K.; Fujita, K.; Tada, K. Experimental and Numerical Study on Warm Hydroforming for T-shape Joint of AZ31 Magnesium Alloy. *J. Chin. Soc. Mech. Eng.* **2010**, *31*, 284–287.
5. Manabe, K.; Ogawa, Y.; Fujita, K.; Tada, K. Wall thickness distribution in Warm hydroforming process for AZ31 Magnesium Alloy Tube. In Proceedings of the International Conference on Materials Processing Technology, Bangkok, Thailand, 5–6 January 2010; pp. 60–63.
6. Hwang, Y.M.; Wang, K.H. Study on y-shape tube hydroforming of magnesium alloys at elevated temperatures. *Int. J. Mater. Form.* **2010**, *3*, 175–178. [CrossRef]
7. He, Z.; Yuan, S.; Liu, G.; Wu, J.; Cha, W. Formability testing of AZ31B magnesium alloy tube at elevated temperature. *J. Mater. Process. Technol.* **2010**, *210*, 877–884. [CrossRef]
8. Furushima, T.; Ikeda, T.; Manabe, K. Deformation and Heat Transfer Analysis for High Speed Dieless Drawing of AZ31 Magnesium Alloy Tubes. *Adv. Mater. Res.* **2012**, *418–420*, 1036–1039. [CrossRef]
9. Furushima, T.; Manabe, K. Workability of AZ31 Magnesium Alloy Tubes in Dieless Drawing Process. *Steel Res. Int.* **2012**, 851–854.
10. Du, P.; Furusawa, S.; Furushima, T. Microstructure and performance of biodegradable magnesium alloy tubes fabricated by local-heating-assisted dieless drawing. *J. Magnes. Alloys* **2020**, *8*, 614–623. [CrossRef]
11. Milenin, A.; Kustra, P.; Furushima, T.; Du, P.; Němeček, J. Design of the laser dieless drawing process of tubes from magnesium alloy using FEM model. *J. Mater. Process. Technol.* **2018**, *262*, 65–74. [CrossRef]
12. Keigler, M.; Bauer, H.; Harrison, D.; Silvia, A. Enhancing the formability of aluminum components via temperature controlled hydroforming. *J. Mater. Process. Technol.* **2005**, *167*, 363–370. [CrossRef]
13. Dong, G.L.; Bi, J.; Du, B.; Chen, X.H.; Zhao, C.C. Research on AA6061 tubular components prepared by combined technology of heat treatment and internal high pressure forming. *J. Mater. Process. Technol.* **2017**, *242*, 126–138. [CrossRef]
14. Maeno, T.; Mori, K.; Fujimoto, K. Development of the hot gas bulging process for aluminum alloy tube using resistance heating. *Key Eng. Mater.* **2009**, *410–411*, 315–323. [CrossRef]
15. Maeno, T.; Mori, K.; Fujimoto, K. Hot gas bulging of sealed aluminum alloy tube using resistance heating. *Manuf. Rev.* **2014**, *1*, 5.
16. Maeno, T.; Mori, K.; Adachi, K. Gas forming of ultra-high strength steel hollow part using air filled into sealed tube and resistance heating. *J. Mater. Process. Technol.* **2014**, *214*, 97–105. [CrossRef]

17. Paul, A.; Strano, M. The influence of process variables on the gas forming and press hardening of steel tubes. *J. Mater. Process. Technol.* **2016**, *228*, 160–169. [CrossRef]
18. Schlemmer, K.L.; Osman, F.H. Differential heating forming of solid and bi-metallic hollow parts. *J. Mater. Process. Technol.* **2005**, *162–163*, 564–569. [CrossRef]
19. Furushima, T.; Hung, N.Q.; Manabe, K.; Sasaki, O. Development of semi-dieless metal bellows forming process. *J. Mater. Process. Technol.* **2013**, *213*, 1406–1411. [CrossRef]
20. Suproadi, S.; Manabe, K. Enhancement of dimensional accuracy of dieless tube-drawing process with vision-based fuzzy control. *J. Mater. Process. Technol.* **2013**, *213*, 905–912. [CrossRef]
21. Asao, H.; Okada, K.; Watanabe, M.; Yonemura, H.; Matsumoto, T.; Umehara, N. Analysis in workability of pipe bending Using High Frequency induction heating. In Proceedings of the 24th International Machine Tool Design and Research Conference (M.T.D.R.), 31 August–1 September 1983; pp. 97–104.
22. Tomizawa, A.; Shimada, N.; Kubota, H.; Okada, N.; Hara, M.; Kuwayama, S. Forming characteristics in three-dimensional hot bending and direct quench process - Development of three-dimensional hot bending and direct quench technology. *J. JSTP* **2015**, *56*, 961–966. [CrossRef]
23. Liu, G.; Zhang, W.; He, Z.; Yuan, S.; Lin, Z. Warm hydroforming of magnesium alloy tube with large expansion ratio within non-uniform temperature field. *Trans. Nonferrous Met. Soc. China* **2012**, *22*, s408–s415. [CrossRef]
24. Yoshihara, S.; Nishimura, H.; Yamamoto, H.; Manabe, K. Formability enhancement in magnesium alloy stamping using a local heating and cooling technique: Circular cup deep drawing process. *J. Mater. Process. Technol.* **2003**, *142*, 579–585. [CrossRef]
25. Manabe, K.; Morishima, T.; Ogawa, Y.; Tada, K.; Murai, T.; Nakagawa, H. Warm Hydroforming Process with Non-Uniform Heating for AZ31 Magnesium Alloy Tube. *Mater. Sci. Forum* **2010**, *654*, 739–742. [CrossRef]
26. Manabe, K.; Morishima, T.; Tada, K.; Mac Donald, B.J. Deformation Behavior of AZ31 Magnesium Alloy Tube in Warm Hydroforming Process with Non-uniform Heating. In Proceeding of the 5th International Conference on Tube Hydroforming (Tube Hydroforming Technology 2011), Noboribetsu, Japan, 24–27 July 2011; pp. 84–87.
27. Manabe, K.; Suetake, M.; Koyama, H.; Yang, M. Hydroforming process optimization of aluminum alloy tube using intelligent control technique. *Int J Mach Tools Manu.* **2006**, *46*, 1207–1211. [CrossRef]
28. Yoshitomi, Y.; Kamohara, H.; Shimaguti, T.; Asao, H.; Nomura, H. An evaluation method of the applicability of the bulge forming for circular tubes. *J. JSTP* **1987**, *28–316*, 432–437.

Article

Description of the Expansion of a Two-Layer Tube: An Analytic Plane-Strain Solution for Arbitrary Pressure-Independent Yield Criterion and Hardening Law

Sergei Alexandrov [1,2], Elena Lyamina [3] and Lihui Lang [1,*]

1 School of Mechanical Engineering and Automation, Beihang University, Beijing 100191, China; sergei_alexandrov@spartak.ru
2 Faculty of Materials Science and Metallurgy Engineering, Federal State Autonomous Educational Institution of Higher Education, South Ural State University (National Research University), 454080 Chelyabinsk, Russia
3 Ishlinsky Institute for Problems in Mechanics RAS, 119526 Moscow, Russia; lyamina@inbox.ru
* Correspondence: lang@buaa.edu.cn; Tel.: +86-10-82316821

Abstract: The main objective of the present paper is to provide a simple analytical solution for describing the expansion of a two-layer tube under plane-strain conditions for its subsequent use in the preliminary design of hydroforming processes. Each layer's constitutive equations are an arbitrary pressure-independent yield criterion, its associated plastic flow rule, and an arbitrary hardening law. The elastic portion of strain is neglected. The method of solution is based on two transformations of space variables. Firstly, a Lagrangian coordinate is introduced instead of the Eulerian radial coordinate. Then, the Lagrangian coordinate is replaced with the equivalent strain. The solution reduces to ordinary integrals that, in general, should be evaluated numerically. However, for two hardening laws of practical importance, these integrals are expressed in terms of special functions. Three geometric parameters for the initial configuration, a constitutive parameter, and two arbitrary functions classify the boundary value problem. Therefore, a detailed parametric analysis of the solution is not feasible. The illustrative example demonstrates the effect of the outer layer's thickness on the pressure applied to the inner radius of the tube.

Keywords: tube hydroforming; two-layer tube; rigid plasticity; arbitrary yield criterion; arbitrary hardening law; analytic solution

Citation: Alexandrov, S.; Lyamina, E.; Lang, L. Description of the Expansion of a Two-Layer Tube: An Analytic Plane-Strain Solution for Arbitrary Pressure-Independent Yield Criterion and Hardening Law. *Metals* **2021**, *11*, 793. https://doi.org/10.3390/met11050793

Academic Editor: Badis Haddag

Received: 9 April 2021
Accepted: 10 May 2021
Published: 14 May 2021

1. Introduction

Tube hydroforming is capable of replacing several traditional manufacturing processes [1]. The products of tube hydroforming processes are widely used in different sectors of the industry [2–6], including the production of micro-parts [7,8]. Several comprehensive reviews on hydroforming technologies are available [9–12], where the advantages and disadvantages of these technologies are discussed in detail.

An important direction of research is to design hydroforming processes. Several methods based on sophisticated numerical modeling have been proposed [1,5,13]. However, it is known from other branches of the mechanics of metal forming processes that simplified methods can be very useful for the preliminary design of metal forming processes. In particular, such methods can provide a reliable initial guess for more sophisticated methods. An example of such simplified methods is the theory of ideal flows [14–16]. The present paper provides a simple analytic solution for a two-layer tube hydroforming process. An advantage of this solution is that it is valid for any pressure-independent yield criterion and any hardening law. Therefore, the solution can be used for parametric analysis and preliminary design of the tube hydroforming process for a large class of materials.

Hydroforming of multi-layer materials is a widely used hydroforming process. The hydroforming technology for producing double-layer spherical vessels was introduced in [17]. This paper includes both experimental and theoretical results. The latter are based on the elastic/plastic finite element method. Paper [18] has applied hydroforming for developing the discrete layer forming of a multi-layer tube. This paper uses an analytic method for finding an optimal loading path to prevent defects in the course of forming. This analytic method has been justified by experiments. The hydraulic bulging test for multi-layer sheets was proposed in [19]. Its theoretical treatment has been based on the finite element method. Various technological aspects of the hydroforming process of multi-layer sheets have been discussed in recent publications [20,21].

The present paper focuses on a new theoretical method for describing two-layer tube hydroforming under plane-strain conditions. It is assumed that each layer is rigid/plastic. No restriction is imposed on the isotropic pressure-independent yield criterion and hardening law. The general solution is analytic. A numerical treatment may be needed for evaluating ordinary integrals.

The success of the method proposed is based on the use of advantageous space variables. In particular, the original formulation of the boundary value problem in Eulerian coordinates is first transformed in the formulation in Lagrangian coordinates. Then, the equivalent strain is used as an independent space variable. In the case of elastic/plastic problems, this change of independent variables has proved advantageous for a class of problems [22–24].

A practical aspect of the solution is that simple solutions are essential for estimating the required forming pressure in tube hydroforming of monometallic and clad tubes [25,26]. Moreover, the solution can be used as a benchmark problem for verifying numerical codes, which is a necessary step before using such codes [27,28].

2. Statement of the Problem

Consider a two-layer tube of initial outer radius R_b and inner radius R_a, subjected to uniform pressure P over the inner radius. The outer radius of the inner layer and the inner radius of the outer layer is R_c (Figure 1a). Each layer is rigid/plastic or hardening. The state of strain is plane. It is natural to adopt a cylindrical coordinate system (r, θ, z), the z-axis of which coincides with the tube's axis of symmetry. Then, the solution is independent of θ. In particular, after any amount of deformation, the outer and inner radii of the tube are b and a, respectively. The radius of the contact surface between the layers is c (Figure 1b). Let σ_r, σ_θ, and σ_z be the normal stresses referred to the cylindrical coordinates. These stresses are the principal stresses. The circumferential velocity vanishes and the radial velocity is denoted as u.

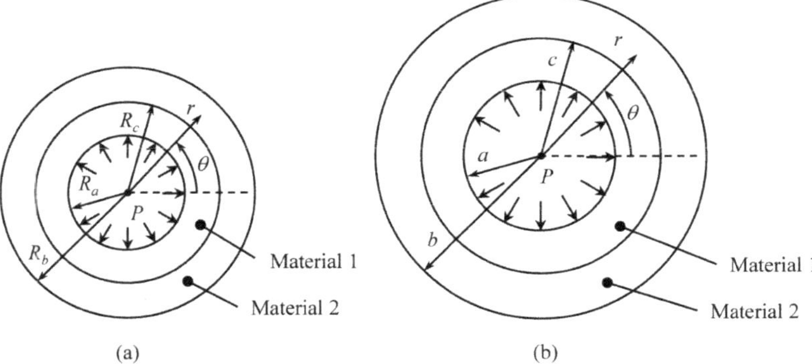

Figure 1. Schematic diagram of the process: (**a**) initial configuration, (**b**) intermediate and final configurations.

In the case under consideration, any pressure-independent yield criterion reduces to

$$|\sigma_r - \sigma_\theta| = \chi\sigma_{eq}. \tag{1}$$

Here, σ_{eq} is a measure of the equivalent stress and χ is constant. Let ξ_r and ξ_θ be the non-zero normal strain rates referred to the cylindrical coordinates. The flow rule associated with the yield criterion (1) reduces to the equation of incompressibility

$$\xi_r + \xi_\theta = 0 \tag{2}$$

and the inequality

$$\xi_r(\sigma_r - \sigma_\theta) > 0. \tag{3}$$

It is assumed that σ_{eq} is a function of the equivalent strain, ε_{eq}. The latter is determined from the equation:

$$\frac{d\varepsilon_{eq}}{dt} = \xi_{eq}. \tag{4}$$

Here, t is the time, d/dt denotes the convected derivative, and ξ_{eq} is the equivalent strain rate. In the case under consideration,

$$\xi_{eq} = \chi|\xi_r| = \chi|\xi_\theta|. \tag{5}$$

Assuming that σ_{eq} is equal to the axial stress in the uniaxial tension test, the von Mises yield criterion follows from Equation (1) at $\chi = 2/\sqrt{3}$ and Tresca's yield criterion at $\chi = 1$.

It is convenient to represent σ_{eq} as $\sigma_{eq} = \sigma_0\Phi(\varepsilon_{eq})$, where σ_0 is the value of the equivalent stress at $\varepsilon_{eq} = 0$ and $\Phi(\varepsilon_{eq})$ is an arbitrary function of its argument, satisfying the conditions $\Phi = 1$ at $\varepsilon_{eq} = 0$ and $d\Phi/d\varepsilon_{eq} \geq 0$ for all ε_{eq}. Then, Equation (1) becomes

$$|\sigma_r - \sigma_\theta| = \chi\sigma_0\Phi(\varepsilon_{eq}). \tag{6}$$

The stress boundary condition is

$$\sigma_r = 0 \tag{7}$$

for $r = b$. The pressure P is determined from the equation

$$P = -\sigma_r \tag{8}$$

where the radial stress is understood to be calculated at $r = a$.

3. General Solution

Since $\xi_r = \partial u/\partial r$ and $\xi_\theta = u/r$, Equation (2) can be immediately integrated to give

$$u = \frac{Ua}{r}. \tag{9}$$

Here, U is the radial velocity at $r = a$. Since the material model is rate-independent, the magnitude of U is immaterial. By definition, $\partial r/\partial t = u$. This equation and Equation (9) combine to give

$$\frac{\partial r}{\partial a} = \frac{a}{r}. \tag{10}$$

It has been taken into account here that $da/dt = U$. Equation (10) can be immediately integrated to give

$$r = \sqrt{R^2 + a^2 - R_a^2}. \tag{11}$$

Here, R is the Lagrangian coordinate such that $r = R$ at $a = R_a$. One can solve Equation (11) for R to obtain

$$R = \sqrt{r^2 - a^2 + R_a^2}. \tag{12}$$

It follows from Equation (10) that

$$\xi_r = \frac{\partial u}{\partial r} = -\frac{Ua}{r^2}. \tag{13}$$

One substitutes Equation (13) into Equation (5) to arrive at

$$\xi_{eq} = \chi \frac{Ua}{r^2}. \tag{14}$$

Equations (9) and (14) combine to give

$$\xi_{eq} = \chi \frac{Ua}{(R^2 + a^2 - R_a^2)}. \tag{15}$$

In Lagrangian coordinates, $d/dt = \partial/\partial t$. Therefore, Equation (15) becomes

$$\frac{\partial \varepsilon_{eq}}{\partial a} = \chi \frac{a}{(R^2 + a^2 - R_a^2)}. \tag{16}$$

It has been taken into account here that $da/dt = U$. One integrates Equation (16) to obtain

$$\varepsilon_{eq} = \frac{\chi}{2} \ln\left(\frac{R^2 + a^2 - R_a^2}{R^2}\right). \tag{17}$$

This solution satisfies the initial condition that $\varepsilon_{eq} = 0$ at $a = R_a$.

The only stress equilibrium equation that is not identically satisfied in the cylindrical coordinates is

$$\frac{\partial \sigma_r}{\partial r} + \frac{\sigma_r - \sigma_\theta}{r} = 0. \tag{18}$$

It is more convenient to rewrite this equation in the Lagrangian coordinates. It follows from Equations (11) and (12) that

$$\frac{\partial R}{\partial r} = \frac{r}{R}. \tag{19}$$

Then, Equation (18) becomes

$$\frac{\partial \sigma_r}{\partial R}\frac{\partial R}{\partial r} + \frac{\sigma_r - \sigma_\theta}{r} = \frac{\partial \sigma_r}{\partial R}\frac{r}{R} + \frac{\sigma_r - \sigma_\theta}{r} = 0. \tag{20}$$

Equations (3) and (13) show that $\sigma_r - \sigma_\theta < 0$. Then, using Equations (6) and (11), one transforms Equation (20) to

$$\frac{\partial \sigma_r}{\sigma_0 \partial R} = \frac{\chi R \Phi(\varepsilon_{eq})}{(R^2 + a^2 - R_a^2)}. \tag{21}$$

It is seen from the general structure of this equation that it is advantageous to use the equivalent strain as the independent variable instead of R. Using Equation (17), one replaces the differentiation with respect to R in Equation (21) with the differentiation with respect to ε_{eq} to attain

$$\frac{\partial \sigma_r}{\sigma_0 \partial \varepsilon_{eq}} = \frac{R^2 \Phi(\varepsilon_{eq})}{(R_a^2 - a^2)}. \tag{22}$$

Moreover, Equation (17) can be solved for R to result in

$$R^2 = \frac{a^2 - R_a^2}{\exp(2\varepsilon_{eq}/\chi) - 1}. \tag{23}$$

Equations (20) and (23) combine to give

$$\frac{\partial \sigma_r}{\sigma_0 \partial \varepsilon_{eq}} = \frac{\Phi(\varepsilon_{eq})}{1 - \exp(2\varepsilon_{eq}/\chi)}. \tag{24}$$

The general solution of this equation is

$$\frac{\sigma_r}{\sigma_0} = \int_{\varepsilon_0}^{\varepsilon_{eq}} \frac{\Phi(\omega)}{1 - \exp(2\omega/\chi)} d\omega + \frac{s_0}{\sigma_0}. \tag{25}$$

Here, ω is a dummy variable of integration, and s_0 is the value of σ_r at $\varepsilon_{eq} = \varepsilon_0$. The circumferential stress is determined from Equations (6) and (25) as

$$\frac{\sigma_\theta}{\sigma_0} = \int_{\varepsilon_0}^{\varepsilon_{eq}} \frac{\Phi(\omega)}{1 - \exp(2\omega/\chi)} d\omega + \chi\Phi(\varepsilon_{eq}) + \frac{s_0}{\sigma_0}. \tag{26}$$

Since ε_{eq} is independent of R at the initial instant, $\partial\varepsilon_{eq}/\partial R = 0$ at $a = R_a$. Therefore, the transformation of Equation (21) into Equation (22) is not justified at the initial instant. As a result, one cannot put $\varepsilon_0 = 0$ in Equation (25) and the following equations that involve this quantity. The solution at the initial instant is not required in the present paper. If one requires such a solution, then it is necessary to return to Equation (21). It is then necessary to put $\varepsilon_{eq} = 0$, which is equivalent to putting $\Phi(\varepsilon_{eq}) = 1$. The resulting equation can be immediately integrated in terms of elementary functions to provide the radial distribution of σ_r.

4. Expansion of a Two-Layer Tube

Throughout this paper's remainder, subscript 1 denotes quantities related to the inner layer and subscript 2 to the outer layer (Figure 1).

Let ε_a, ε_b, and ε_c be the values of the equivalent strain at $R = R_a$, $R = R_b$, and $R = R_c$, respectively. Then, it follows from Equation (17) that

$$\varepsilon_a = \chi \ln\left(\frac{a}{R_a}\right), \quad \varepsilon_b = \frac{\chi}{2} \ln\left(\frac{R_b^2 + a^2 - R_a^2}{R_b^2}\right), \text{ and } \varepsilon_c = \frac{\chi}{2} \ln\left(\frac{R_c^2 + a^2 - R_a^2}{R_c^2}\right). \tag{27}$$

The solution to Equation (25) satisfies the boundary condition of Equation (7) if $s = s_0$ and $\varepsilon_0 = \varepsilon_b$. Then,

$$\frac{\sigma_r^{(2)}}{\sigma_0^{(2)}} = \int_{\varepsilon_b}^{\varepsilon_{eq}} \frac{\Phi^{(2)}(\omega)}{1 - \exp(2\omega/\chi)} d\omega. \tag{28}$$

Let σ_c be the value of the radial stress at $R = R_c$. It follows from Equation (28) that

$$\frac{\sigma_c}{\sigma_0^{(2)}} = \int_{\varepsilon_b}^{\varepsilon_c} \frac{\Phi^{(2)}(\omega)}{1 - \exp(2\omega/\chi)} d\omega. \tag{29}$$

The radial stress must be continuous across the bi-material interface. Therefore, $\sigma_r^{(1)} = \sigma_c$ at $R = R_c$. The solution to Equation (25) satisfying this condition is

$$\frac{\sigma_r^{(1)}}{\sigma_0^{(2)}} = k \int_{\varepsilon_c}^{\varepsilon_{eq}} \frac{\Phi^{(1)}(\omega)}{1 - \exp(2\omega/\chi)} d\omega + \frac{\sigma_c}{\sigma_0^{(2)}} \tag{30}$$

where $k = \sigma_0^{(1)}/\sigma_0^{(2)}$. Equations (29) and (30) combine to give

$$\frac{\sigma_r^{(1)}}{\sigma_0^{(2)}} = k\int_{\varepsilon_c}^{\varepsilon_{eq}} \frac{\Phi^{(1)}(\omega)}{1-\exp(2\omega/\chi)}d\omega + \int_{\varepsilon_b}^{\varepsilon_c} \frac{\Phi^{(2)}(\omega)}{1-\exp(2\omega/\chi)}d\omega \tag{31}$$

The pressure of the inner radius is determined from Equations (8) and (31) as

$$\frac{P}{\sigma_0^{(2)}} = -k\int_{\varepsilon_c}^{\varepsilon_{eq}} \frac{\Phi^{(1)}(\omega)}{1-\exp(2\omega/\chi)}d\omega - \int_{\varepsilon_b}^{\varepsilon_c} \frac{\Phi^{(2)}(\omega)}{1-\exp(2\omega/\chi)}d\omega \tag{32}$$

Together with Equations (17) and (26), the solution above supplies the dependence of the radial and circumferential stresses on the Lagrangian coordinate in parametric form, with the equivalent strain being the parameter. One can use Equation (12) to find the distribution of these stresses along the r-axis.

In general, the integral in Equation (25) should be evaluated numerically. However, two hardening laws of practical importance allow for the evaluation of this integral in terms of special functions. In the case of linear hardening, $\Phi(\varepsilon_{eq}) = 1 + \beta\varepsilon_{eq}$. Using this function, one finds

$$\int_{\varepsilon_0}^{\varepsilon_{eq}} \frac{\Phi(\omega)}{1-\exp(2\omega/\chi)}d\omega = \frac{\chi}{4}\left\{ \begin{array}{c} \beta\chi \mathrm{Li}_2\left[\exp\left(-\frac{2\varepsilon_{eq}}{\chi}\right)\right] - \beta\chi \mathrm{Li}_2\left[\exp\left(-\frac{2\varepsilon_0}{\chi}\right)\right] - \\ 2(1+\beta\varepsilon_{eq})\ln\left[1-\exp\left(-\frac{2\varepsilon_{eq}}{\chi}\right)\right] + 2(1+\beta\varepsilon_0)\ln\left[1-\exp\left(-\frac{2\varepsilon_0}{\chi}\right)\right] \end{array} \right\}. \tag{33}$$

Here, $\mathrm{Li}_2(\varepsilon_{eq})$ is the dilogarithm function. In the case of Voce's hardening law, $\Phi(\varepsilon_{eq}) = 1 + (\beta - 1)[1 - \exp(-n\varepsilon_{eq})]$. Using this function, one finds

$$\int_{\varepsilon_0}^{\varepsilon_{eq}} \frac{\Phi(\omega)}{1-\exp(2\omega/\chi)}d\omega = \beta(\varepsilon_{eq} - \varepsilon_0) - \frac{1}{2}\beta\chi\ln\left[\frac{1-\exp(2\varepsilon_{eq}/\chi)}{1-\exp(2\varepsilon_0/\chi)}\right] + \frac{(\beta-1)}{n}\left[\exp(-n\varepsilon_{eq})\Lambda(\varepsilon_{eq}) - \exp(-n\varepsilon_0)\Lambda(\varepsilon_0)\right]. \tag{34}$$

Here, $\Lambda(x) \equiv 2F1[1, -n\chi/2, 1 - n\chi/2, exp(2\varepsilon eq/\chi)]$ is the hypergeometric function.

5. Experimental Verification and Illustrative Examples

Three geometric parameters for the initial configuration, the value of χ, and two arbitrary functions classify the boundary value problem. Therefore, a detailed parametric analysis of the solution is not feasible. However, the solution is very simple for any given set of initial data. The numerical results below focus on the maximum value of the internal pressure, $P_{\max}$. Simple solutions for this quantity are essential for estimating the required forming pressure in tube hydroforming of monometallic and clad tubes [25,26]. The solution above certainly belongs to this class of solutions.

Experimental data on hydroforming of clad tubes are presented in [26]. The outer tube is made of aluminum alloy A1060-O and the inner tube from copper alloy C1020TS-O. The initial outer radius of the clad tube is 20 mm, and the initial total thickness of the two layers is 1.5 mm. The experiments have been carried out for several ratios of the outer tube's thickness to the inner tube's thickness. The stress–strain curves of the copper and aluminum alloys have been represented as

$$\sigma_{eq} = 518\varepsilon_{eq}^{0.45} \quad \text{and} \quad \sigma_{eq} = 144\varepsilon_{eq}^{0.25}, \tag{35}$$

respectively. Here, the equivalent stress is expressed in MPa. The equations in Equation (35) are not compatible with the restrictions imposed on the function $\Phi(\varepsilon_{eq})$ before Equation (6). Therefore, Equation (35) is replaced with Ludwik's hardening law

$$\sigma_{eq} = 31.3 + 496\varepsilon_{eq}^{0.507} \quad \text{and} \quad \sigma_{eq} = 32.2 + 115.7\varepsilon_{eq}^{0.37}. \tag{36}$$

Here, the first equation corresponds to the copper alloy and the second to the aluminum alloy. The difference between the laws in Equations (35) and (36) is negligible, and is revealed only at small strains (Figure 2). The solid curves correspond to Equation (35) and the broken curves to Equation (36). Using Equation (36) and assuming that $\chi = 2/\sqrt{3}$, one can find the function involved in Equation (6) as

$$\Phi^{(1)}(\varepsilon_{eq}) = 1 + 15.85\varepsilon_{eq}^{0.507} \quad \text{and} \quad \Phi^{(2)}(\varepsilon_{eq}) = 1 + 3.6\varepsilon_{eq}^{0.37}. \tag{37}$$

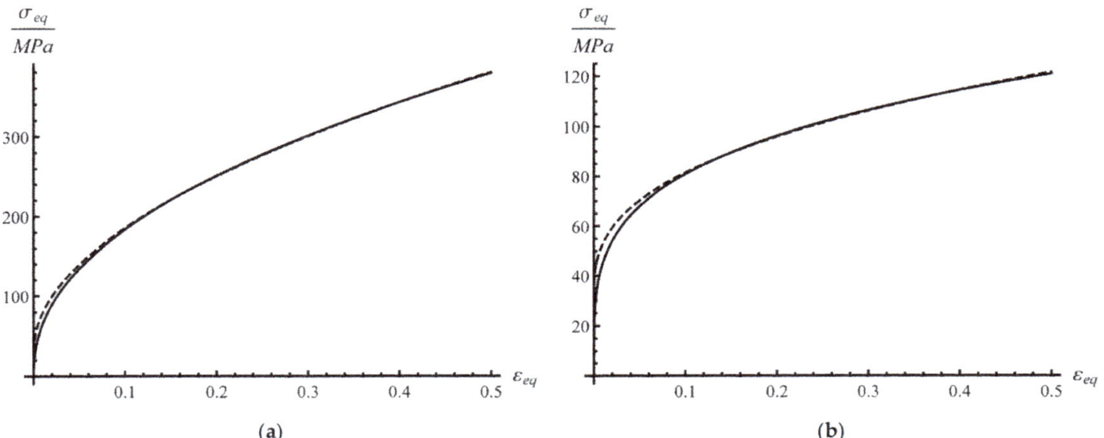

Figure 2. Stress–strain curves according to Equations (35) and (36): (**a**) copper alloy, (**b**) aluminum alloy. The solid curves correspond to Equation (35) and the broken curves to Equation (36).

Moreover, $\sigma_0^{(1)} = 31.3$ MPa and $\sigma_0^{(2)} = 32.2$ MPa. The volume fraction of the copper alloy is determined as

$$\lambda = \left(\frac{R_c^2 - R_a^2}{R_b^2 - R_a^2}\right) \times 100\%. \tag{38}$$

Since R_a and R_b are fixed in [10], λ is controlled by R_c. It is seen from the experimental data depicted in Figure 10 in [10] that $P_{\max}$ is practically a linear function of λ. This function can be interpolated as

$$P_{\max} = 6 + 9.5\left(\frac{\lambda}{100\%}\right). \tag{39}$$

Here, $P_{\max}$ is expressed in MPa. One can substitute Equation (37) into Equation (32) for calculating P as a function of a. A local maximum of this function is found numerically. A comparison of the experimental data from [26] and the theoretical solution is shown in Figure 3. The solid line represents Equation (39) and the broken line is the theoretical solution found using Equation (32). It is seen from this figure that the theoretical solution is quite accurate.

As another example, a tube made of Al–Li alloy (Material 1 in Figure 1) and 5A06 aluminum alloy (Material 2 in Figure 1) is considered. Paper [20] provides the mechanical properties of these materials. In our nomenclature, $\sigma_0^{(1)} = 77.7$ MPa and $\sigma_0^{(2)} = 155$ MPa. Moreover,

$$\Phi^{(1)} = 1 + 5.08\varepsilon_{eq}^{0.28} \quad \text{and} \quad \Phi^{(1)} = 1 + 4.4\varepsilon_{eq}^{0.3}. \tag{40}$$

The initial configuration is determined by

$$\frac{R_c}{R_a} = 1 + \frac{t_1}{R_a} \quad \text{and} \quad \frac{R_b}{R_a} = 1 + \frac{t_1 + t_2}{R_a}. \tag{41}$$

Here, t_1 is the initial thickness of the inner layer and t_2 is the initial thickness of the outer layer. In all calculations, $t_1 = 1.8\,\text{mm}$, $R_a = 100\,\text{mm}$, and $\chi = 2/\sqrt{3}$. Figure 4 depicts the variation of the inner pressure with the inner radius of the tube for three values of t_2. The value of P increases with t_2. All three curves attain a local maximum at a certain value of a. The value of $P_{\max}$ can be found numerically with no difficulty.

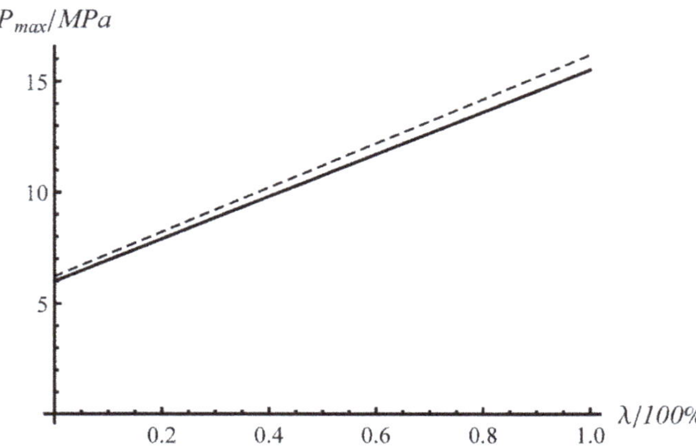

Figure 3. Comparison of the experimental data from [26], and the theoretical solution resulting from Equation (32). The solid line represents the experimental data and the broken line the theoretical solution found using Equation (32).

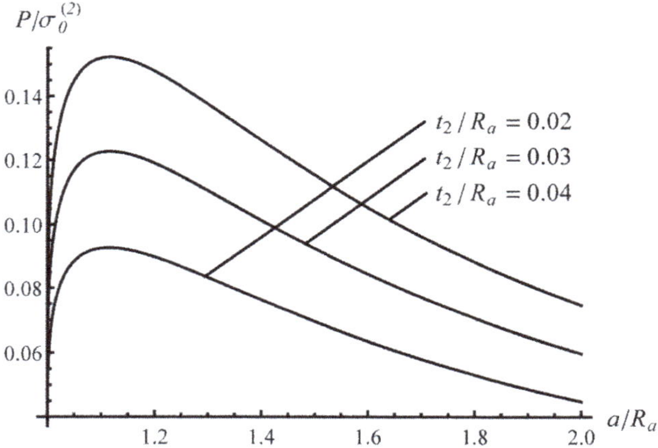

Figure 4. Dependence of the internal pressure on the inner radius for several initial thicknesses of the outer layer.

6. Conclusions

The solution presented describes the plane-strain expansion of a two-layer tube for tube hydroforming applications. The solution has been reduced to evaluating the integral in Equation (25). If the maximum value of the pressure is required, then a numerical technique is necessary. No restriction on the pressure-independent yield criterion and hardening law has been applied. Therefore, the solution is advantageous for the preliminary design of the tube hydroforming process. It is of special importance because many parameters and functions classify the boundary value problem. The solution's accuracy is verified by comparing it to the experimental data provided in [26] (Figure 3). As another example, the effect of the thickness of the outer layer on the pressure applied over the inner radius of the tube has been investigated using the data provided in [20] (Figure 4). It is seen from this figure that its magnitude attains a local maximum at a certain stage of the process. It is probably because of competition between the change of geometric parameters and hardening. Simple solutions for P's maximum value are essential for estimating the required forming pressure in tube hydroforming of clad tubes [26].

The solution is exact. Therefore, it can be used for validating numerical solutions, which is a necessary step before their use in applications [27].

The solution in Section 3 is valid for any value of a. Therefore, it can be used to analyze and design hydroforging processes introduced in [29].

The solution in Section 4 is for two-layer tubes. The line of reasoning in this section shows that extending this solution to multi-layer tubes requires adding equations similar to Equation (31) for each interface, which is a relatively simple task.

Author Contributions: Formal analysis, S.A. and E.L.; supervision, L.L.; conceptualization, L.L.; writing—original draft, S.A. and E.L. All authors have read and agreed to the published version of the manuscript.

Funding: This research was made possible by the project AAAA-A20-120011690136-2 (Russian Ministry of Science and Education).

Institutional Review Board Statement: Not applicable.

Informed Consent Statement: Not applicable.

Data Availability Statement: Not applicable.

Conflicts of Interest: The authors declare no conflict of interest.

Nomenclature

a	inner radius of the tube after any amount of deformation
b	outer radius of the tube after any amount of deformation
c	radius of the interface between the layers after any amount of deformation
k	parameter introduced after Equation (30)
P	pressure over the inner radius of the tube
P_{max}	maximum inner pressure in the course of the hydroforming process
R	Lagrangian coordinate
R_a	Lagrangian coordinate of the inner radius of the tube
R_b	Lagrangian coordinate of the outer radius of the tube
R_c	Lagrangian coordinate of the interface between the layers
(r, θ, z)	cylindrical coordinate system
t	Time
t_1	initial thickness of the inner layer
t_2	initial thickness of the outer layer
U	radial velocity at the inner radius of the tube
u	radial velocity
ε_{eq}	equivalent strain
ξ_{eq}	equivalent strain rate
ξ_r and ξ_θ	radial and circumferential strain rates, respectively
σ_{eq}	equivalent stress
σ_r, σ_θ and σ_z	radial, circumferential, and axial stresses, respectively
σ_0	initial yield stress in tension
Φ	arbitrary function of the equivalent strain introduced in Equation (6)
χ	parameter introduced in Equation (1)

References

1. Fiorentino, A.; Ginestra, P.S.; Attanasio, A.; Ceretti, E. Numerical optimization of the blank dimensions in tube hydroforming using line-search and bisection methods. *Materials* **2020**, *13*, 945. [CrossRef] [PubMed]
2. Kocańda, A.; Sadłowska, H. Automotive component development by means of hydroforming. *Archiv. Civ. Mech. Eng.* **2008**, *8*, 55–72. [CrossRef]
3. Han, S.W.; Woo, Y.Y.; Hwang, T.W.; Oh, I.Y.; Moon, Y.H. Tailor layered tube hydroforming for fabricating tubular parts with dissimilar thickness. *Int. J. Mach. Tools Manuf.* **2018**, *138*, 51–65. [CrossRef]
4. Saboori, M.; Gholipour, J.; Champliaud, H.; Wanjara, P.; Gakwaya, A.; Savoie, J. Prediction of burst pressure in multistage tube hydroforming of aerospace alloys. *ASME J. Eng. Gas Turbines Power* **2016**, *138*, 082101. [CrossRef] [PubMed]
5. Sadłowska, H.; Kochański, A.; Czapla, M. Application of the numerical model to design the geometry of a unit tool in the innovative rth hydroforming technology. *Materials* **2020**, *13*, 5427. [CrossRef] [PubMed]
6. *Micro-Manufacturing Technologies and Their Applications: A Theoretical and Practical Guide*, 1st ed.; Fassi, I.; Shipley, D. (Eds.) Springer International Publishing: Cham, Switzerland, 2017; p. 301. [CrossRef]
7. Ngaile, G.; Lowrie, J. New micro tube hydroforming system based on floating die assembly concept. *ASME J. Micro Nano-Manuf.* **2014**, *2*, 41004. [CrossRef]
8. Hartl, H.; Schiefer, H.; Chlynin, A. Evaluation of experimental and numerical investigations into micro-hydroforming of platinum tubes for an industrial application. *Manuf. Rev.* **2014**, *1*, 17. [CrossRef]
9. Koç, M.; Altan, T. Overall review of the tube hydroforming (THF) technology. *J. Mater. Process. Technol.* **2001**, *108*, 384–393. [CrossRef]
10. Bell, C.; Corney, J.; Zuelli, N.; Savings, D. A state of the art review of hydroforming technology: Its applications, research areas, history, and future in manufacturing. *Int. J. Mater. Form.* **2019**, *13*, 789–828. [CrossRef]
11. Hartl, C. Review on advances in metal micro-tube forming. *Metals* **2019**, *9*, 542. [CrossRef]
12. Reddy, P.V.; Reddy, B.V.; Ramulu, P.J. Evolution of hydroforming technologies and its applications—A review. *J. Adv. Manuf. Syst.* **2020**, *19*, 737–780. [CrossRef]
13. Hwang, Y.-M.; Pham, H.-N.; Tsui, H.-S.R. Investigation of punch shape and loading path design in hydro-flanging processes of aluminum alloy tubes. *Metals* **2021**, *11*, 636. [CrossRef]
14. Chung, K.; Richmond, O. Ideal forming—II. Sheet forming with optimum deformation. *Int. J. Mech. Sci.* **1992**, *34*, 617–633. [CrossRef]
15. Barlat, F.; Chung, K.; Richmond, O. Anisotropic plastic potentials for polycrystals and application to the design of optimum blank shapes in sheet forming. *Metall. Mater. Trans. A* **1994**, *25*, 1209–1216. [CrossRef]
16. Chung, K.; Alexandrov, S. Ideal flow in plasticity. *ASME. Appl. Mech. Rev.* **2007**, *60*, 316–335. [CrossRef]
17. Zhang, S.H.; Wang, Z.R. Research into the dieless hydroforming of double layer spherical vessels. *Int. J. Pres. Ves. Pip.* **1994**, *60*, 145–149. [CrossRef]
18. Kim, S.Y.; Joo, B.D.; Shin, S.; Van Tyne, C.J.; Moon, Y.H. Discrete layer hydroforming of three-layered tubes. *Int. J. Mach. Tools Manuf.* **2013**, *68*, 56–62. [CrossRef]
19. Liu, S.; Lang, L.; Guan, S.; Zeng, Y. Investigation into composites property effect on the forming limits of multi-layer hybrid sheets using hydroforming technology. *Appl. Compos. Mater.* **2019**, *26*, 205–217. [CrossRef]
20. Zhou, B.J.; Xu, Y.C.; Zhang, Z.C. Research on the selection principle of upper sheet in double-layer sheets hydroforming. *Int. J. Adv. Manuf. Technol.* **2020**, *109*, 1663–1669. [CrossRef]
21. Blala, H.; Lang, L.; Khan, S.; Li, L.; Alexandrov, S. An analysis of process parameters in the hydroforming of a hemispherical dome made of fiber metal laminate. *Appl. Compos. Mater.* **2021**. [CrossRef]
22. Durban, D.; Baruch, M. Analysis of an elasto-plastic thick walled sphere loaded by internal and external pressure. *Int. J. Non-Linear Mech.* **1997**, *12*, 9–21. [CrossRef]
23. Alexandrov, S.; Hwang, Y.-M. Plane strain bending with isotropic strain hardening at large strains. *Trans. ASME J. Appl. Mech.* **2010**, *77*. [CrossRef]
24. Alexandrov, S.; Pirumov, A.; Jeng, Y.-R. Expansion/contraction of a spherical elastic/plastic shell revisited. *Cont. Mech. Therm.* **2015**, *27*, 483–494. [CrossRef]
25. Manabe, K.; Nishimura, H. Influence of material properties in forming of tubes. *Bander Bleche Rohre* **1983**, *9*, 266–269.
26. Mori, S.; Manabe, K.-I.; Nishimura, H. Hydraulic bulge forming of clad thin-walled tubes. *Adv. Tech. Plast.* **1990**, *3*, 1549–1554.
27. Roberts, S.M.; Hall, F.; Van Bael, A.; Hartley, P.; Pillinger, I.; Sturgess, E.N.; Van Houtte, P.; Aernoudt, E. Benchmark tests for 3-D, elasto-plastic, finite-element codes for the modelling of metal forming processes. *J. Mater. Process. Technol.* **1992**, *34*, 61–68. [CrossRef]
28. Abali, B.E.; Reich, F.A. Verification of deforming polarized structure computation by using a closed-form solution. *Contin. Mech. Thermodyn.* **2020**, *32*, 693–708. [CrossRef]
29. Alzahrani, B.; Ngaile, G. Preliminary investigation of the process capabilities of hydroforging. *Materials* **2016**, *9*, 40. [CrossRef] [PubMed]

MDPI

Article

Local One-Sided Rubber Bulging Test to Measure Various Strain Paths of Metal Tube

Hidenori Yoshimura [1,*], Kana Nakahara [2] and Masaaki Otsu [3]

[1] Faculty of Engineering and Design, Kagawa University, Takamatsu 761-0396, Japan
[2] Graduate School of Engineering, Kagawa University, Takamatsu 761-0396, Japan; s19g522@stu.kagawa-u.ac.jp
[3] Faculty of Engineering, University of Fukui, Fukui 910-8507, Japan; otsu@u-fukui.ac.jp
* Correspondence: yoshimura.hidenori@kagawa-u.ac.jp; Tel.: +81-87-864-2345

Abstract: We proposed a local one-sided rubber bulging method of metal tubes to evaluate various strain paths at an aimed portion and measured the forming limit strains of metal tubes at the place of the occurrence of necking under biaxial deformation. Using this method, since rubber is used to give pressure from the inner side of the tube, no sealing mechanisms were necessary unlike during hydraulic pressure bulging. An opening was prepared in front of the die to locally bulge a tube at only the evaluation portion. To change the restriction conditions of the bulged region for biaxial deformation at the opening, a round or square cutout, or a slit was introduced. The test was conducted using a universal compression test machine and simple dies rather than a dedicated machine. Considering the experimental results, it was confirmed that the strain path was varied by changing the position and size of slits and cutouts. Using either a cutout or a slit, the strain path in the side of the metal tubes can be either equi-biaxial tension or simple tension, respectively. Additionally, by changing the size of the cuts or slits, the strain path can be varied.

Keywords: tube forming; tube bulging test; formability test; forming limit; biaxial strain; local rubber bulging; cutout shape; slit length

Citation: Yoshimura, H.; Nakahara, K.; Otsu, M. Local One-Sided Rubber Bulging Test to Measure Various Strain Paths of Metal Tube. *Metals* **2021**, *11*, 751. https://doi.org/10.3390/met11050751

Academic Editor: Umberto Prisco

Received: 25 March 2021
Accepted: 30 April 2021
Published: 2 May 2021

1. Introduction

Recently, the application of tube-formed products has been utilized in various hollow transport components and mechanical structures to improve their rigidity and strength and to reduce their production cost. When the cross-sectional areas of solid and hollow materials are the same, the moment of inertia of the hollow-like tube is larger than that of the solid. Regarding the case of producing hollow parts, a welding margin is necessary when forming from sheet metal but unnecessary from a tube. Thus, the mass of the parts can be reduced, and the specific rigidity and specific strength will increase. Regarding the latter, because some workings such as bulging and piercing can be performed simultaneously by a tube-forming machine, production costs can be reduced. Concerning tube forming, tube materials are subjected to complex and severe deformation and, thus, some forming defects such as cracking and buckling often occur. To avoid such forming defects, the formability of the tube materials should be evaluated appropriately. Although trial forming tests require both time and cost, finite element method (FEM) simulation can reduce both of them. To obtain satisfactory simulated results during the forming processes, trustworthy input data should be obtained by accurate measurement of the material properties of the flow stress, the forming limit, and the fracture limit. Since tubes have distributions of initial thickness and hardness in the circumferential direction that is caused by the manufacturing method, input data for the simulations should be obtained, but not from sheet metal before forming into a tube, rather from a tube [1,2]. Metal tubes tend to have more anisotropy of strength because they are made from pre-finished products by the additive working process. The purpose of this research is to establish an easy biaxial test method for metal tubes to enable the construction of a forming limit diagram.

Yoshida has reported on the examination of standard test methods such as Japanese Industrial Standards (JIS) and International Organization for Standardization (ISO) and other methods [3]. Generally, mechanical properties tests for tubes are classified into axial and circumferential directions. There are JIS Z 2241 in the axial tensile test with which it is possible to obtain the ultimate tensile strength, yield point, breaking elongation, etc. Conversely, regarding the circumferential direction, the ring tension (ISO 8496), flatness (ISO 8492), push-spread (ISO 8493), flange extension (ISO 8494), ring spread (ISO 8495) tests, etc. are standardized. However, the purpose of them is to detect defects in the tube material such as a defective weld. Moreover, they are only soundness evaluation tests of the pre-service tube and cannot evaluate tube formability.

Therefore, various advanced researches have been conducted on the evaluation test of the tube formability. Regarding the axial direction, Yoshida et al. developed the tube tensile test using image processing for measurement of the flow stress and forming limit. Also, the influence of triaxiality on the fractured strain change around the weld seam [4,5] was investigated. Concerning the circumferential direction, Yoshimura et al. reported the ring tensile test method which can evaluate the flow stress and forming limit strain of the welded tube's base metal [6,7]. Sokolowski, Muammer, and Hwang proposed methods to evaluate flow stress using a free bulging test which does not need to cut a specimen from the tube [8–10]. Additionally, Manabe et al. proposed the conical flaring test in which the tube was expanded until cracks occurred at the end of the tube, and also investigated the effects of test conditions, mechanical properties of the material, and the apex angle of the conical tool on the deformation behavior [11,12]. Also, using a conical flaring test, they identified the parameters of the ductile fracture criterion proposed by Oyane [13]. Manabe and Yoshida compared a conical flaring test to three kinds of bulging test and reported that the flaring test results are in relatively good agreement with the result of the rubber bulging test [14].

It is well-known that the ratios of the circumferential strain and axial strain influence the forming limit of the tube similar to the sheet forming. Therefore, several evaluation methods for the forming limit on the biaxial deformation of the tube have been proposed. Most of them are methods which apply the free bulge test. Guo et al. improved the accuracy of the evaluation of the forming limit of the tube under hydraulic free bulging without axial feeding using a combination of Oyane's ductile fracture criterion with the M-K theory [15].

Generally, free-bulging tests with internal pressure and axial feeding are classified into two kinds, depending on axial feeding; one has only compressive axial feeding, the other has compressive and tensile axial feeding. Many investigations are concerned with the former to obtain only the left-hand side of the forming limit diagram (FLD). Hwang et al. constructed a FLD of AA6011 using a free bulge test machine which clamps and pushes both ends of the tube [16], for example. Kim proposed analytical and numerical analyses of the forming limit in the free bulge test with axial compressive feeding, and it was shown that the results agreed with the experimental results [17]. Song et al. also executed a series of free bulge tests on an FLD of a tube with a similar experimental apparatus and investigated the effect of the flow stress on the FLD [18]. Kuwabara et al. conducted simple hydraulic bulging tests in which ends of the tube were not restricted and examined the breakage expansion ratio and the forming limit of the circumferential strain. Moreover, they applied the tests on three types of tubes(1)as-roll, (2)as-roll and normalized, and (3)cold-drawn and normalized, and reported in detail about the relation of uneven initial tube thickness distribution and uneven hardness distribution to the fracture position [19].

To obtain the strain paths, including the right side of the FLD by the free bulge test, it is necessary to apply axial tension to both ends of the pipe. The difficulty is in clamping both ends of the tube without leakage of the internal pressure medium. Kuwabara et al. developed a hydraulic multiaxial tube expansion testing machine that can measure material properties along a predetermined strain and stress path by controlling the axial load and the internal hydraulic pressure, and can evaluate the mechanical properties of tubes and plates, yield surface, and so on, until large strains cause failure [20]. During this current research, a test method for obtaining the forming limit stress curve (FLSC) has been proposed. During the subsequent research of the former study, Kuwabara et al. developed the test method which enables continuous measurement of large-strain biaxial stress–strain curves without resticking strain gages [21]. Korkolis and Kyriakides investigated the inflation and burst of Al-6260-T4 tubes under combined tensile or compressive axial loads with internal pressure and observed that localized wall thinning, and burst can be very sensitive to the constitutive description employed for the material [22].

Added to the free bulge test, by controlling the internal pressure and axial feeding, other kinds of material property evaluation tests also have been proposed. As a test method designating the evaluated area in a tube, Shirayori et al. proposed a partial hydraulic bulge test method in which only a designated part of a tube was expanded [23] and described the deformation behavior on the wall thickness and the bulged rate under multiple axial loading conditions [24]. Chen et al. proposed a new theoretical model to predict the FLD for a seamed tube. Regarding the same report, they obtained the left-hand side FLD using a classical free hydroforming tool set, and the right-hand side FLD via the novel device to simultaneously give the lateral compression force and the internal pressure to control the material flow [25]. Furthermore, Lin et al. proposed a novel experimental device by partially bulging a tube at the upper elliptical opening to evaluate the right side FLD and could obtain the forming limit strains at the weld line and the opposite region of the tube [26]. Magrinho et al. obtained multiple strain paths in both sides of the FLD using three kinds of tests: a uniaxial tensile test of a specimen cutout from a tube, a tube expansion test of the tube end with a rigid punch, and a rubber pressured free bulge test using a digital image correlation system (DIC) [27]. Furthermore, Magrinho et al. also proposed a methodology to determine the forming limits by buckling, necking, and fracture during an external thin-walled tube inversion [28].

Since welded steel tubes, which are frequently used for parts of automobiles, have a different hardness between the weld and the base metal caused by the thermal treatment, it is necessary to evaluate the weld and the base metal, respectively. However, there is no test of the tube to evaluate a designated location, except those by Shirayori et al. [23] and Lin et al. [26]. Most biaxial tube formability evaluation tests require expensive special equipment for hydraulic loading, sealing, and axial loading, excluding a conical flaring test and rubber bulge test. Furthermore, it is important to check the linearity of the strain path because the forming limit curve (FLC) is changed depending on the strain path. Chen et al. and Lin et al. measured strain paths using each test method [25,26]. However, these methods have problems caused by stopping the test, such as the effect of unloading, the length of the test time, and damage to the specimen. Therefore, a non-contact strain measurement method is required as a method that can be continuously and quickly tested and measured with a single test tube.

During this research, to obtain a forming limit under various biaxial deformations of a tube, we propose a local one-sided rubber bulging test. Using this test method, a tube is locally bulged to specify the evaluation part, for example, the base material except for the weld and heat affected zone (HAZ). Since the pressure medium is rubber, the test machine is inexpensive. Although the final goal of this research is measuring the FLD of the tube, this report focuses on the possibility of measuring various proportional strain paths in the right and left sides of the strain space. To realize the various strain paths, we propose that a cutout or slit is preprocessed on the tube specimen. It is possible for noncontact while continually measuring the strain path using image processing because the solid rubber pressure medium has little risk to damage the camera. Thus, the test can be simply and continuously conducted until the fracture limit. This test method was applied for a pure aluminum tube (JIS A 1070TD), and we confirmed the linearity of the obtained strain path and the wide range of the strain ratio of the strain paths.

2. Materials and Methods

2.1. Overview of the Local Rubber Bulging Test

During the proposed local one-sided rubber bulging test shown in Figure 1, the outer side of a tube is restrained with a die (a container), and silicone rubber that is inserted inside the tube is compressed axially by a punch to provide the tube internal pressure. Only the local surface at the opening of the die is bulged and strain on the bulged surface is measured. Silicone rubber can be compressed easily with a punch using a universal testing machine. Regarding the case of hydraulic pressure, the pressure medium leaks easily, despite the small expansion of the tube diameter in the clearance between the tube and the die. Conversely, in the proposed method, silicone rubber can be easily sealed using urethane rubber, which is harder than silicone rubber. Thus, complicated and any expensive equipment, including a hydraulic pressure pump and sealing mechanisms, are unnecessary in this method. The circumferential and axial strains at the top of the bulged surface, which is the center of the die's opening, are measured using an image processing system.

Multiple strain paths are required to construct the FLD of the tube material. Magrinho et al. have reported that the free rubber pressure-bulging tests in which the movement of the tube ends is fixed can obtain multiple strain paths only on the right side of the strain space (from plane strain tension to equi-biaxial tension) by changing the bulged length [25]. Using the proposed test method, it is expected to measure not only the right-side strain paths but also the left-hand strain paths. Various deformations can be obtained by making a cutout or slits on the tube specimens. Figure 2 shows three kinds of specimens before and after the test: no cutout, cutout, and slit. No-cutout is an unprocessed tubular specimen, and the strain path on the blue line of the plane strain tension will be measured. Cutout is a preprocessed tubular specimen cutout on the opposite side to the bulged surface, and strain paths in the right hand of the No-cutout will be measured in the green space shown in Figure 2. A slit is added to a tubular specimen on the same side as the bulging, and the strain paths in the left hand of the strain paths of the No-cutout will be measured in the yellow space in Figure 2.

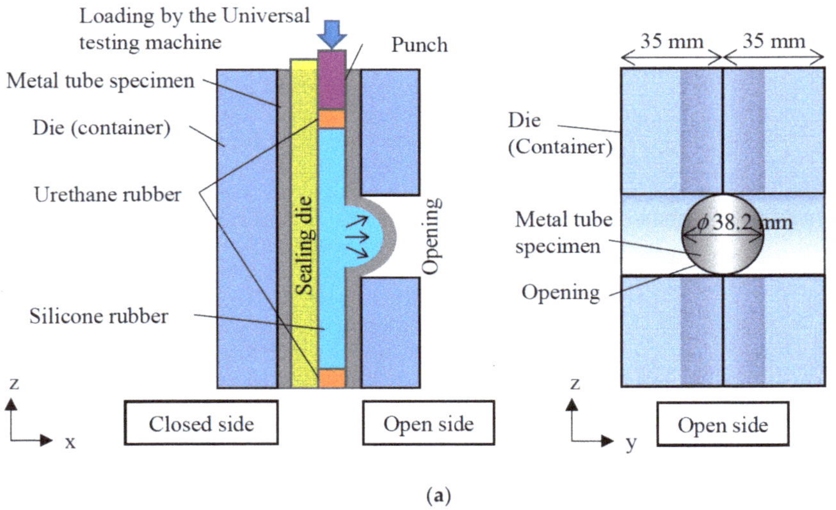

(a)

Figure 1. *Cont.*

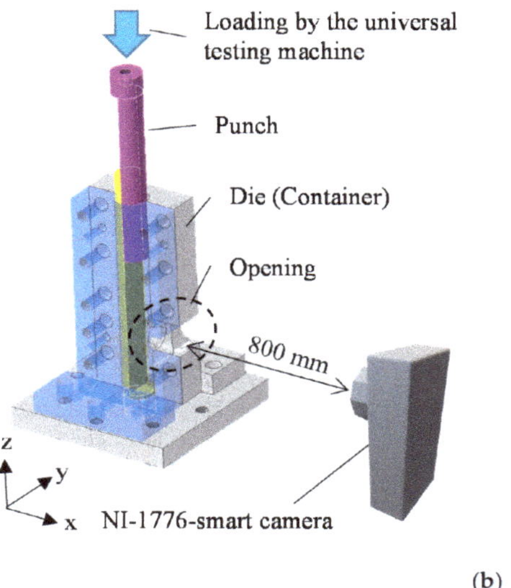

(b)

Figure 1. Illustration of local one-sided rubber bulging test: (**a**) Cross-sectional view; (**b**) General view.

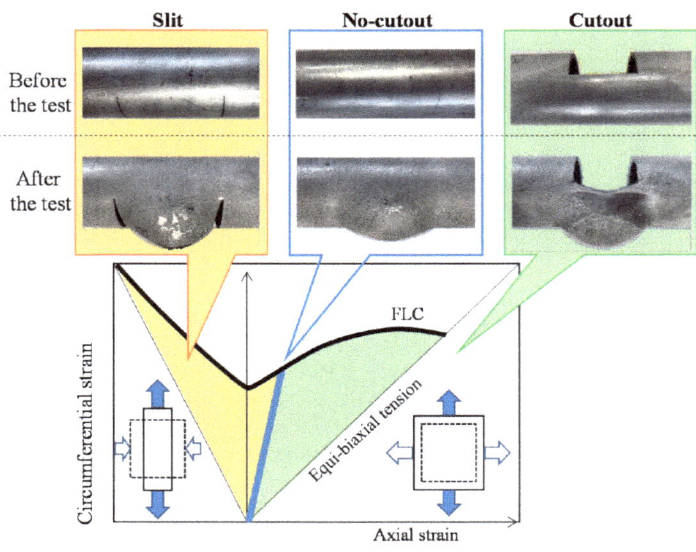

Figure 2. Change of strain path by introducing a cutout or slits to the tube specimen.

During this test method, since a local bulged region can be selected, the bulged area is set to exclude a singular region, for example, the welded or heat-affected zone of the welded steel tube. Besides the welded tube, the other metal tubes also have thickness and hardness distributions in the circumferential direction due to the manufacturing process. Although the weak point, which is the thinner or softer region, may be broken, generally such weak points cannot be avoided when designing hollow products. Therefore, in this paper, it is explained how to know beforehand where the weak position of a metal tube is using preliminary tests and then by measuring the strain path around the position.

2.2. Specimen, Rubber Pressure Medium and Test Equipment

The proposed test method was applied to the round pure aluminum tube, JIS: A1070TD, with an outer diameter D_0 = 38.0 mm, and a nominal thickness t = 2.0 mm. Since this pure aluminum tube was made using porthole extrusion, it did not have the heat affect caused by welding. Also, the initial thickness distribution was less than ±0.035 mm and seemed very small. Prior to the test, the specimens used were annealed for 1 h at 400 °C. The average and deviation of the hardness was 25/HV0.3 and less than ±2.8%, respectively.

Silicone KE1417 made by Shin-Etsu Chemical Ltd. (Tokyo, Japan) was used as a pressure medium, and a 3.5 wt.% curing agent was mixed with the liquid state silicone to make a semi-cylindrical shape with a diameter DG = 33.5 mm and a length LG = 170 mm. The length of the semi-cylindrical rubber was sufficient so the end of the rubber did not reach the die opening when compressed by the punch. The semi-cylindrical rubber was used to save the rubber by the use of a semi-cylindrical sealing die, as shown in Figure 1. Naturally, cylindrical rubber can be used, as well. Due to the solid pressure medium, leakage of rubber from the cutout and the slit did not occur except at the opening of the die. The sealing was very easy. Figure 3 shows the results of the uniaxial compression test for two cylindrical silicone rubbers, N1 and N2, which are different production lots. Since rubber often is treated as an incompressible material [29], the strain on the rubber in the proposed rubber bulging test may be minimal when estimating it roughly from the compression ratio of it in the vicinity of the die opening. Moreover, the flow stress on the rubber is less than that on the tube metal. Thus, we assessed that the difference between rubber lots can be ignored.

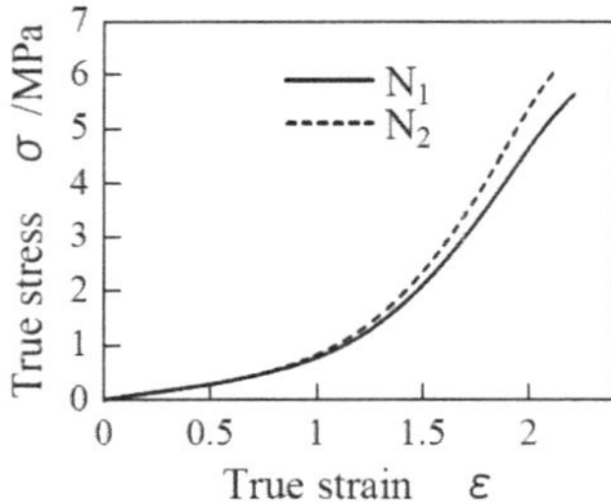

Figure 3. Stress-strain curve of rubber.

Figure 4 shows the used separate die. The hole of ϕ 38.2 mm, which contacts the specimen, is dug to a T-shape by a milling machine with a ball-end mill. The corner roundness on the separated surface is R = 5 mm, so the material moves to the opening easily. Although the ends of the tube can move in the rotation axis direction, the lower tube end is fixed to affect the asymmetrical deformation at the opening due to the tube rotation.

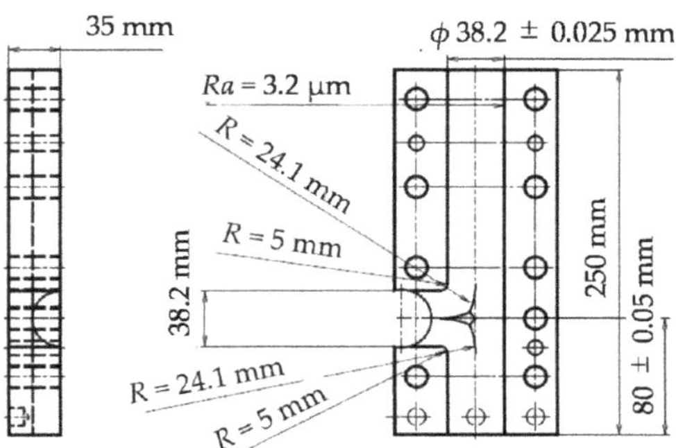

Figure 4. Dimensions of separate die (container).

2.3. The Test Procedure

Although the used pure aluminum tubes seem quite homogeneous on the thickness and hardness distributions in comparison with the welded steel tube, the distributions could not be ignored due to the manufacturing process of hot porthole extrusion. Actually, even after the annealing treatment, the A1070TD specimens used have the effect of four joints in the circumferential direction of the deformation behavior. It was confirmed that the deformation behavior differs between the joint region and others by Matsuoka et al. [1,2]. Since it is difficult to exclude the locations that are easily fractured in tube forming, it is necessary to selectively evaluate the weak location. Regarding the reproducibility of the proposed test method, the fracture point must be at the center of the opening in all tests.

The procedure of the proposed test method is shown in Figures 5 and 6. Since tubes often have a weak position in the circumferential direction due to its production method, we realistically must check the forming limit of the weak position. Thus, we conducted the preliminary test to specify the weak position of the tube specimen. During the main test, the weak position was set at the center of the die opening.

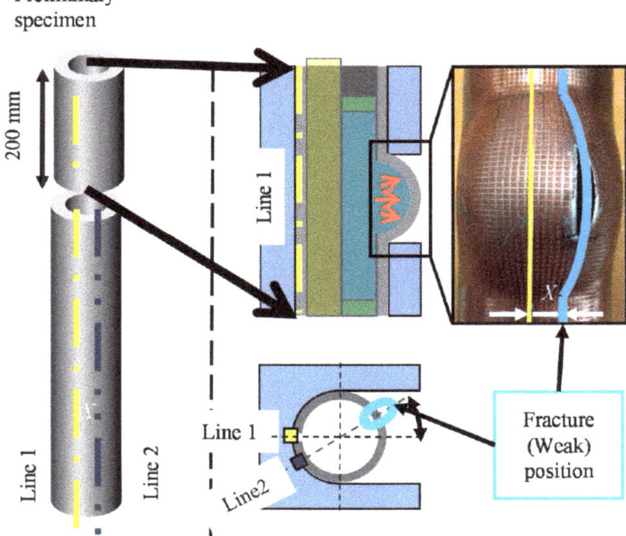

Figure 5. Overview of the preliminary test to determine the weak position.

Prior to cutting out the test material from the delivered 4000 mm tube, one marking line (line 1) that was parallel to the tube axial direction was drawn on the long pipe using a scriber. A specimen was cut out and grid lines were drawn at 180° on the opposite side of line 1 to measure the circumferential position of the fracture. The specimen length was 200 mm, and the grid lines were given using a laser-processing machine and painting. The preliminary test specimen was placed into a die to set line 1 in the 180° opposite position of the opening. Then, a local rubber pressure one-sided bulge test was conducted. Shown in Figure 5, we regarded the circumferential position of the fracture as a weak position, and the 180°opposite side position to the weak position as the position of Line 2. Axial marking line 2 is drawn at the 180° opposite side of this weak point on the residual long tube. Shown in Figure 6, the main test specimen of the 200 mm tube was cut out from the residual tube, and line 2 was set to 180° on the opposite side of the center of the opening of the die. Using this preliminary test method, the center of the bulged surface was fractured in all main tests, and the position always can be at the center of the opening, as captured by the imaging camera with a monocular lens. The reproducibility of the proposed test method was confirmed.

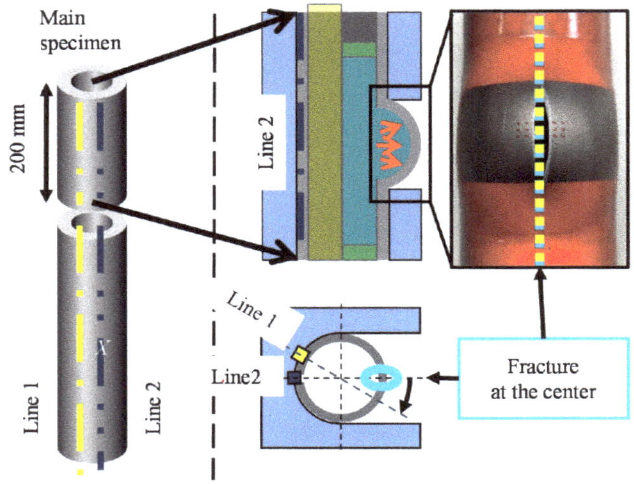

Figure 6. Overview of the main test.

To reduce friction over the contact surfaces of the sealing tool and silicone rubber with the test specimen, machine oil containing molybdenum disulfide was applied as a lubricant. The silicone rubber was inserted into the tube, and then urethane rubber for sealing was pushed from the upper side of it by a punch. Autograph AG-IS50kN (Shimadzu Corp., Kyoto, Japan) was used as a universal testing machine to push the punch, and the punch feed speed was 10 mm/min. The possibility to obtain several biaxial strain paths using the proposed testing method is examined in this paper.

2.4. The Measurement Method of Strain Path by Image Processing with a Monocular Camera

Prior to the main test, twelve red marking points for image processing were added to the specimens at the center of the die opening, as shown in Figure 7. During this research, red resin was coated over the pipe surface using a spray and the region, except for the marking points, was etched using a laser-processing machine. The strain path was acquired by measuring the centroid of each marking point using image processing. The attached red round points were 0.8 mm in diameter, and 12 points were arranged at 1.2 mm intervals. The camera was a NI-1776C smart camera (National Instrument Corp., Austin, TX, USA) with a monocular lens, and National Instruments LabVIEW (National Instrument Corp., Austin, TX, USA) was used as the recording software. The measurement data interval was 8 frames/s. The camera was placed vertical to the test material for the measurement, so the object distance between the lens and the test material was about 800 mm, as shown in Figure 1. NI Vision Assistant was used as the image processing software (National Instrument Corp., Austin, TX, USA), and it calculated each centroid of each frame. Using this method, we obtained strain paths by determining the tube axial and circumferential strains. To obtain a forming limit at the occurrence of necking, strain components were measured near the fracture position, as shown in Figure 8. Although the deformed surface was curved, the accuracy, even by a monocular camera, was sufficient because the measured region at the center of the opening was small and vertical to the shooting axis.

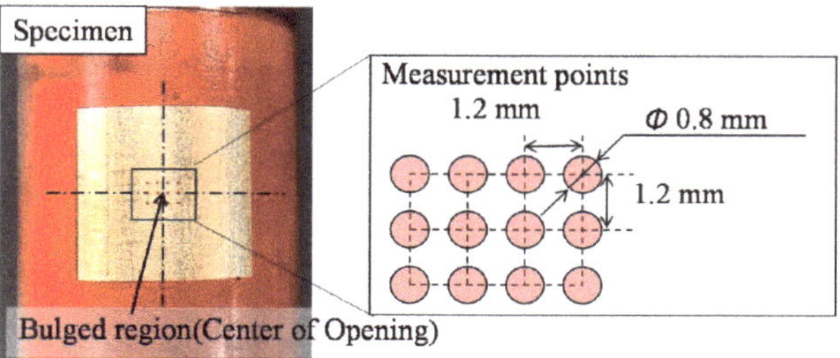

Figure 7. Size and shape of marking points for measurement by image processing.

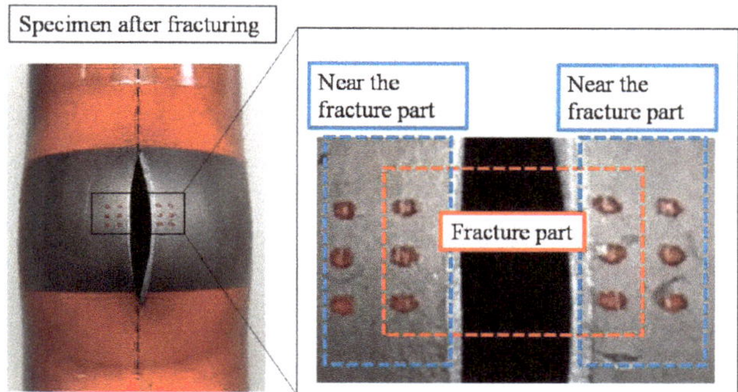

Figure 8. Appearance of measurement points on deformed tube.

2.5. Cutout and Slit to Change a Strain Path

Previously mentioned, it is supposed that cutouts and slits added to the specimen change the strain path. Thus, we examined the effect of cutout and slit shapes on the strain paths. The cutout shapes, rectangular, I-type, and round, shown in Figures 9 and 10, were examined. Cutout and slit were easily made with a general machine such as a milling machine or an electric discharge machine. Since cracks from the periphery, or corners of them, did not occur, there was no damage to the processed surface roughness, the roundness of the cutout corners, or the slit edges.

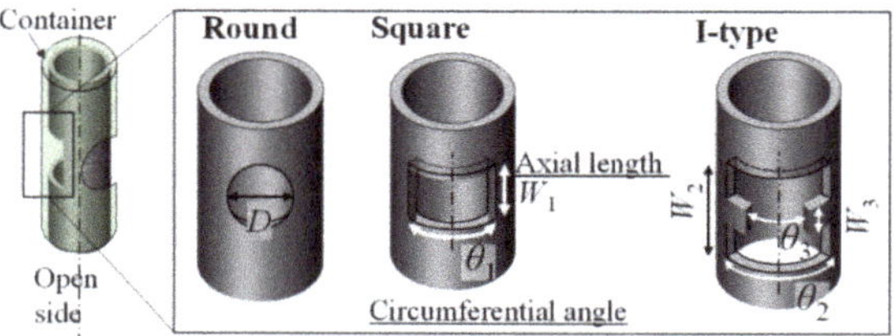

Figure 9. Variations of cutouts shape and position.

The cutout was made on the opposite side of the die opening, as shown in Figure 9. Regarding the case of the round cutout, the diameters D were 10 and 30 mm. Regarding the case of the rectangular cutout, the angle θ_1 in the circumferential direction and the length W_1 in the axial direction were changed between θ_1 = 60–140° and W_1 = 10–60 mm, respectively. The size of the cutout was limited because the deformed cutout ends reached the die opening during the test and the rubber might have leaked from the opening. Therefore, an I-type cutout was designed to prevent this leak of rubber by keeping the material near the opening. Concerning this paper, only one pattern of I-type was examined, and the cutout dimensions were θ_2 =170°, θ_3 = 100° and the lengths were W_2 = 40 mm and W_3 = 20 mm, respectively. The round cutout was opened with a drill and its diameter was changed from 10 to 30 mm. Shown in Figure 10, the slit was made in the specimen at the opening side at the ±24 mm tube axial distant positions from the center of the opening. The length of the slit, θ_4, also changed from 90 to 180° in the circumferential direction. The wire diameter of the electric discharge machine was 0.2 mm.

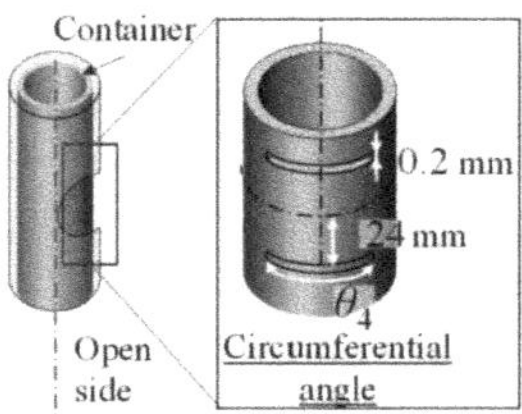

Figure 10. Size and position of slits.

3. Results and Discussion

3.1. Measuring Precision of Strain by a Monocular Camera

We obtained strain paths of the specimen using the proposed method. Figure 11 shows the strain path of a specimen with no cutout, and a forming limit strain measured by a microscope. Comparing the final strain on the strain path using image processing and the forming limit strain measured by a microscope, the average discrepancy of the circumferential strain is 0.026, and that of the axial strain is 0.020. It is considered that the discrepancy of 6–8% was a calculation error of the image processing caused by the marker' peeling or deforming after the necking. Actually, the value measured by a microscope was on the strain path obtained using image processing. Thus, we regarded that the proposed method, using image processing, could be used as a strain measurement method. Moreover, we determined the object distance was 800 mm and the marker distance and the measurement error were about several percent different due to the pixel number of the camera. The object distance, however, could be 250–300 mm if the diaphragm of the lens used was larger using a defused lighting method. It is estimated that a less than 0.3% error can be achieved by this camera.

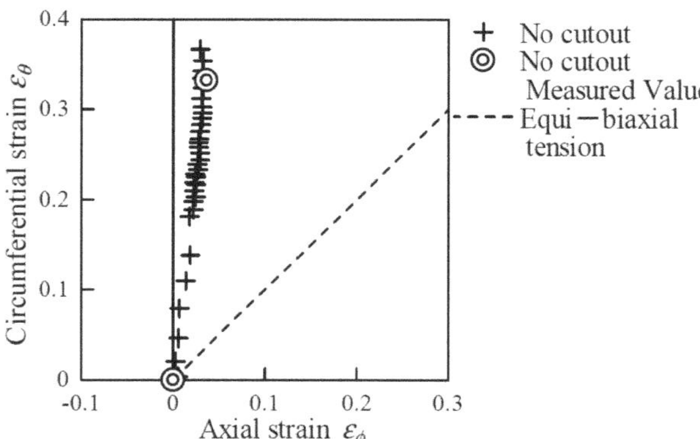

Figure 11. Strain path measured by image processing and forming limit strain measured by microscope.

Next, we verified whether the obtained strain paths were proportional loading states or not. Concerning this paper, the strain ratio is defined as the axial strain divided by the circumferential strain at the necking, which is measured near the fracture. The linear approximation of each strain path was conducted. The average value of the R^2 representing the errors were 0.95 for no cutout, 0.92 for cutout, and 0.89 for slit. Therefore, we can confirm that the obtained strain paths are almost close to the proportional loading state. Different from hydraulic bulging, the inner pressure distribution for rubber bulging may be nonuniform. Flow stress in the porthole extrusion tube is inhomogeneous. We have not established the numerical analysis model yet due to the difficulty of their treatments, and the reason for this proportional loading state deformation could not be clarified.

3.2. Cutout and Slit

During the tube bulge hydroforming, the strain path is often between circumferential uniaxial tension and equal biaxial tension, and it covers the strain ratio between −0.5 and 1.0. Shown in Figure 11, the strain ratio was 0.080 for the specimen without the cutout. Figure 12 shows the results for round cutouts. The strain ratios of D = 10 mm and 30 mm are 0.18 and 0.24, respectively. When the round cutout diameter D is larger, the closer the strain paths are to the equi-biaxial tension. However, even for a cutout diameter of D = 30 mm, the strain ratio is small. The residual material around the die's opening abounds and it is considered that the effect of the cutout on the material flow to the opening makes it smaller.

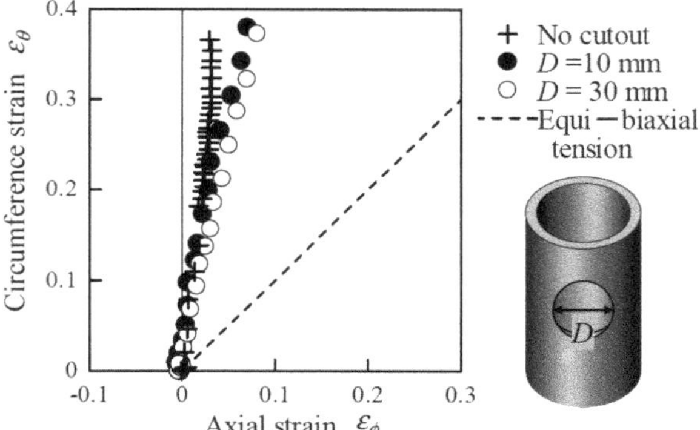

Figure 12. Strain paths obtained by specimens with round cutouts.

Figure 13 shows the strain paths of specimens with square cutouts. Figure 13a shows the change in length W at the circumferential angle of θ_1 = 140°; when W_1 = 10, 30, and 60 mm, the strain ratios are 0.13, 0.21, and 0.48, respectively. Thus, the larger W_1 is, the closer the strain paths are to the equi-biaxial tension. Viewing Figure 13b, the length W_1 = 30 mm and the angle θ_1 in the circumferential direction is changed. Occurring at θ_1 = 60, 140°, the strain ratios are 0.11 and 0.21, respectively. It was found that the larger θ_1 was, the closer it was to the equi-biaxial tension, as well. Therefore, it is supposed that as the cutout area becomes larger, the material around the cutout easily moves to the die's opening in the circumferential direction, as shown in Figure 2 and, as a result, the effect of the circumferential deformation is smaller than that of the tube axial.

Figure 14 shows the strain paths of specimens with an I-type cutout. Although the square cutout of more than approximately θ_1 =140° tends to cause rubber leakage, the cutout area can increase at the corner in the circumferential direction without rubber leakage using the I-type cutout. Considering the case of θ_2 = 170°, the strain ratio was 0.67, which was larger than the 0.48 obtained by a square cutout of θ_1 = 140°. It is expected that the strain ratio can be closer to the equi-biaxial tension via optimization of the cutout shape, for example, θ_3 or W_3.

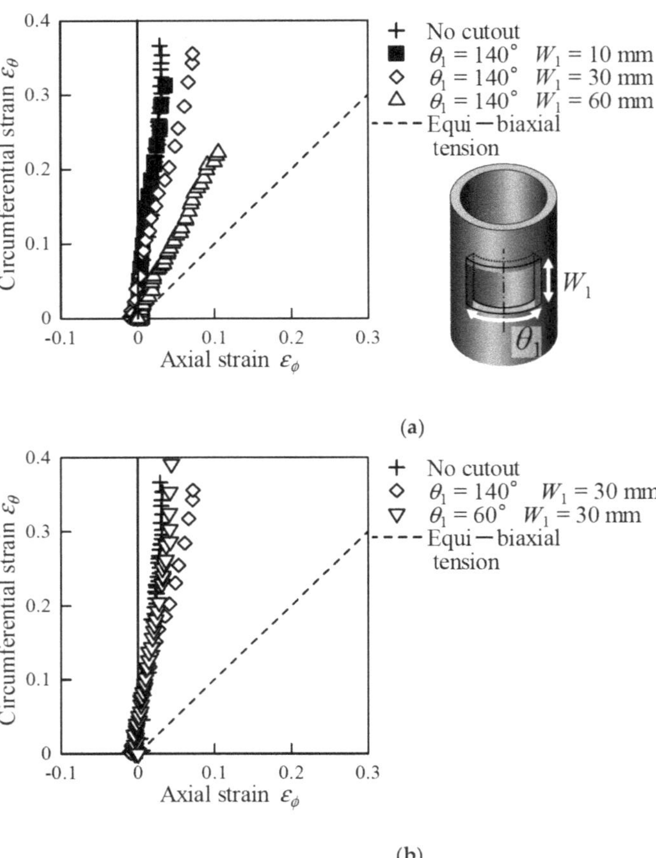

Figure 13. Strain paths of specimen with and without a square cutout; (**a**) different axial length W_1 and circumferential angle θ_1 of 140° (**b**) different circumferential angle θ_1 and axial length W_1 of 30 mm.

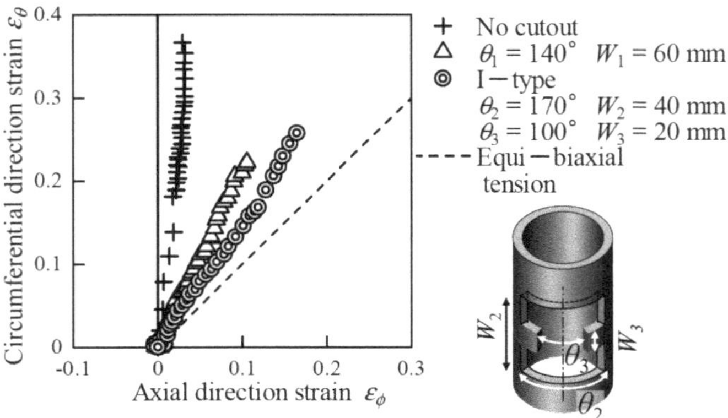

Figure 14. Strain paths of specimen with an I-type cutout.

Figure 15 shows the strain paths of specimens with slits. The strain ratios defined by this paper for slit sizes of θ_4 = 90°, 120°, and 150° are –0.12, −0.11, and –0.14, respectively.

Via giving two slits on the tube, the strain ratio is closer to the uniaxial tension's value, −0.5, in comparison with the specimen without a cutout and slit. The material near the slit seems to be moved in the tube axial direction, as shown in Figure 2. Although the larger θ_4 is, the closer the strain paths look to uniaxial tension, the change is very small. Although the region between the two slits is circumferentially elongated like the uniaxial tension test, the circumferential ends of the region are connected to the tube. The distance between the two slits is 48 mm and the circumferential length of the region is short, even if θ_4 = 150°. Therefore, the circumferential ends of the region are restricted to not being shrunk in the tube axial direction. Consequently, it is thought that the effect of the slit length on the strain path is slight.

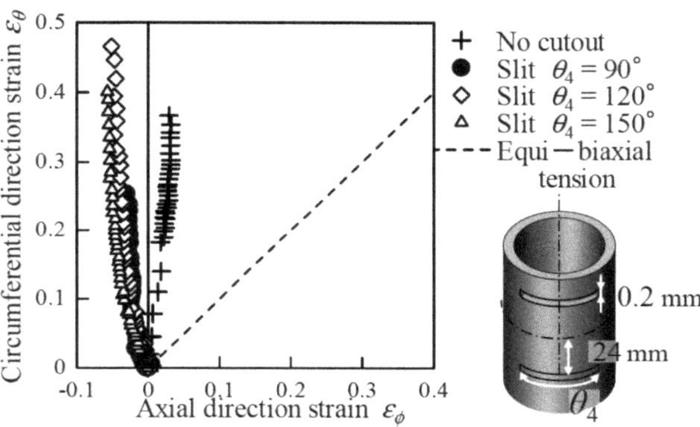

Figure 15. Strain paths of specimen with slits.

Regarding cases of a cutout, the end of each strain path is different from the biaxial test results of the metal sheet, as well. Concerning this paper, the strain ratio is defined by an average of the strain calculated from the distance between two markers, except for the fractured position. The distance is subjected to the effect of the local circumferential elongation at the fracture and it is guessed that despite the near stain path, the periods of the strain path end may be quite different. Evidently, although this phenomenon occurs in the case of the metal sheet, the tubes produced using porthole extrusion have a strong inhomogeneity in the circumferential direction and the deviation of the strain on both sides of the fracture is very large. To obtain the forming limit for the metal tubes, it is desired that the deformation behavior, and how to decide the forming limit, will be clarified by continuous research using a 3-D DIC system.

4. Conclusions

A local one-sided rubber bulging testing method was proposed to measure proportional strain paths in the right and left side of the strain space. To obtain the various strain paths in the biaxial deformation of the tube axial and circumferential directions, a cutout was preprocessed in the tubal specimen at the opposite side of the bulged surface, and slits also were made beforehand in the vicinity of the bulged surface in the tube axial direction. The size and shape of them were changed and we examined the effect of these on the strain paths. The testing method was applied to an annealed pure aluminum tube of A1070TD, and we confirmed that all measured strain paths almost were proportional. Logically, using the proposed testing method, some strain paths with strain ratios from −0.14 to 0.67 were obtained using tubular specimens. A larger cutout resulted in a strain path that was closer to the equi-biaxial strain path, and a larger slit resulted in a strain path that was closer to the uniaxial strain path. Although, at present, the forming limit on each strain path is roughly specified from imaging processing, the specification method is the next research stage.

Author Contributions: Conceptualization, H.Y.; methodology, H.Y.; validation, H.Y., K.N. and M.O.; formal analysis, K.N.; investigation, H.Y. and K.N.; resources, H.Y. and M.O.; data curation, K.N.; writing—original draft preparation, K.N.; writing—review and editing, H.Y. and M.O.; visualization, K.N.; supervision, H.Y.; project administration, H.Y.; funding acquisition, H.Y. All authors have read and agreed to the published version of the manuscript.

Funding: This research received no external funding.

Conflicts of Interest: The authors declare no conflict of interest.

References

1. Matsuoka, H.; Neumann, L.; Hamano, H.; Sakaguchi, M. Effect of microstructures on tensile properties of a 6061 extruded alloy containing weld parts. *J. Jpn. Inst. Light Met.* **2002**, *52*, 155–160. (In Japanese) [CrossRef]
2. Eda, H.; Sakaguchi, M.; Maehara, H.; Sukimoto, M.; Fuchizawa, S. Effects of production conditions on free bulge formability of 6063 aluminum alloy tubes. *J. Jpn. Inst. Light Met.* **1993**, *43*, 438–443. (In Japanese) [CrossRef]
3. Yoshida, Y. Tensile test and evaluation tests of tube and pipe. *J. Jpn. Soc. Technol. Plast.* **2010**, *51*, 313–317. (In Japanese) [CrossRef]
4. Yoshida, Y.; Kume, S.; Yukawa, N.; Ishikawa, T.; Manabe, K. Tube Tension Test Using Image Analysis. In Proceedings of the 148th the Iron and Steel Institute of Japan Meeting, 17–5, Akita, Japan, 28–30 September 2004; p. 1054. (In Japanese)
5. Yoshida, Y.; Yukawa, N.; Ishikawa, T. Evaluation of Fracture Property of Tube by Tensile Test Using Image Analysis. In Proceedings of the 4th International Conference on Tube Hydroforming (Tubehydro 2009), Kaohsiung, Taiwan, 6–9 September 2009; pp. 73–78.
6. Yoshimura, H.; Mihara, Y. Applicability of Mechanical Property of Steel Tubes by Ring Tensile Test. In Proceedings of the 166th the Iron and Steel Institute of Japan Meeting, 26–2, Kanazawa, Japan, 17–19 September 2013; pp. 611–614. (In Japanese)
7. Yoshimura, H.; Tajima, M.; Mihara, Y. Estimation of Ductile Fracture of Tube Material by Ring Tensile Test. In Proceedings of the 5th International Conference on Tube Hydroforming (Tubehydro 2011), Noboribetsu, Japan, 24–27 July 2011; pp. 203–206.
8. Sokolowski, T.; Gerke, K.; Ahmetoglu, M.; Altan, T. Evaluation of tube formability and material characteristics: Hydraulic bulge testing of tubes. *J. Mater. Process. Technol.* **2000**, *98*, 34–40. [CrossRef]
9. Koç, M.; Aue-U-Lan, Y.; Altan, T. On the characteristics of tubular materials for hydroforming—Experimentation and analysis. *Int. J. Mach. Tools Manuf.* **2001**, *41*, 761–772. [CrossRef]
10. Hwang, Y.M.; Lin, Y.K.; Altan, T. Evaluation of tubular materials by a hydraulic bulge test. *Int. J. Mach. Tools Manuf.* **2007**, *47*, 343–351. [CrossRef]
11. Manabe, K.; Yoshida, Y. Effect of Lubricants in the Flaring Test of Steel Tubes. In Proceedings of the 148th the Iron and Steel Institute of Japan Meeting, 17–5, Akita, Japan, 28–30 September 2004; p. 1053. (In Japanese)
12. Manabe, K.; Nakata, K. 2736 Effect of Tube-end Finishing Process on Deformability of Steel Tubes by Flaring Test. In Proceedings of the Mechanical Engineering Congress (MECJ-07), Osaka, Japan, 9–12 September 2007; pp. 633–634. [CrossRef]
13. Oyane, M. On criteria for ductile fracture (Minor special issue on plastic working and materials characteristics). *J. Jpn. Soc. Mech. Eng.* **1972**, *75*, 596–601. (In Japanese) [CrossRef]
14. Manabe, K.; Yoshida, Y. Evaluation of Hydroformability of Steel Pipes by Conical Flaring Test. In Proceedings of the 3rd International Conference on Tube Hydroforming (Tubehydro 2007), Harbin, China, 14–17 November 2007; pp. 39–45.
15. Guo, X.; Ma, F.; Guo, Q.; Luo, X.; Kim, N.; Jin, K. A calculating method of tube constants of ductile fracture criteria in tube free bulging process based on M-K theory. *Int. J. Mech. Sci.* **2017**, *128–129*, 140–146. [CrossRef]
16. Hwang, Y.M.; Lin, Y.K.; Chuang, H.C. Forming limit diagrams of tubular materials by bulge tests. *J. Mater. Process. Technol.* **2009**, *209*, 5024–5034. [CrossRef]
17. Kim, J.; Kim, S.W.; Song, W.J.; Kang, B.S. Analytical and numerical approach to prediction of forming limit in tube hydroforming. *Int. J. Mech. Sci.* **2005**, *47*, 1023–1037. [CrossRef]
18. Song, W.J.; Heo, S.C.; Ku, T.W.; Kim, J.; Kang, B.S. Evaluation of effect of flow stress characteristics of tubular material on forming limit in tube hydroforming process. *Int. J. Mach. Tools Manuf.* **2010**, *50*, 753–764. [CrossRef]
19. Kuwabara, T.; Moriguchi, K. Formability Evaluation of Steel Tubes Using a Simplified Hydraulic Bulging Test. In Proceedings of the 148th the Iron and Steel Institute of Japan Meeting, 17–5, Akita, Japan, 28–30 September 2004; p. 1051. (In Japanese)
20. Yoshida, K.; Kuwabara, T.; Takahashi, S. Stress Based Forming Limit Criterion for Aluminum Alloy Tubes. In Proceedings of the 2003 Annual Meeting of the JSME/MMD (M&M2003), Toyama, Japan, 24–26 September 2003; pp. 821–822. (In Japanese) [CrossRef]
21. Sugawara, F.; Kuwabara, T. Development of a multiaxial tube expansion testing machine that enables the continuous measurement of large-strain biaxial stress-strain curves of sheet metals. *J. Jpn. Soc. Technol. Plast.* **2013**, *54*, 57–63. (In Japanese) [CrossRef]
22. Korkolis, Y.P.; Kyriakides, S. Inflation and burst of anisotropic aluminum tubes for hydroforming applications. *Int. J. Plast.* **2008**, *24*, 509–543. [CrossRef]
23. Shirayori, A. Effect of Circumferential Initial Imperfection on Hydraulic Bulging of ERW Steel Tubes. In Proceedings of the 5th International Conference on Tube Hydroforming (Tubehydro 2011), Noboribetsu, Japan, 24–27 July 2011; pp. 179–182.
24. Shirayori, A.; Oshima, R.; Narazaki, M. Effect of Axial Feeding in Partial Hydraulic Bulging of ERW steel tube. In Proceedings of the 62nd Japanese Joint Conference for the Technology of Plasticity, Toyohashi, Japan, 26–29 October 2011; pp. 165–166. (In Japanese)

25. Chen, X.; Yu, Z.; Hou, B.; Li, S.; Lin, Z. A theoretical and experimental study on forming limit diagram for a seamed tube hydroforming. *J. Mater. Process. Technol.* **2011**, *211*, 2012–2021. [CrossRef]
26. Li, S.; Chen, X.; Kong, Q.; Yu, Z.; Lin, Z. Study on formability of tube hydroforming through elliptical die inserts. *J. Mater. Process. Technol.* **2012**, *212*, 1916–1924. [CrossRef]
27. Magrinho, J.P.; Silva, M.B.; Centeno, G.; Moedas, F.; Vallellano, C.; Martins, P.A.F. On the determination of forming limits in thin-walled tubes. *Int. J. Mech. Sci.* **2019**, *155*, 381–391. [CrossRef]
28. Magrinho, J.P.; Centeno, G.; Silva, M.B.; Vallellano, C.; Martins, P.A.F. On the formability limits of thin-walled tube inversion using different die fillet radii. *Thin-Walled Struct.* **2019**, *144*, 106328. [CrossRef]
29. Gent, A.N. A new constitutive relation for rubber. *Rubber Chem. Technol.* **1996**, *69*, 59–61. [CrossRef]

Article

Investigation of Punch Shape and Loading Path Design in Hydro-Flanging Processes of Aluminum Alloy Tubes

Yeong-Maw Hwang *, Hong-Nhan Pham and Hiu-Shan Rachel Tsui

Department of Mechanical and Electro-Mechanical Engineering, National Sun Yat-Sen University, Lien-Hai Rd., Kaohsiung 804, Taiwan; d033020006@nsysu.edu.tw (H.-N.P.); m093020096@nsysu.edu.tw (H.-S.R.T.)
* Correspondence: ymhwang@mail.nsysu.edu.tw; Tel.: +886-7-525-2000

Abstract: Hydro-joining is composed of hydro-piercing, hole flanging and nut-inlaying processes. In this study, a new hydro-flanging process combining hydro-piercing and hydro-flanging is proposed. An internal pressured fluid is used as the supporting medium instead of a rigid die. Three kinds of punch head shapes are designed to explore the thickness distribution of the flanged tube and the fluid leakage effects between the punch head and the flanged tube in the hydro-flanging process. A finite element code DEFORM 3D is used to simulate the tube material deformation behavior and to investigate the formability of the hydro-flanging processes of aluminum alloy tubes. The effects of various forming parameters, such as punch shapes, internal pressure, die hole diameter, etc., on the hydro-flanged tube thickness distributions are discussed. Hydro-flanging experiments are also carried out. The die hole radius is designed to make the maximum internal forming pressure needed smaller than 70 MPa, so that a general hydraulic power unit can be used to implement the proposed hole flanging experiments. The flanged thickness distributions are compared with simulation results to verify the validity of the proposed models and the designed punch head shapes.

Keywords: tube hydroforming; hydro-flanging; punch head shape; finite element analysis; alumimum alloy

Citation: Hwang, Y.-M.; Pham, H.-N.; Tsui, H.-S.R. Investigation of Punch Shape and Loading Path Design in Hydro-Flanging Processes of Aluminum Alloy Tubes. *Metals* **2021**, *11*, 636. https://doi.org/10.3390/met11040636

Academic Editor: Umberto Prisco

Received: 15 March 2021
Accepted: 12 April 2021
Published: 13 April 2021

1. Introduction

Nowadays, energy saving and carbon dioxide reduction have become important issues in the world, especially in aerospace and transportation fields. Tube hydroforming (THF) processes have been applied to manufacture lightweight parts in various fields, such as bicycles, automobiles and aerospace industries. Compared with conventional metal forming processes, tube hydroforming has some merits, such as reductions in workpiece cost, tool cost and product weight. THF can also improve structural stability and increase strength and stiffness of the formed parts, such as front and rear axles, exhaust system components, body frames, etc. [1,2] Recently, tube hydroforming processes sometimes incorporate piercing, flanging, or joining processes to become hydro-piercing, hydro-flanging, or hydro-joining, which are more efficient compared with a single process and can reduce the total weight of the final product.

Some research concerning hydro-piercing and hydro-flanging processes of sheets or tubes have been presented. For example, Fracz et al. [3] investigated the effect of punch geometry on the sheet thickness distribution during hole-flanging process. Three different punch geometries: cylindrical (flat-bottomed), hemispherical and conical, were used in the experiment, as well as in numerical simulation. The results of experimental investigations were compared with the FE simulation results. Kacem et al. [4] used a conical punch to characterize and predict numerically the limits of the hole flanging process arising from material failure for two different aluminum sheets. Then, a fracture criterion based on local strain measures in tension has been identified for both materials. Finally, numerical predictions of the strain limits obtained from successful parts were compared to experimental results. Liu et al. [5] proposed a new hybrid technology of hole hydro-piercing-flanging processes. The influence of punch shape on geometrical profile and quality of holes was investigated by experiments and simulations using a punch of different shapes at the transition zone. The results showed that the geometrical dimensions of the roll-over depth, straight-ring zone height and thickness distributions varied with punch shapes. Finally, the mechanism of hydro-piercing-flanging process affected by the punch shape was clarified from the stress and strain distributions by the finite element analysis.

Thipprakmas et al. [6] proposed a fine blanked-hole (FB-hole) flanging process to investigate the deformation of a sheet by the finite element method. The FE simulation results of the flanged shapes were compared with experimental results and good agreement was found. The results verified that the FB-hole flanging process resulted in better-flanged shapes and flangeability than those obtained by the conventional-hole flanging process. Mizumura et al. [7] investigated the influence of internal pressure on hydro-burring after hydro-piercing of steel tubes. They found internal pressures during hydro-burring have large effects on the formability of burring processes. The hydroformed component can be joined to another part with the thread tapped at the hydro-burring portion. However, it is difficult to tap thin-ring parted tubes. Therefore, they also developed a new nut-inlaying method in a hydroformed component, by which thin-ring parted hydroformed components can also be joined to other parts using this nut-inlaying method. Mizumura et al. [8] developed a method of hydro-burring, wherein a hole made by hydro-piercing was expanded while the internal pressure was maintained. It became possible to join formed parts to others with bolts by tapping screw threads to the hole or inlaying a nut there. Choi et al. [9] analyzed the tube deformation behavior surrounding a hole produced by a hydro-piercing process. They investigated both experimentally and analytically the relationship between the deformation radius and the roll-over under different punch diameters and internal pressures.

One of the present authors has published a series of works related to tube hydroforming processes [10–12]. For example, Hwang and Wu [10] proposed a compound forming process including crushing, hydroforming and calibration, for hydroforming a rectangular cross-sectional tube of aluminum alloys A6061-O. Using the proposed compound forming processes and an appropriate loading path, a uniform thickness distribution in the formed tube, a lower internal pressure and a smaller clamping force were obtained compared with the tradition hydroforming process. Hwang et al. [11] used finite element simulations and experiments to investigate the effects of punch shape and various parameters such as punch strokes, internal pressures, etc., on the pierced hole surface characteristics in a tube hydro-piercing process of SPFC590Y carbon steel tubes. The deformation mechanisms for obtaining a better surface characteristic were also discussed. Hwang and Tsai [12] proposed a tube hydroforming process with a novel movable die concept and loading path design to manufacture irregular bellows with small thinning ratios in the formed product. Two kinds of feeding types were proposed to make the maximal thinning ratio in the formed bellows as small as possible. A finite element simulation software "DEFORM 3D" was used to analyze the plastic deformation of the tube within the die cavity using the proposed movable die design. Using the movable die design with an appropriate die gap width, the internal forming pressure needed can be reduced to only one-sixth of the internal pressure needed without the movable die design.

Park et al. [13] proposed an advanced sealing system to prevent fluid leakage during the hydroforming process. The advanced sealing system was composed of a die spring, cylindrical sleeve and punch with end fillet. With the proposed sealing system, the circumference of the tube end became more tightly sealed by the axial pressure between the punch and sleeve and the axial pressure increased with increasing axial feeding. The feasibility of the proposed sealing system was experimentally confirmed by hydroforming a non-axisymmetric complicated part. Yu et al. [14] introduced a quantitative method to evaluate effects of crack size and material anisotropy on edge stretchability with an index of effective failure strain ratio during punching of sheet metals. Numerical studies were conducted to investigate the interaction effect of cracks and anisotropy on the edge stretchability during hole stretching. The results showed that cracks are prone to appearing along the direction with the lowest r-value within the sheet plate. Punching and hole-expansion experiments using Dual Phase steel were conducted to validate the conclusions. Kumar et al. [15] used finite element simulations and experiments to investigate the effect of punch head profiles on deformation behavior of AA5052 alloy sheet in stretch-flanging processes. Six different punch geometries were used. The simulative and experimental results showed that the circumferential strain, radial strain and punching load are minimum with a hemispherical punch profile as compared to other punch profiles.

Material behavior modellings influence the design of processes, tools and the final products in metal forming or machining processes. Dixit et al. [16] presented a comprehensive review of various approaches of material behavior modellings. Metal forming processes, traditional machining processes and non-traditional machining processes were all considered in this review paper. Different material models were compared with respect to their suitability for the design of processes, tools and products. Del Pozo et al. [17] proposed a methodological scheme for a reduction of both the try-out and lead-time of complex dies. From the finite element simulation of the press/tool deflection during the stamping process, the best design of high-cost dies/punches was recommended. Fernández-Abia et al. [18] proposed a mechanistic model for cutting force prediction. The effect of the edge force due to the rounded cutting edge was also considered in their model. In addition, a set of machining tests were carried out to obtain the specific force coefficients expressions for austenitic stainless steels using the mechanistic approach at high cutting speeds. The results were validated by comparing the values estimated by the model with the ones obtained by experimentation.

The present authors have proposed a punch design concept in hydro-reaming and hydro-flanging of aluminum alloy tube [19]. In this study, a new hybrid forming process combining hydro-piercing and hydro-flanging of a tube is proposed. The formability of a round hole flanging process of aluminum alloy A6063 tubes is investigated numerically and experimentally. The effects of punch geometries, die hole radius and internal pressure on the formability of the hydro-flanging process and the tube thickness distributions at the ring zone after hydro-flanging are discussed. The forming conditions for obtaining sound product with a ring zone over 3 mm thick and without oil leakage at the interface between the tube and punch are also explored.

2. Finite Element Modelling

2.1. Geometric Configurations during Hydro-Flanging

The specimens for the tensile test were cut using wire-cutting directly from the aluminum alloy tube in the longitudinal direction with the ASTM standard dimensions [20]. Then, tensile tests were conducted at constant strain rates of 0.1 and 0.01 s^{-1} at room temperature using an INSTRON universal testing machine. After the tensile tests, the recorded loads and elongations of the specimens were transferred into true stresses and true strains, respectively, as shown in Figure 1. The yielding stress with the 0.2% offset method is about 40 MPa. At small strains, the two curves almost overlap, whereas at large strains, there is about a 2% difference in the flow stresses. The flow stresses at the two different strain rates are adopted in the finite element simulations.

The geometric configurations between the tube, punch and die before a hydro-flanging process are shown in Figure 2. The outer surface of a tube is supported by a die with a hole, through which a punch is moved forward to implement piering and flanging processes. The inside of the tube is filled with a high pressure fluid, acted as a supporting media to increase the process formability and make the formed product shape and dimensions meet the requirements as can as possible.

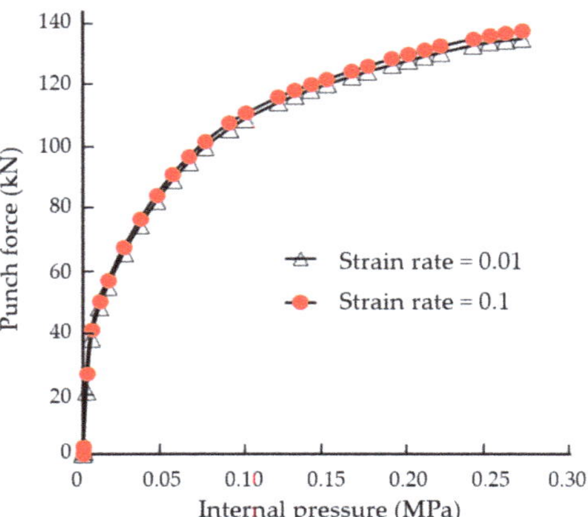

Figure 1. Stress-strain curves of aluminum alloy tube A6063.

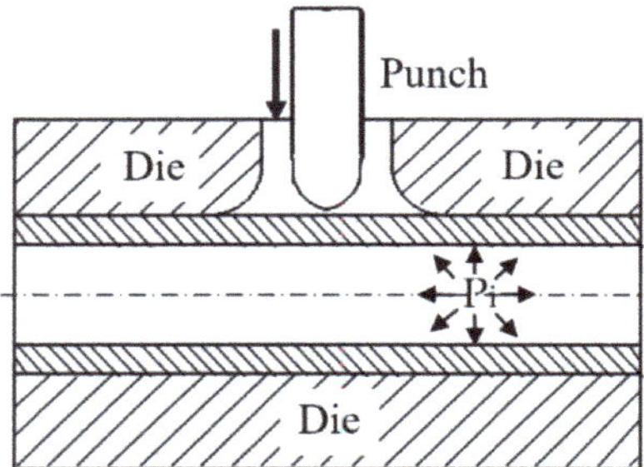

Figure 2. Geometric configurations of tube, die and punch in hydro-flanging.

The whole hydro-flanging process is divided into two stages: (1) free bulging and (2) hole flanging, as shown in Figure 3a,b, respectively, where r_h and r_p are the radii of the die hole and the punch head, respectively. P_i is the internal pressure, R is the die fillet radius, t is tube thickness and H_p is the height of the punch head. In Figure 3a, as the tube is bulged upward with a distance H_b in z direction, the punch starts to move downward to implement hole flanging stage, as shown in Figure 3b. t_0, t_1 and t_2 are the thicknesses of the flanging region at positions of s_0, s_1 and s_2, respectively, which correspond to z coordinates of z_0, z_1 and z_2, respectively. If the punch fillet radius ρ (shown in Figure 3a) and other forming conditions are not appropriately set, the thickness t_0 at the tip of the punch head (zone A) becomes close to that at zone B and the tube breakage probably occurs in zone B. The punch head shape is an important factor influencing the position of tube breakage greatly. The punch head aspect ratio is defined as H_p/r_p, which is related with the punch fillet radius ρ. As rp is set as 3.45 mm, then three different punch heights of H_p = 4, 4.5 and 5 mm will correspond to three different punch fillet radii of ρ = 4.04, 4.68 and 5.41 mm, respectively.

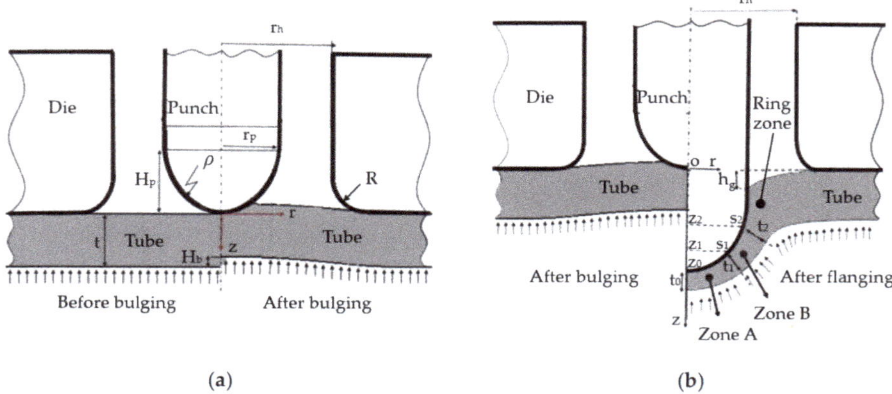

Figure 3. Schematic diagrams in different stages. (**a**) Free bulging stage; (**b**) Hole flanging stage.

2.2. Finite Element Simulations in Hydro-Flanging

Table 1 shows the forming conditions used in the finite element simulations in hole hydro-flanging processes. The aluminum alloy A6063 tube is set as an elasto-plastic material, whereas die and punch are set as rigid body in the finite element simulations. The tube outer diameter d_0 is 50.8 mm and the initial thickness t is 3 mm. Three punch heights of 4, 4.5 and 5 mm are designed. Three die hole radii r_h are 3.5 mm, 5 mm and 6.5 mm. Three internal pressures of 30, 50 and 70 MPa are selected. The forming temperature is set as 25 °C. The die fillet radius is set as 0.2 mm.

The loading paths of internal pressure and punch positions for free bulging and flanging stages are shown in Figure 4. At stage 1, the internal pressure is increased from 0 to P_{i1}. At stage 2, the punch starts to move downward to implement flanging stage with punch speed of 1 mm/s.

Table 1. Forming conditions used in finite element simulations.

Object	Dimension or Property
Tube material	A6063-T0 (Elasto-plastic)
Tube outer diameter d0 (mm)	50.8
Tube initial thickness t (mm)	3
Punch material	Rigid
Punch radius rp (mm)	3.45
Punch head height Hp (mm)	4.5
Die material	Rigid
Die inner diameter (mm)	50.8
Die hole radius rh (mm)	3.5
Internal pressure P_i (MPa)	50
Temperature (°C)	25

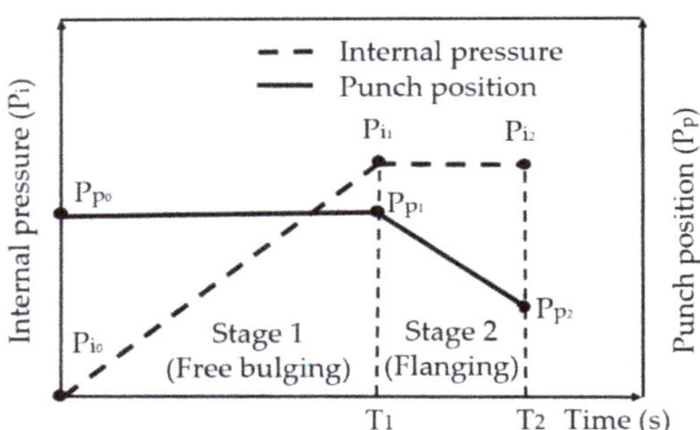

Figure 4. Loading path for internal pressurization and punch position.

An implicit and static FE code "DEFORM 3D" was adopted to analyze the plastic deformation pattern of an aluminum alloy tube during hydro-flanging processes. The finite element code is based on the flow formulation approach using an updated Lagrange procedure. At first, the geometries of the objects are constructed using a commercial software Solidworks. Then, DEFORM 3D is used to implement the simulation of hydro-flanging processes. The punch and die are regarded as rigid bodies and the tube is elasto-plastic. The flow stresses of the tube material are considered as a function of strain and strain rate. The stress-strain curves at strain rates of 0.1 and 0.01 s^{-1} shown in Figure 1 were input in the finite element modelling. The coulomb friction model was adopted at the interfaces between the tube and the punch head and the die. For a dry friction interface, the friction coefficients are usually assumed as 0.1–0.2. Because of different surface roughness at the punch head and the die, friction coefficients of 0.15 and 0.2 were assumed for the interfaces in contact with the punch and tube, respectively. The constant shear friction model is usually adopted at the contact surfaces in warm forming or hot forming processes for avoiding overestimate in the friction stresses. The length of the tube is 60 mm. Due to symmetry on the left-right and front-back side, only one quarter of the objects was adopted in the simulations to save the simulation time. One quarter of the tube having totally about 50,000 elements is divided into three zones, which have different element size ratios, set as 0.2, 0.4 and 1 for zones I, II and III, respectively, as shown in Figure 5a. The tube material in zone I just below the punch head will undergo severest plastic deformation; thus, a smallest element size ratio was set in this region. It is known that there are over ten layers of meshes in zones I and II in the thickness direction of the tube. The mesh configurations before and after hydro-flanging are shown in Figure 5b,c, respectively. The effects of various forming conditions on the thickness distributions at the flanging region and the punch force will be discussed in the next section.

Convergence analyses for simulation results were implemented to understand the ranges of the simulation result errors. Finite element simulations with different total element numbers were conducted. The effects of total element number on the maximal punch force during hydro-flanging processes are shown in Figure 6. Clearly, the relative differences in the maximal punch forces decrease to within 0.5% as the total element number increases to 40,000 elements. Accordingly, approximately 50,000 tetrahedron elements were set for the tube object in the following finite element simulations.

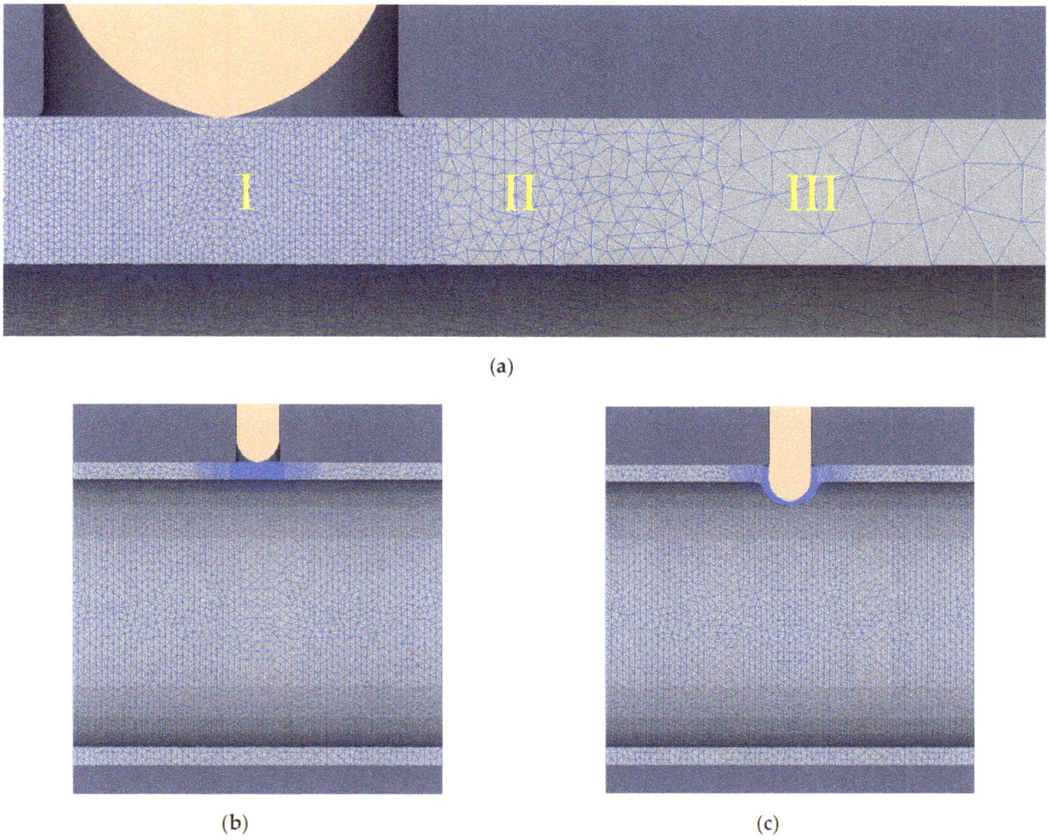

(a)

(b) (c)

Figure 5. Mesh configurations before and after hydro-flanging. (**a**) Element size ratios of 0.2, 0.4 and 1 for zones I, II and III, respectively; (**b**) Before forming; (**c**) After forming.

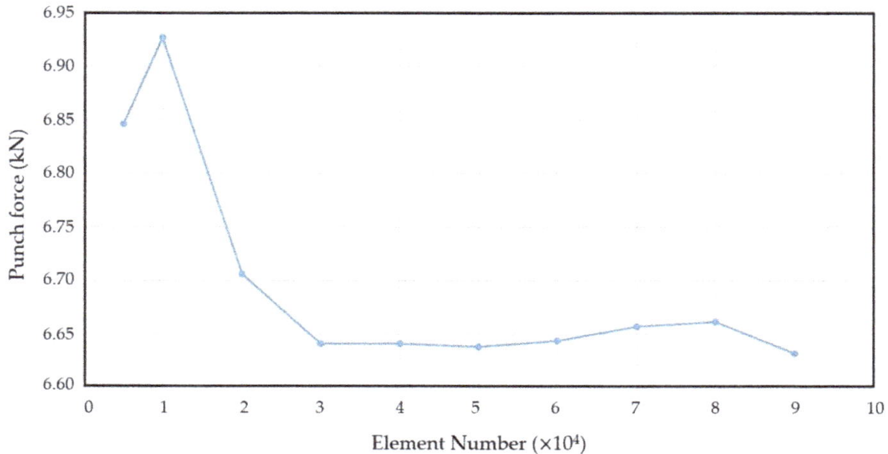

Figure 6. Effects of total element number on maximal punch forces.

3. Simulation Results and Discussion

3.1. Thickness Distributions at Flanging Region

It is important to obtain a thick and long ring zone after hydro-flanging processes for subsequent nut-laying processes. In order to achieve this objective, the tube is first bulged up in the free bulging stage with a distance of H_b to fill the hole vacancy, which can increase the thickness at the ring zone after the flanging stage. Secondly, the forming conditions in the flanging stage, especially the punch shape is designed to make the thickness distributions at the flanging region with $t_2 > t_1 > t_0$. In this way, the fracture position of the tube occurs at the bottom of the punch head, which can make the ring zone become thicker and longer eventually. If the thickness distribution is $t_2 > t_0 > t_1$, or $t_0 > t_2 > t_1$, fracture will probably occur in zone B and the fluid sealing function between the punch head and flanging region may fail. For the case of $t_2 > t_1 > t_0$ at the flanging region, fracture probably occurs at the bottom of the punch head (zone A) and due to the elastic recovery of the tube at the flanging region and the internal pressure compression, the flanging region can stick to the punch head surface and prevent liquid leakage for an effective sealing function.

Figure 7a,b show the simulation results of the thickness distributions with a short and long punch head heights, respectively, at a high internal pressure of P_i = 70 MPa and a stroke of 6.5 mm. The die hole radius is r_h = 3.5 mm, slightly larger than the punch radius r_p = 3.45 mm. From Figure 7a, it can be seen that the thickness of flanging region is quite uniform and there is a small gap or roll-over of h_g = 0.4 mm. Fracture probably occurs at zone B, as the punch moves downward further. That is because the downward drawing force from the surface friction between the tube and punch is quite large, which makes the material close to the lower punch head is pulled downward by the punch. Meanwhile, the tube material close to the upper punch head is pushed up by the high internal pressure. These two effects cause the material in zone B to elongate and, finally, fracture occurs at zone B. Figure 7b shows the effects of a longer punch (H_p = 5 mm) on the thickness distribution at the flanging region. Clearly, the thickness t_0 at the bottom of flanging region is thinnest and even necking occurred there. The thickness t_2 close to the upper punch head is much larger than t_0. If the punch moves downward further, fracturing will happen at the bottom of the punch head (zone A). That is because a sharp punch head makes the tube material more easily to flow upward and make the tube thicker close to the upper punch head. Accordingly, the thickness ratio of t_2/t_0 is an important factor for predicting the occurrence of fracturing within zone A or zone B. On the other hand, in Figure 7b, the gap or roll-over is almost zero, because of the high internal pressure pushing the tube upward to contact with the die tightly. A longer ring zone with a smaller roll-over is beneficial for subsequent nut-inlaying processes.

Table 2 shows thickness distributions at the flanging region for different punch heights and internal pressures. Figure 8 shows the effects of punch height H_p and internal pressure P_i on the thickness ratio t_2/t_0 at the flanging region at a punch stroke of 5 mm. As stated earlier, the thickness ratio of t_2/t_0 is the larger the better. A larger t_2/t_0 can make the fracture occurring at the bottom of the flanging region, have a good sealing effect and obtain a longer ring zone. From Figure 8, it is known that the thickness ratios t_2/t_0 obtained with a larger punch height (an acute punch head) are larger than those using a smaller punch height (a blunt punch head), which means an acute punch head is helpful for obtaining a longer ring zone. As H_p = 5 mm, a larger internal pressure can increase the thickness ratio, whereas, as H_p = 4 mm, a larger internal pressure decreases the thickness ratio. It can be concluded that among the 9 cases, H_p = 5 mm and P_i = 70 MPa are better conditions to obtain a thick and long ring zone after hydro-flanging. The mechanism for an acute or sharp punch head that can obtain a better ring zone is explained in the following figures.

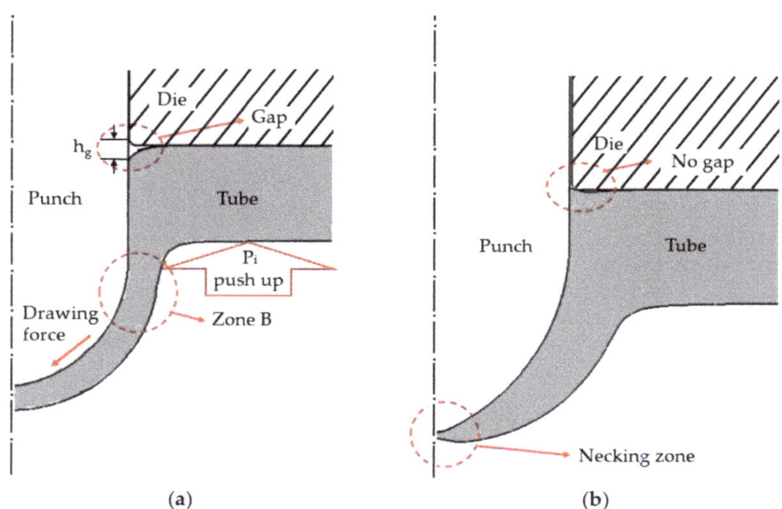

Figure 7. Hydro-flanging results for different punch head heights at P_i = 70 MPa. (**a**) Hp = 4 mm; (**b**) Hp = 5 mm.

From Figure 8, it is known that the internal pressure slightly influences the thickness ratio, whereas the punch head height influences the thickness ratio significantly. In order to obtain the desired thickness ratio or deforming shape of the flanging region, punch shape design is more effective than internal pressure design. The relationships between the forming conditions and the thickness ratio or the gap height are quite complicated. For example, in Figure 8, the tendency of the effects of the internal pressure on the thickness ratio is different for different punch head heights. From the simulation results obtained in this paper, it is difficult to propose an empirical equation to present the relationships between the forming conditions and the thickness ratio or the gap height. In the future, a more detailed plan for selections of various forming parameters will be carried out to obtain more detailed simulation results and then some empirical equations can be proposed.

Table 2. Thickness distributions at flanging region for different punch head heights and internal pressures.

Punch Height (H_p)	Internal Pressure (P_i)	t_2 (Z_2 = 2.5 mm)	t_1 (Z_1 = 3.75 mm)	t_0 (Z_0 = 5 mm)	t_2/t_0
4 mm	30 MPa	1.918	1.750	1.549	1.238
	50 MPa	1.686	1.627	1.420	1.187
	70 MPa	1.523	1.608	1.456	1.046
4.5 mm	30 MPa	2.076	1.664	1.206	1.721
	50 MPa	1.842	1.600	1.100	1.675
	70 MPa	1.760	1.563	1.031	1.707
5 mm	30 MPa	2.117	1.669	0.990	2.138
	50 MPa	1.932	1.597	0.876	2.205
	70 MPa	1.846	1.545	0.810	2.279

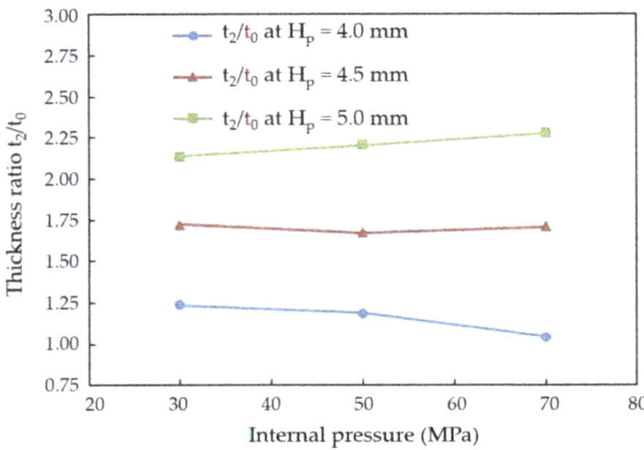

Figure 8. Effects of punch height and internal pressure on thickness ratio at zone B.

Figure 9 shows the effects of internal pressures on the roll-over or gap height with a blunt punch head of H_p = 4 mm and a large die hole radius of r_h = 6.5 mm. In Figure 9a, at a lower internal pressure, a thicker flanging region and a larger gap h_g = 1.0 are obtained. Meanwhile, at a higher internal pressure in Figure 9c, a thinner flanging region and a negative gap height h_g = −0.44 mm are obtained. At a lower internal pressure, the tube material is pulled downward by the blunt punch head, which results in a thicker flanging region and a positive gap height. At a higher internal pressure, the tube material is bulged upward, which results in a thinner flanging region and a negative gap height. At an intermediate internal pressure of P_i = 50 MPa, shown in Figure 9b, an almost zero gap height is obtained, which geometry is beneficial for subsequent inlaying joining processes.

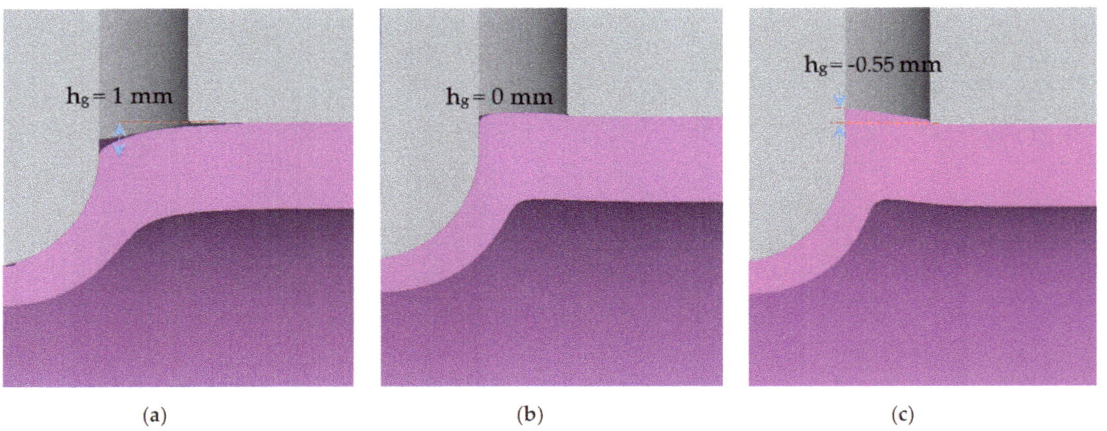

Figure 9. Effects of internal pressure on gap height with larger die hole radius. (**a**) P_i = 30 MPa; (**b**) P_i = 50 MPa; (**c**) P_i = 70 MPa.

Figure 10 shows the effects of die hole radius on the roll-over or gap height with a blunt punch head of H_p = 4 mm and an internal pressure of P_i = 50 MPa. It can be seen from the figure that as the die hole radius r_h increases, much more tube materials can be raised by the internal pressure to the hole cavity of the die and a smaller roll-over or gap is obtained. However, if the internal pressure is larger than 50 MPa, the gap height hg may become negative, which is not beneficial for subsequent inlaying processes. The target of this hydro-flanging process is obtaining a zero-gap height (h_g = 0) and making the ring zone as thick as possible.

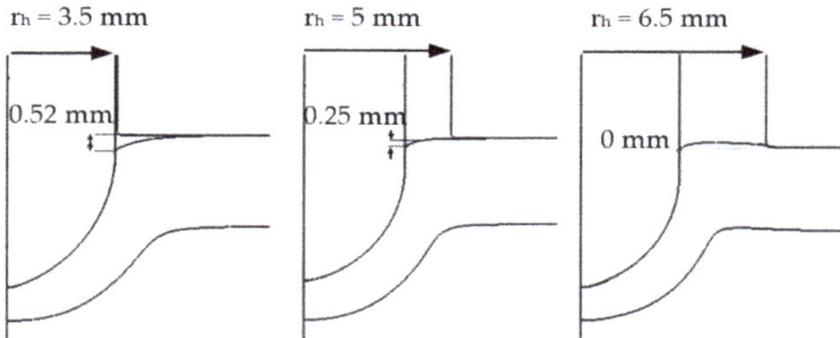

Figure 10. Effects of die hole radius on gap height under P_i = 50 MPa.

3.2. Punch Load Variations

Figure 11 shows the punch load variations with different punch height H_p at P_i = 50 MPa and r_h = 5 mm. The punch load is determined by the punch contact area and the tube thickness at the flanging region. At the early stage, the punch contact area increases with the punch stroke; thus, the punch load increases with the stroke. As the stroke reaches about 4~5 mm, the whole punch head surface is in contact with the tube. After that, the tube at the flanging region continuously becomes thinner. Accordingly, the load begins to decrease at the late stage. For a long punch head (larger H_p), the projected area is smaller than that obtained with a short punch head (smaller H_p) at the same stroke, thus, a smaller punch load is obtained with a larger H_p. The stroke corresponding to the peak value with a larger H_p occurs later, that is because the whole contact surface at the punch head reaches the maximum at a later moment compared with a smaller H_p. At the final stage, the fracture point with H_p = 5 mm occurs at the bottom of the flanging region (zone A), whereas the fracture point with H_p = 4.5 mm occurs at the middle of the flanging region (zone B). The average thickness with H_p = 5 mm is thicker than that with H_p = 4.5 mm, thus, the punch load at the fracture point with H_p = 5 mm is slightly larger than that with H_p = 4.5 mm.

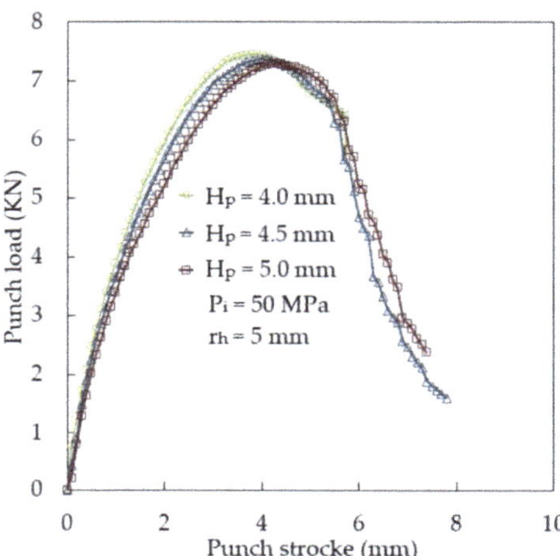

Figure 11. Punch load variations for different punch heights.

Figure 12 shows the punch load variations for different internal pressures P_i at H_p = 4.5 mm and r_h = 5 mm. Clearly, the punch load increases with the internal pressure, because a larger punch load is needed to counterbalance the internal pressure imposed on the tube inner surface. At a higher pressure, more tube material is pushed up and more tube material is in contact with the punch head; thus, the punch load reaches its peak value earlier than that at a lower pressure.

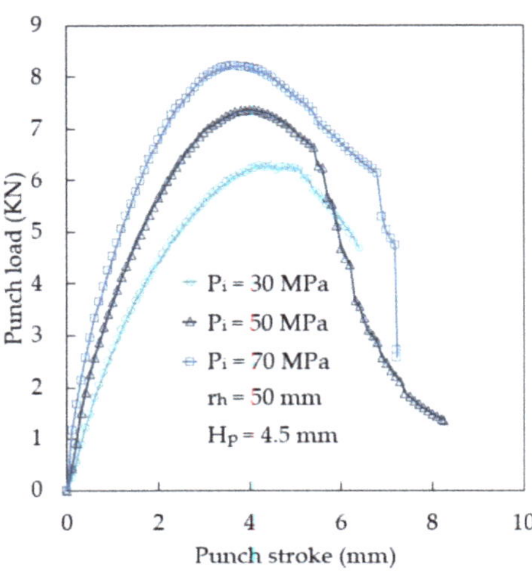

Figure 12. Punch load variations for different internal pressures.

4. Experiments of Hydro-Flanging of Aluminum Alloy Tubes

Aluminum alloy tubes A6063 were annealed before hydro-flanging experiments. The annealing temperature was kept at 420 °C for 3 h, then reduced at a cooling rate of 30 °C/h until 260 °C and, finally, cooled down to room temperature naturally. Figure 13 is the cross-sectional drawing of the die set for hydro-flanging experiments. Table 3 is the list of the components used in the die set. A tube (1) was first put into the die cavity and then 2 nuts (9) was screwed to pull two mandrels (4) backward. Two rubber rings (6) were squeezed by the mandrels (4) and two steel rings (5) to dilate laterally to achieve oil sealing functions.

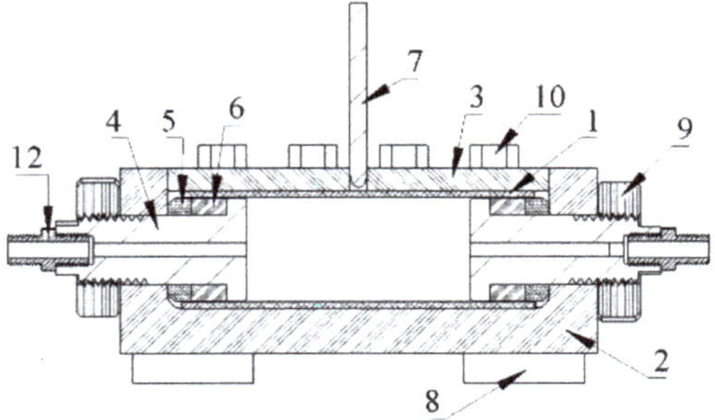

Figure 13. Cross sectional drawing of die set for hydro-flanging.

Table 3. List of die components.

No.	Component	No.	Component
1	Aluminum tube	7	Punch
2	Lower Die	8	Positioning pin
3	Upper Die	9	Nut (M30)
4	Mandrel	10	Screw (M12)
5	Steel ring	11	Screw (M6)
6	Rubber ring	12	Hose connector

Figure 14 shows the main apparatus used in the hydro-flanging experiments. A universal testing machine was used to control the punch movement. A hydraulic system was used to supply oil source and a booster cylinder was designed to make the output pressure reach as high as 100 MPa. The die set, shown in Figure 15, was positioned on the working table of the universal test machine. After the tube was pressurized to the set value, the punch controlled by the universal testing machine started to move downward to implement hydro-flanging experiments. The main purpose of hydro-flanging experiments is to validate the finite element modelling and the punch head shape design concept. Thus, only different internal pressures and punch head shapes were planned for the experiments. The dimensions of the tube, punch and die, such as the tube diameter, punch radius and die hole radius, were so chosen that the maximal internal pressure needed is smaller than 70 MPa, which is the safe capacity of the hydraulic power system used.

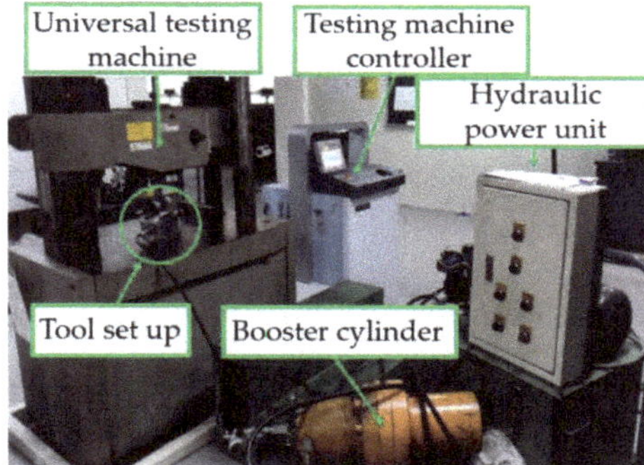

Figure 14. Main apparatus for hydro-flanging experiments.

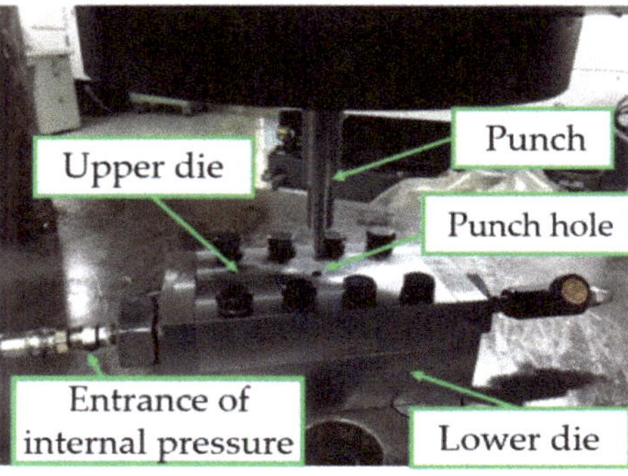

Figure 15. Appearance of die set and punch.

4.1. Experimental Results of Hydro-Flanging Processes

Table 4 shows the appearance of the hydro-flanged tubes and cross-sections at flanged regions under different internal pressures. The punch head height is H_p = 4 mm, the punch head radius is r_p = 3.45 mm and die hole radius is r_h = 3.5 mm. In case 1 (P_i = 0 MPa), fracture occurred at the bottom of the flanging region (zone A) and the aluminum alloy tube was pierced through by the punch with a large roll-over at a stroke of 15.6 mm. As the internal pressure increases, the roll-over or the gap height hg decreases and the thickness at the ring zone increases. However, at P_i = 50 MPa and a stroke of 8 mm in case 4, necking occurs in zone B. As the internal pressure increases to 70 MPa in case 5, the flanging region broken in zone B separates completely from the ring zone. Generally speaking, the results in case 5 with a smaller roll-over and a about 3.4 mm-thick ring zone are beneficial for the subsequent inlaying process.

Table 4. Appearances of hydro-flanged tubes and cross-sections at flanged regions under different internal pressures.

Forming Conditions	Product Appearance	Cross-Sectional Appearance
Case 1 Pi = 0 MPa Stroke = 15.6 mm	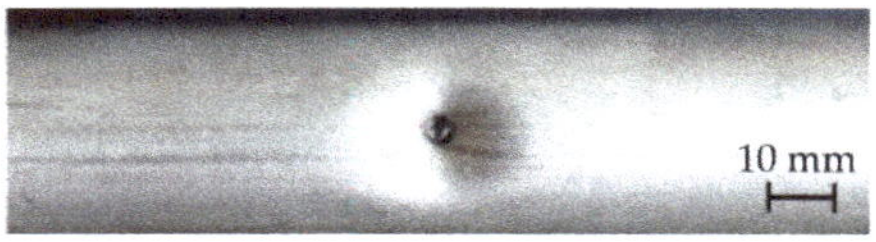 10 mm	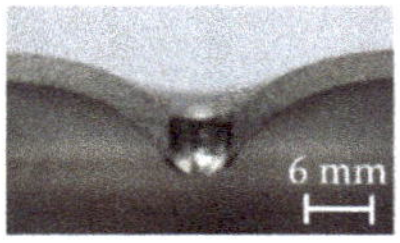 6 mm
Case 2 Pi = 30 MPa Stroke = 7 mm	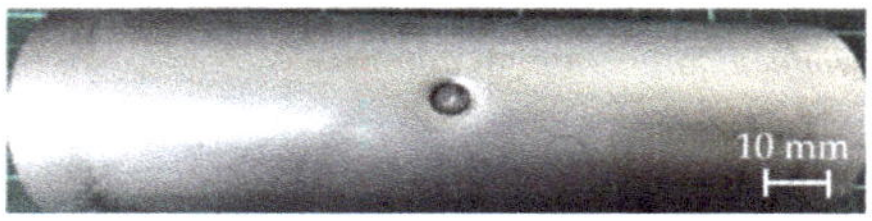 10 mm	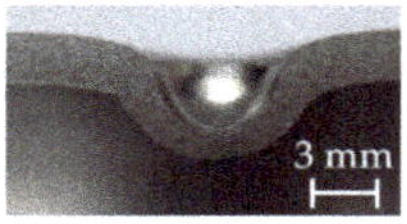 3 mm
Case 3 Pi = 50 MPa Stroke = 7 mm	 10 mm	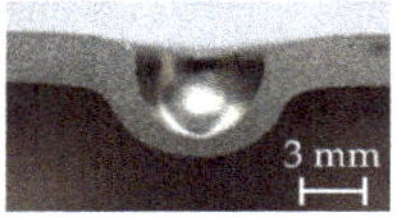 3 mm
Case 4 Pi = 50 MPa Stroke = 8 mm	 10 mm	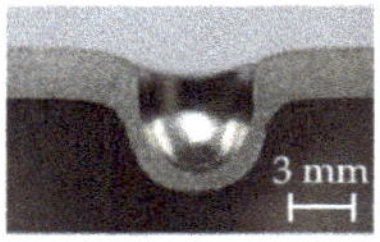 3 mm
Case 5 Pi = 70 MPa Stroke = 8 mm	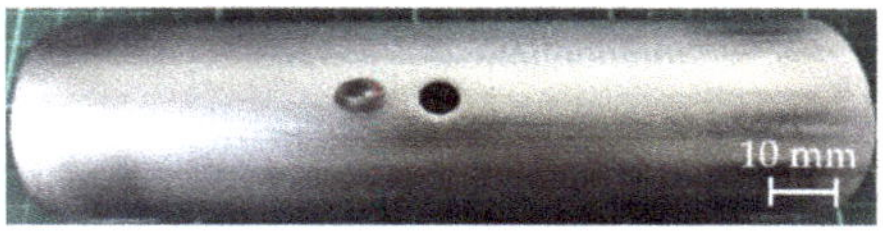 10 mm	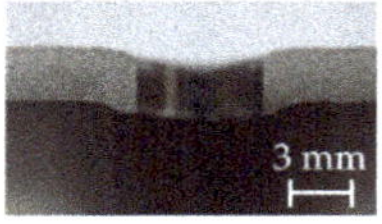 3 mm

Figure 16 shows the experimental punch load variations for different internal pressures under H_p = 4 mm and r_h = 3.5 mm. It can be seen that the punch load increases significantly with the increase of internal pressure and the maximal load occurs earlier at a higher pressure. As the internal pressure is 70 MPa, the maximum punch load occurring at a stroke of 5.3 mm is about 11.4 kN. As the internal pressure is 50 MPa, the maximum punch load occurring at a stroke of 6.2 mm is about 9.1 kN. For P_i = 30 MPa, at a stroke of 7 mm, the punch load has not reached the maximal value yet.

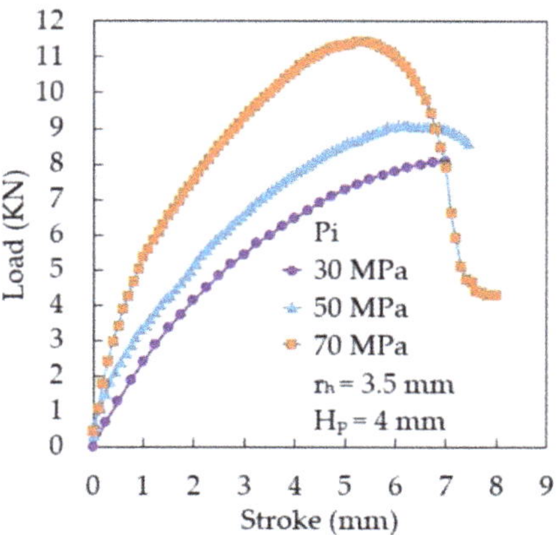

Figure 16. Experimental punch load variations for different internal pressures.

4.2. *Comparisons of Simulative and Experimental Results*

Table 5 shows the comparisons of thickness distributions at the flanging region between simulations and experiments. The punch height is fixed as H_p = 4 mm, the die hole radius is r_h = 3.5 mm and the stroke is 7 mm. It can be seen from the table that as the internal pressure increases, both the simulative and experimental thicknesses decrease. Generally, the differences between the simulation and experimental thicknesses were within 20%. The predicted fracture or necking positions were consistent with the simulation results. For P_i = 30 MPa, if the punch moves downward further, fracture occurs in zone B as shown in Figure 17. For P_i = 50 MPa, necking occurs in zone B as shown in Figure 18. The Normalized Cockcroft and Latham (NCL) ductile fracture criterion [21] was used to determine if the tube material reaches the fracture condition and the corresponding element is removed. The critical damage value was obtained by comparing the elongation and loading curve between the FE simulations and the real tensile tests. The obtained critical damage value for aluminum alloy A6063 is 0.63. The simulation result in Figure 17b showing tube fracture at the upper flanging region was obtained according to the NCL criterion adopted in "DEFORM3D" software.

Figure 19 shows the comparisons of the simulative and experimental punch load variations during hydro-flanging processes under H_p = 4 mm and r_h = 3.5 mm. The coefficient of friction used in the finite element simulations was μ = 0.3, because a lot of new surface of the aluminum alloy tube generated at the flanging region during the hydro-flanging processes. It can be seen that the punch load increases with increasing internal pressures. For internal pressures of P_i = 30 MPa and 50 MPa, the simulative maximum loads are almost the same as the experimental values. For internal pressure of P_i = 70 MPa, some deviation of about 10% exists between the simulative and experimental maximal punch loads. For the maximal loads, generally the simulative values are close to the experimental data within an error of 10%. However, the forming moments or punch strokes corresponding to the maximal loads in the experiments occurred slightly later compared with the simulations. The reasons for the delay are probably because of the elastic deformation of the die and the compressibility of the hydraulic oil. During the finite element simulations, the punch and die were all assumed as rigid bodies and the compressibility of fluid media was not considered. During hydro-flanging, the load usually increases at the early stage and decreases at the late stage. A maximal value occurs as the whole punch head surface is in contact with the tube, as shown in Figures 11 and 12. For internal pressure of P_i = 30 MPa in the experiments, because of the delayed response of the load variation, the late stage for the decreasing period did not appear on the curve for the experiments with P_i = 30 MPa.

Table 5. Comparisons of thickness distributions at flanging region between simulation and experiment.

H_p = 4 mm		Simulation		Experiment		Difference	
r_h	Z	30 MPa	50 MPa	30 MPa	50 MPa	30 MPa	50 MPa
3.5 mm	t_2 (z_2 = 2.5 mm)	1.65	1.22	1.79	1.58	−7.82%	−22.7%
	t_1 (z_1 = 3.75 mm)	1.43	1.27	1.51	1.41	−5.29%	−9.93%
	t_0 (z_0 = 5 mm)	1.13	1.12	1.26	1.20	−10.3%	−6.67%
	t_2/t_0	1.46	1.089	1.421	1.345	2.74%	−19.0%

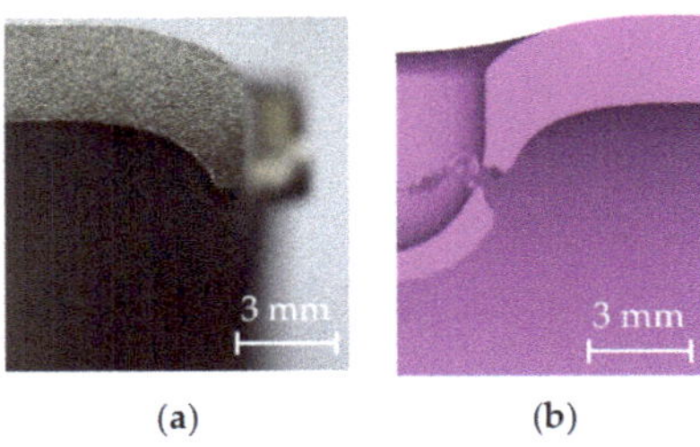

Figure 17. Cross-sectional configurations after hydro-flanging at P_i = 30 MPa. (**a**) Experiment; (**b**) Simulation.

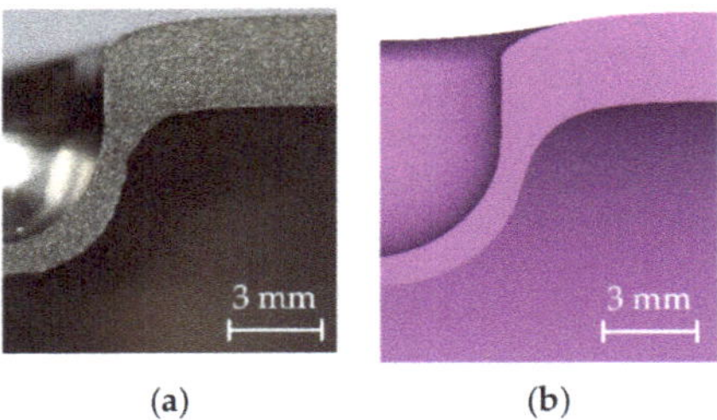

Figure 18. Cross-sectional configurations after hydro-flanging at P_i = 50 MPa. (**a**) Experiment; (**b**) Simulation.

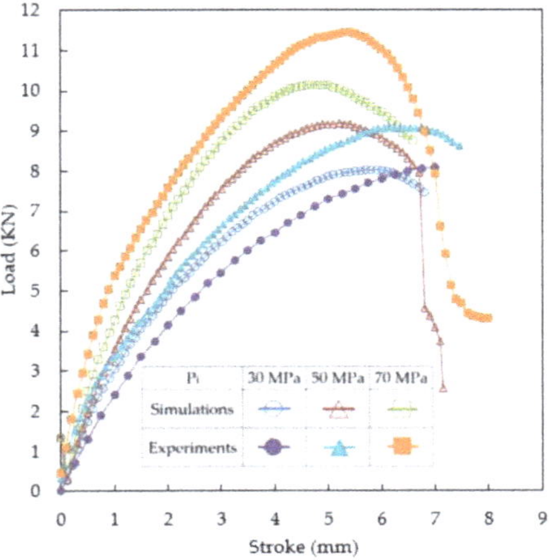

Figure 19. Simulative and experimental punch load variations during hydro-flanging.

5. Conclusions

In this study, a new hydro-flanging process was proposed to investigate the plastic deformation pattern of an aluminum alloy tube at the flanging region. A series of finite element simulations and experiments of hydro-flanging processes were carried out and some conclusions can be drawn as below:

(1) As the internal pressure increased, the thickness of the tube at the bottom of the punch head became thinner. At a higher internal pressure, the tube was bulged up and, hence, the thickness close to the ring zone became thicker and that under the punch head became thinner.
(2) A short blunt punch head made the thickness distribution more uniform and the thickness t_2/t_0 ratio was close to unity, which was more likely to result in fracture in zone B. If a long punch head was used instead, t_2/t_0 became larger, fracture would probably occur at zone A at the bottom of the punch head.
(3) The change of punch hole radius r_h at a lower pressure did not affect the thicknesses at the flanging region; however, at a higher pressure of 70 MPa, the thicknesses decreased obviously as r_h increased.
(4) The simulation results of thickness distributions and maximum punch loads were generally the same as the experimental results. Under a high internal pressure and a short blunt punch head, fracture was more likely to occur in zone B.

In this paper, the thickness distributions at the flanging region affected by the punch head shapes and internal pressures or loading paths were systematically discussed. In the future, bonding forces or holding forces of the inlaid nut from the ring zone for various hydro-flanged ring zone shapes during a nut hydro-inlaying process will be investigated.

Author Contributions: Conceptualization, methodology, research supervision and writing—review and editing were conducted by Y.-M.H. Simulations, experiments, were conducted by H.-N.P. Validation and writing—original draft preparation were completed by H.-S.R.T. All authors have read and agreed to the published version of the manuscript.

Funding: This research received a funding from the Ministry of Science and Technology of the Republic of China under Grant no. MOST 106-2221-E-110-029-MY3.

Institutional Review Board Statement: Not applicable.

Informed Consent Statement: Not applicable.

Data Availability Statement: Not applicable.

Acknowledgments: The authors would like to extend their thanks to the Ministry of Science and Technology of the Republic of China under Grant no. MOST 106-2221-E-110-029-MY3. The advice and financial support of MOST are greatly acknowledged.

Conflicts of Interest: The authors declare no conflict of interest.

References

1. Dohmann, F.; Hartl, C. Tube hydroforming—Research and practical application. *J. Mater. Process. Technol.* **1997**, *71*, 174. [CrossRef]
2. Ahmetoglu, M.; Altan, T. Tube hydroforming: State-of-the-art and future trends. *J. Mater. Process. Technol.* **2000**, *98*, 25. [CrossRef]
3. Fracz, W.; Stachowicz, F.; Trzepiecinski, T. Investigations of thickness distribution in hole expanding of thin steel sheets. *Arch. Civ. Mech. Eng.* **2012**, *12*, 279–283. [CrossRef]
4. Kacem, A.; Krichen, A.; Manach, P.Y.; Thuillier, S.; Yoon, J.W. Failure prediction in the hole-flanging process of aluminum alloys. *Eng. Fract. Mech.* **2013**, *99*, 261–265. [CrossRef]
5. Liu, W.; Hao, J.; Liu, G.; Gao, G.; Yuan, S. Influence of punch shape on geometrical profile and quality of hole piercing-flanging under high pressure. *Int. J. Adv. Manuf. Technol.* **2016**, *86*, 1253–1262. [CrossRef]
6. Thipprakmas, S.; Jin, M.; Murakawa, M. Study on flanged shapes in fineblanked-hole flanging process using finite element method. *J. Mater. Process. Technol.* **2007**, *193*, 128–133. [CrossRef]
7. Mizumura, M.; Sato, K.; Kuriyama, Y. Development of nut-inlaying technique in hydroformed component by hydro-burring. *Mater. Trans.* **2012**, *53*, 801–806. [CrossRef]
8. Mizumura, M.; Sato, K.; Suehiro, M. Development of new hydroforming methods. *Nippon Steel Tech. Rep.* 2013. Available online: https://www.nipponsteel.com/en/tech/report/nsc/pdf/103-07.pdf (accessed on 15 March 2021).

9. Choi, S.K.; Kim, W.T.; Moon, Y.H. Analysis of deformation surrounding a hole produced by tube hydro-piercing. *J. Eng. Manuf.* **2004**, *218*, 1091–1097. [CrossRef]
10. Hwang, Y.M.; Wu, R.K. Process and loading path design for hydraulic compound forming of rectangular tubes. *Int. J. Adv. Manuf. Technol.* **2017**, *91*, 2135–2142. [CrossRef]
11. Hwang, Y.M.; Dai, W.H.; Chen, C.C. Investigation of punch shape design in tube hydro-piercing processes. *Int. J. Adv. Manuf. Technol.* **2020**, *110*, 2211–2220. [CrossRef]
12. Hwang, Y.M.; Tsai, Y.J. Movable die and loading path design in tube hydroforming of irregular bellows. *Metals* **2020**, *10*, 1518. [CrossRef]
13. Park, J.Y.; Han, S.W.; Jeong, H.S.; Cho, J.R.; Moon, Y.H. Advanced sealing system to prevent leakage in hydroforming. *J. Mater. Process. Tech.* **2017**, *247*, 103–110. [CrossRef]
14. Yu, X.Y.; Chen, J.; Chen, J.S. Interaction Effect of Cracks and Anisotropic Influence on Degradation of Edge Stretchability in Hole-expansion of Advanced High Strength Steel. *Int. J. Mech. Sci.* **2016**, *105*, 348–359. [CrossRef]
15. Kumar, S.; Ahmed, M.; Panthi, S.K. Effect of Punch Profile on Deformation Behaviour of AA5052 Sheet in Stretch Flanging Process. *Arch. Civil Mech. Eng.* **2020**, *20*, 1–17. [CrossRef]
16. Dixit, U.S.; Joshi, S.N.; Davim, J.P. Incorporation of material behavior in modeling of metal forming and machining processes: A review. *Mater. Des.* **2011**, *32*, 3655–3670. [CrossRef]
17. Del Pozo, D.; López de Lacalle, L.N.; López, J.M.; Hernández, A. Prediction of press/die deformation for an accurate manufacturing of drawing dies. *Int. J. Adv. Manuf. Technol.* **2008**, *37*, 649–656. [CrossRef]
18. Fernández-Abia, A.I.; Barreiro, J.; López de Lacalle, L.N.; Martínez-Pellitero, S. Behavior of austenitic stainless steels at high speed turning using specific force coefficients. *Int. J. Adv. Manuf. Technol.* **2012**, *62*, 505–515. [CrossRef]
19. Hwang, Y.M.; Pham, H.N. Study of Hydro-Reaming and Flanging of Aluminum Tubes. Tube Hydroforming Technology. In Proceedings of the 9th International Conference on Tube Hydroforming, Kaohsiung, Taiwan, 18–21 November 2019; pp. 124–129.
20. ASTM. *Standard Test Methods for Tension Testing of Metallic Materials*; E8/E8M-15; ASTM International: West Conshohocken, PA, USA, 2015; pp. 10–11.
21. Oh, S.I.; Chen, C.C.; Kobayashi, S. Ductile fracture in axisymmetric extrusion and drawing. *ASME J. Eng. Ind.* **1979**, *101*, 36–44. [CrossRef]

Article

Tube Drawing Process with Diameter Expansion for Effectively Reducing Thickness

Shohei Kajikawa [1,*], Hikaru Kawaguchi [1], Takashi Kuboki [1], Isamu Akasaka [2], Yuzo Terashita [2] and Masayoshi Akiyama [3]

1. Department of Mechanical and Intelligent Systems Engineering, The University of Electro-Communications, 1-5-1 Chofu Gaoka, Chofu-shi, Tokyo 182-8585, Japan; kawaguchi@mt.mce.uec.ac.jp (H.K.); kuboki@mce.uec.ac.jp (T.K.)
2. Miyazaki Machinery Systems Co., Ltd., 1 Nii, Kaizuka-shi, Osaka 597-8588, Japan; Akasaka@miyazakijp.com (I.A.); Terashita@miyazakijp.com (Y.T.)
3. Akiyama Mechanical Engineering Consulting, 2-7-306 Tanaka Sekiden-cho, Sakyo-ku, Kyoto-shi, Kyoto 606-8203, Japan; masayoshiakiyama@ac.auone-net.jp

* Correspondence: s.kajikawa@uec.ac.jp; Tel.: +81-42-443-5294

Received: 31 October 2020; Accepted: 4 December 2020; Published: 6 December 2020

Abstract: The present paper describes a tube drawing method with diameter expansion, which is herein referred to as "expansion drawing", for effectively producing thin-walled tube. In the proposed method, the tube end is flared by pushing a plug into the tube, and the tube is then expanded by drawing the plug in the axial direction while the flared end is chucked. The forming characteristics and effectiveness of the proposed method were investigated through a series of finite element method (FEM) analyses and experiments. As a result of FEM analysis, the expansion drawing effectively reduced the tube thickness with a smaller axial load when compared with the conventional method. According to the experimental results, the thin-walled tube was produced successfully by the expansion drawing. Maximum thickness reduction ratios for a carbon steel (STKM13C) and an aluminum alloy (AA1070) were 0.15 and 0.29 when the maximum expansion ratios were 0.23 and 0.31, respectively. The above results suggest that the proposed expansion drawing method is effective for producing thin-walled tubes.

Keywords: drawing; flaring; tube expansion; plug drawing; thickness reduction

1. Introduction

Thin-walled tubes, which are used for various machine components, contribute to the reduction in size and weight of various machines. The thin-walled tube is manufactured from rather thick-walled raw tubes by multi-pass drawing, which is the conventional cold working forming process of tubes, that determines many properties of the tube, such as residual stress [1], dimensional precision [2,3], surface integrity [4], and strength [5]. However, many drawing passes are needed for manufacturing very-thin-walled tubes. This is because the tube tends to fracture due to large axial load in the drawing process when the thickness reduction is too large for one drawing pass. The production cost also increases with the increase in the number of drawing passes.

A number of technologies were proposed for manufacturing thin-walled tubes. The spinning process is effective for greatly reducing the tube thickness [6,7]. However, its productivity is low due to low feed rate of the process. Pilger rolling is another method for reducing the diameter and thickness of the tube, with an area reduction of more than 80% [8]. However, the productivity is also low because the tube is deformed incrementally by reciprocal movement of a pair of rolls. Therefore, an effective processing method, which realizes both effective thickness reduction and good productivity, should be developed. Tube expansion process is considered to be one of the methods for reducing wall thickness. The thickness reduces due to stretching in the hoop direction because the material deforms with constant volume. In general, tubes can be partially expanded by hydro forming or press forming for producing components with various shapes [9,10]. In addition, the tube expansion is also applied for joining tubes to other components [11], and for forming a well-bore tube [12]. Therefore, various fundamental studies about the tube expansion have been conducted numerically and experimentally [13–16]. On the other hand, there is little research about the production of thin-walled tube by expanding the whole length of the tube.

The present paper proposes a tube drawing method with diameter expansion for effectively manufacturing thin-walled tube. Figure 1 shows a schematic diagram of the proposed method. In the first step, the tube end is flared by pushing the plug in the axial direction, as shown in Figure 1a. In the second step, the tube was expanded by pushing the plug while the flared end is chucked, as shown in Figure 1b. Figure 2 compares the stress state of the tube wall during drawing with diameter shrinkage or expansion. In the expansion drawing, the tube wall stretches biaxially in the axial and hoop directions, and then a negative deviatoric stress in the thickness direction becomes larger than that in the conventional drawing with diameter shrinkage. Therefore, the thickness should be effectively reduced by resorting to a small drawing load.

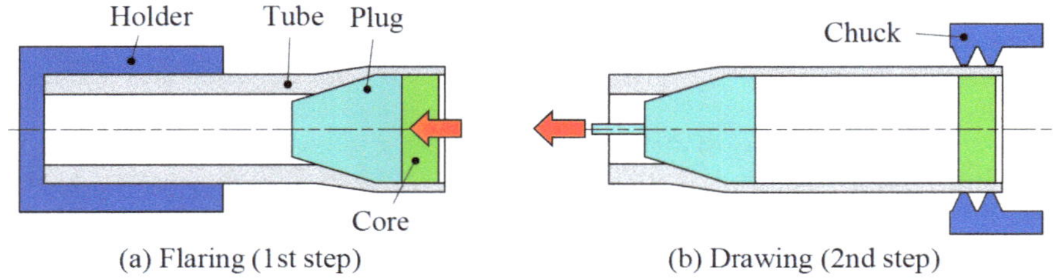

Figure 1. Schematic diagram of drawing process with diameter expansion.

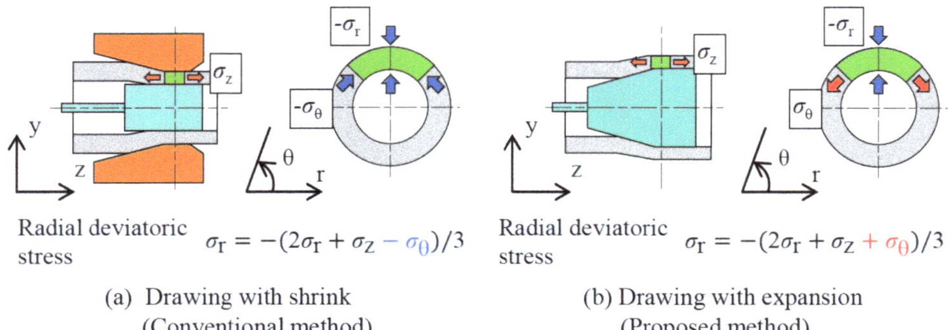

Figure 2. Stress state of tube wall during drawing with shrink and drawing with expansion.

The present study fundamentally investigated the forming characteristics and the effectiveness of the proposed method through a series of finite element method (FEM) analyses and experiments. First, the tube flaring was conducted, and the friction coefficient, μ, was estimated by comparing the FEM and experimental results. Based on the results for the tube flaring, FEM analysis of the expansion drawing was carried out to investigate the forming characteristics and the deformation mechanism, and the proposed method was compared with the conventional method. Finally, the forming limit, thickness reduction, ovality, and the thickness deviation were evaluated through a series of experiments.

2. Materials and Methods

2.1. Method for Estimating Frictional Property by Tube Flaring

The friction coefficient μ between the tube and the expansion plug was investigated by comparison of the experimental and FEM result in the flaring process. Figure 3 shows a schematic diagram of the tube flaring experiment. One tube end was flared by pushing a conical plug, while the other end was fixed by a holder. The flaring ratio κ_f was defined by the equation

$$\kappa_f = \frac{d_{if} - d_{i0}}{d_{i0}} \quad (1)$$

where d_{i0} is the initial inner diameter of the tube. Typical forming defects in this experiment were crack and buckling, as shown in Figure 3c. The maximum flaring ratio κ_{f_max}, which indicates the flaring limit, was defined as the maximum value of κ_f before the buckling or the crack occurrence. In the experiment, a hydraulic universal testing machine was used to press the plug into the tube while the load was measured. The inner diameter d_{if} of the flared end was measured every time the plug was pressed 2 mm in the axial direction, and the maximum value of d_{if} before defect occurrence was used to calculate the maximum flaring ratio κ_{f_max}.

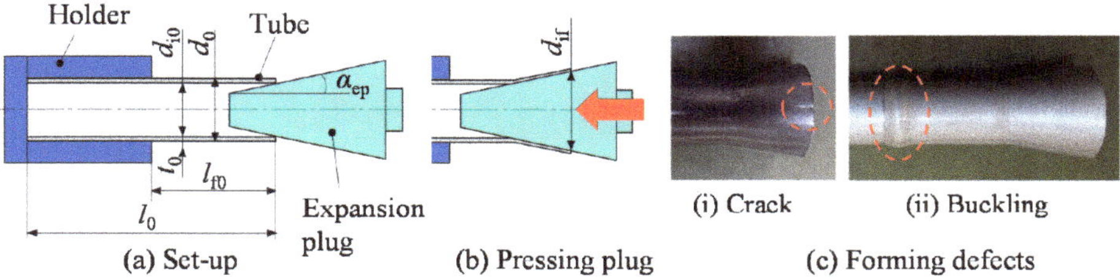

Figure 3. Schematic diagram of flaring experiment.

Figure 4 shows a schematic diagram of the FEM model for the tube flaring (FL-2D-Impl). An elastic-plastic analysis was carried out using the commercial code "ELFEN" which was developed by Rockfield Software Limited, Swansea, UK. The model is a two-dimensional axisymmetric model. An implicit scheme and four-node rectangular elements were adopted. The F-bar method was applied to the element to overcome volumetric locking [17]. In the analysis of the tube flaring, one tube end was flared by the plug being pressed until buckling occurred or until the plug reached the fixed portion, while the other end and the outer surface were fixed at the holder portion.

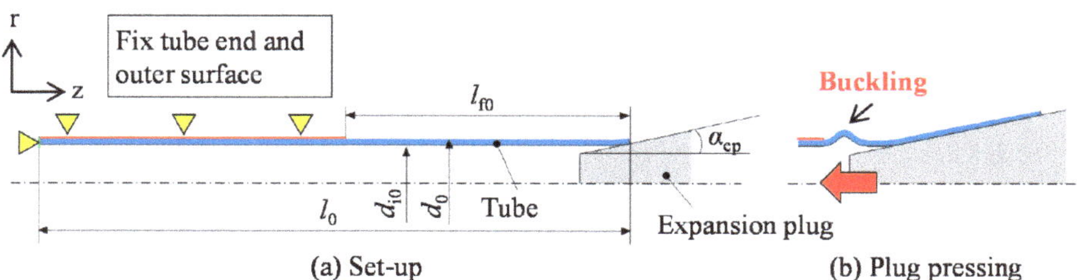

Figure 4. Two-dimensional implicit model for flaring (FL-2D-Impl) for FEM analysis.

Table 1 shows the process conditions for the tube flaring in the FEM and the experiment. Two tube materials were tested: a carbon steel STKM13C (Japan Industrial Standard, JIS) and an aluminum alloy

AA1070. Table 2 shows these material properties used for FEM analysis. The Hooke's law and the von Mises yield criterion were adopted. The constraints were determined by the penalty function method. Figure 5 shows effective stress σ_{eq} vs. effective plastic strain ε_{eq} curves of the tubes. Here, the σ_{eq}–ε_{eq} curves were obtained by tensile tests, and these curves were approximated by the Swift equation. FEM analysis was carried out under various friction coefficients μ, which ranged from 0.1 to 0.3. Here, μ between the tube and the plug was estimated by comparing the experimental flaring load P with that of the analysis for various μ.

Table 1. Process conditions for tube flaring.

Tube	Material	Carbon Steel (STKM13C) Aluminum Alloy (AA1070)
	Initial diameter d_0 (mm)	30
	Initial thickness t_0 (mm)	2
	Initial length l_0 (mm)	200
	Initial flared length l_{f0} (mm)	100
	Element size for flaring region (FEM)	8 elements in thickness 0.5 mm/div along the z axis
Expansion plug	Material	SKD11 (Experiment) Rigid (FEM)
	Half angle α_{ep} (°)	12
Lubricant (Experiment)		Stamping and Forming Lubricant G-3344 (Nihon Kohsakuyu Co., Ltd., Minato-ku, Japan)
Friction coefficient μ (FEM)		0.1–0.3

Table 2. Material properties of tube materials (FEM).

Property	STKM13C	AA1070
Young's modulus (GPa)	200	66
Poisson ratio	0.3	0.34
Yield stress (MPa)	637	88
Swift equation $\sigma_{eq} = F(\varepsilon_{eq} + \varepsilon_0)^n$	$\sigma_{eq} = 725(\varepsilon_{eq} + 0.0032)^{0.023}$	$\sigma_{eq} = 121(\varepsilon_{eq} + 0.0013)^{0.048}$

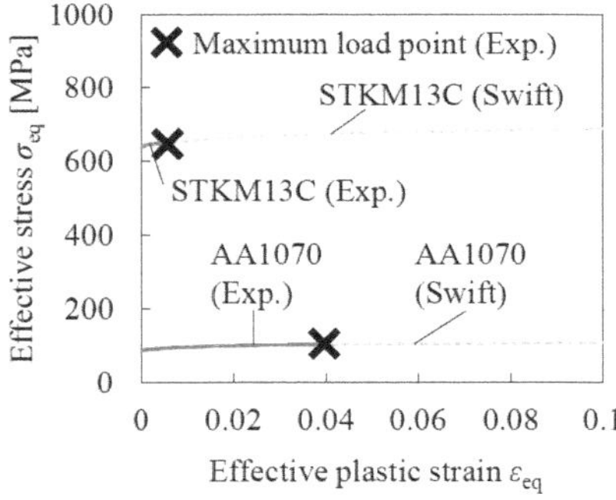

Figure 5. Effective stress σ_{eq} vs. effective plastic strain ε_{eq} curves of STKM13C and AA1070.

2.2. Expansion Drawing Method and Conditions

FEM analysis of the drawing process was carried out in order to investigate the effectiveness of the proposed method over the conventional method. In addition, expansion drawing experiments were performed and the results were compared with those from FEM analysis. Figure 6 shows schematic diagrams of the FEM models for the drawing processes. Two-dimensional axisymmetric models, which are shown in Figure 6a,b, were used to compare the characteristics of the proposed method to those of the conventional drawing method with diameter shrinkage. An implicit scheme and four-node rectangular elements were adopted for the two-dimensional models. The F-bar method was applied to the element to overcome volumetric locking [9]. In expansion drawing analysis using the model (FD-2D-Impl), the expansion plug was displaced in the axial direction while fixing one tube end, as shown in Figure 6a. Analysis of plug drawing, which is the conventional method, was carried out using the model (PD-2D-Impl), as shown in Figure 6b. In this model, both the die and the plug were displaced in the axial direction, promoting the tube thickness reduction by ironing between the die and the plug. The target thickness g was controlled by changing the plug diameter d_p while the die diameter d_d was constant. In addition, a three-dimensional model (FD-3D-Expl), which is shown in Figure 6c, was used to consider the effect of initial thickness deviation in expansion drawing. An explicit scheme and eight-node hexahedral elements were adopted. The thickness deviation was reproduced by changing the center of axis of the outer surface in the x direction as compared to that of the inner surface.

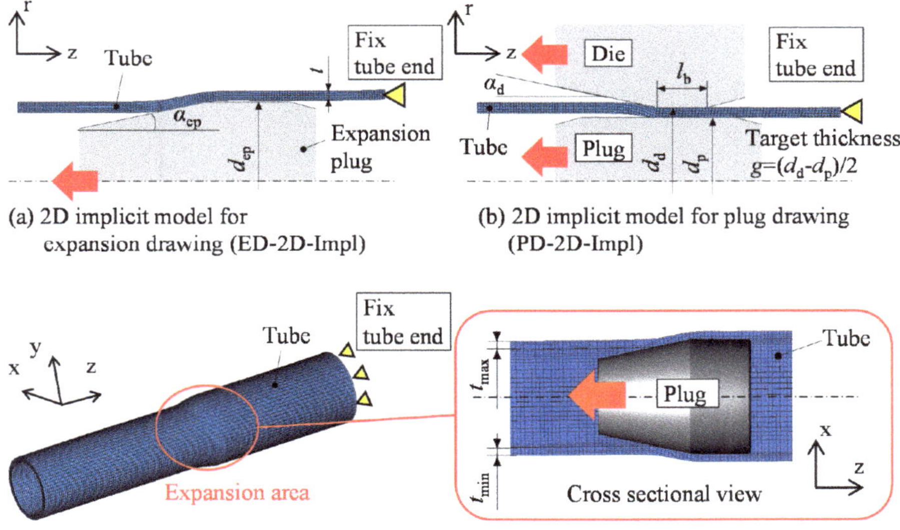

Figure 6. FEM models for drawing processes.

Figure 7 shows a schematic diagram of the experiment for the expansion drawing. Tubes having ends that were previously flared were used in this experiment. The flared portion was chucked, and the plug was drawn in the tube axial direction using a hydraulic cylinder. Expansion ratio κ_d, thickness change ratio η, ovality ratio ω, and thickness deviation ratio λ were evaluated. Here, κ_d, η, ω, and λ were defined by the following equations:

$$\kappa_d = \frac{d_{ep} - d_{i0}}{d_{i0}} \quad (2)$$

$$\eta = \frac{t - t_0}{t_0} \quad (3)$$

$$\omega = \frac{d_{max} - d_{min}}{d_{ave}} \quad (4)$$

$$\lambda = \frac{t_{max} - t_{min}}{t_{ave}} \tag{5}$$

where d_{ep} is the plug diameter, d_{i0} is the initial tube inner diameter, and t_0 and t are the tube thicknesses before and after the drawing, respectively. Here, d_{max}, t_{max}, d_{min}, t_{min}, d_{ave}, and t_{ave} are, respectively, the maximum, minimum, and averaged values of the diameter and the thickness in the hoop direction at one cross section. Figure 8 shows the measurement position of t in this experiment. t was measured at 45° intervals in the hoop direction and 10 mm intervals in the axial direction by using an ultrasonic thickness gauge "ECHOMETER 1061" which was produced by Nihon Matech Corporation, Shinjuku-ku, Japan.

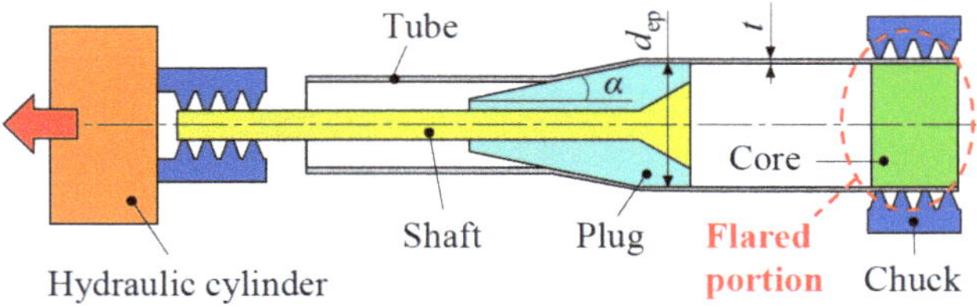

Figure 7. Schematic diagram of expansion drawing experiment.

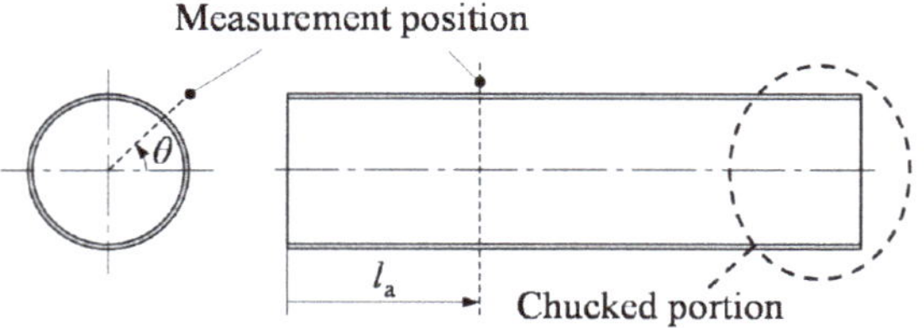

Figure 8. Measurement position of thickness t in drawn tube.

Table 3 shows the conditions of the drawing processes. The materials properties were the same as those in the flaring analysis, which were shown in Table 2 and Figure 5. In expansion drawing, the plug diameter, d_{ep}, was varied in order to control the tube thickness t. In the plug drawing, t was controlled by changing the plug diameter, d_p, while the die diameter, d_d, was constant. The friction coefficient, μ, was set to 0.1 or 0.25 based on the result of flaring.

Table 3. Process conditions for the expansion and plug drawing processes.

Tube (FEM and experiment)	Material		Carbon Steel (STKM13C) Aluminum Alloy (AA1070)
	Initial diameter d_0 (mm)		30
	Initial thickness t_0 (mm)		2, 4
	Initial length l_0 (mm)	Exp.	200
		2D FEM	160
		3D FEM	170
	Element size for flaring region (FEM)	2D FEM	0.5 mm/div along the z axis 8 elements in thickness
		3D FEM	1 mm/div along the z axis 4 elements in thickness 64 elements in hoop

Table 3. *Cont.*

Plug for expansion drawing (FEM and experiment)	Material		SKD11 (Experiment) Rigid (FEM)
	Half angle α_{ep} (°)		12
	Diameter d_{ep} (mm) (Expansion ratio κ_d)		24–38 (0.07–0.46)
Die for plug drawing (FEM)	Material		Rigid
	Half angle α_d (°)		12
	Diameter d_d (mm)		28
	Bearing length l_b (mm)		9
Plug for plug drawing (FEM)	Material		Rigid
	Diameter d_p (mm) (Target thickness g (mm))	t_0 = 2.0 mm	24–25 (1.5–2.0)
		t_0 = 4.0 mm	20–22 (3.0–4.0)
Lubricant (Experiment)			Stamping and Forming Lubricant G-3344 (Nihon Kohsakuyu Co., Ltd., Minato-ku, Japan)
Friction coefficient μ (FEM)			0.1 (STKM13C) 0.25 (AA1070)

3. Results and Discussions

3.1. Identification of Friction Coefficient

Figure 9 shows a load-stroke diagram of the tube flaring. In the experiment, the load P increased with an increase in the stroke S, and P drastically decreased when the forming defect occurred. For both materials, STKM13C and AA1070, the maximum flaring ratio κ_{f_max} was 0.59. The forming defect of STKM13C was cracking, which is shown in Figure 3c(i), while the forming defect of AA1070 was buckling, which is shown in Figure 3c(ii). The load P vs. stroke S curve of FEM showed the same tendency as that of the experiment, and P increased with the increase in the friction coefficient μ. The experimental P vs. S curves fit the FEM P vs.S curves for μ = 0.1 and 0.25, in the cases STKM13C and AA1070, respectively. In the following FEM analysis of the drawing process, μ was set to 0.1 and 0.25 for STKM13C and AA1070, respectively.

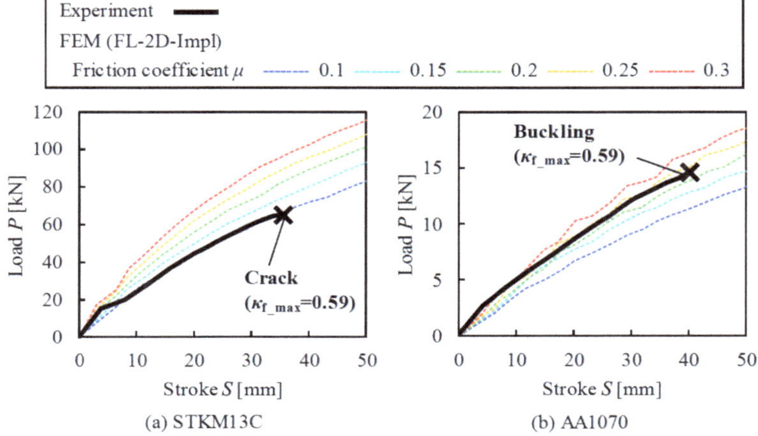

Figure 9. Load P vs. stroke S diagram for tube flaring.

3.2. Investigation of Thickness Reduction Efficiency by FEM Analysis

FEM analysis with two dimensional models, which are shown in Figure 6a,b, was conducted to investigate the forming characteristics and the effectiveness of the expansion drawing, compared to the conventional drawing process with the diameter shrinkage. In this investigation, the material properties of STKM13C were adopted. Figure 10 shows the effect of the expansion ratio κ_d on the thickness reduction ratio γ, evaluated numerically and theoretically (assuming an axial strain $\varepsilon_z = 0$). Here, γ is equal to the negative value of the thickness change ratio η (i.e., $\gamma = -\eta$). Moreover, γ of FEM is the averaged value in a steady region of the drawing. Theoretical value of processed tube thickness $t_{\varepsilon z=0}$ was calculated from the following equation, which assumes that the cross-sectional area does not change, and the inner diameter expand to the plug diameter d_{ep}.

$$t_{\varepsilon z=0} = \frac{\sqrt{4t_0(d_0-t_0) + d_{ep}^2} - d_{ep}}{2} \tag{6}$$

Theoretical value of γ was calculated using $t_{\varepsilon z=0}$. γ increased with the increase in κ_d, similarly to the theoretical value. Therefore, the tube mainly elongated in the hoop direction in this process. The effect of the initial tube thickness t_0 on γ was small. The axial strain ε_z was considered to be small, and either the elongation or the shrinkage depended on the forming conditions, such as κ_d and t_0. For example, when κ_d was over 0.15 and $t_0 = 2$ mm, the value of γ obtained by FEM, was smaller than its theoretical value due to a shrank in the axial direction; while for $t_0 = 4$ mm, the value of γ obtained by FEM was larger than the theoretical value due to an elongation in the axial direction.

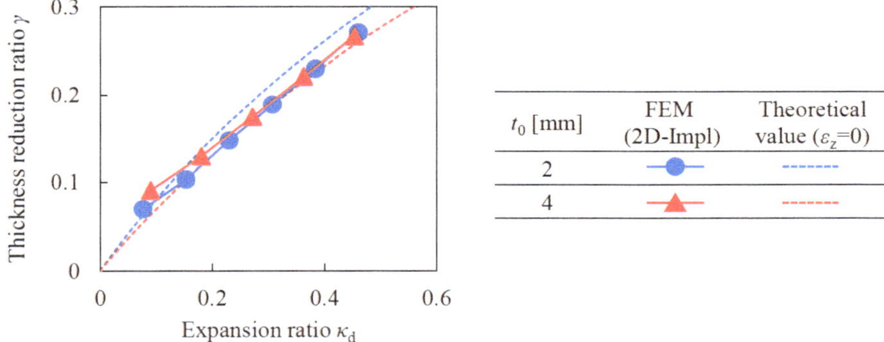

Figure 10. Effect of expansion ratio κ_d on thickness reduction ratio γ (STKM13C, 2D FEM).

Figure 11 shows the relationship between the thickness reduction ratio γ and the drawing load P. Here, γ of the expansion drawing was larger compared to that of the plug drawing. The result suggests that the expansion drawing requires a lower load, in comparison with the plug drawing, to reach the same thickness reduction. On the other hand, γ drastically increased when P reached a large value, namely, 96 and 114 kN for $t_0 = 2$ and 4 mm, respectively, in the case of the plug drawing. Necking is considered to occur easily in this load region, leading to eventual tube fracture. A drastic increase of γ was not observed in the expansion drawing. Therefore, expansion drawing has the potential to improve the thickness reduction limit.

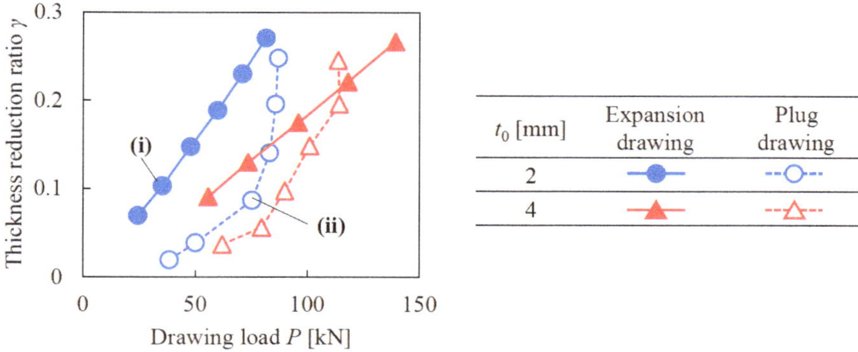

Figure 11. Relationship between thickness reduction ratio γ and drawing load P (STKM13C, 2D FEM).

Figure 12 shows the relationship between the effective plastic strain ε_{eq} and the thickness reduction ratio γ. ε_{eq} increased with the increase in γ. This means that the tube work-hardened with the thickness reduction. ε_{eq} changed by the drawing method, although the effect of the tube initial thickness t_0 on ε_{eq} was small. For the expansion drawing, ε_{eq} was smaller than that of the plug drawing when γ was small, and ε_{eq} came close to that of the plug drawing with the increase in γ.

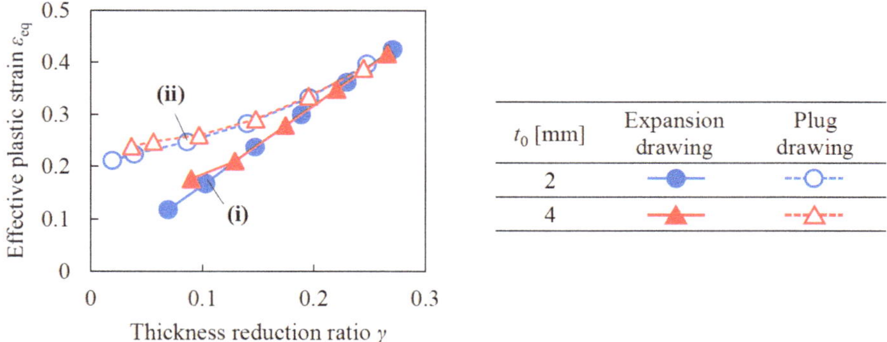

Figure 12. Relationship between effective plastic strain ε_{eq} and thickness reduction ratio γ (STKM13C, 2D FEM).

Figure 13 shows the distribution of deviatoric stress and strain for each drawing process under the condition that the thickness reduction ratio γ was approximately 0.1, which are indicated by (i) and (ii) in Figures 11 and 12. In the case of plug drawing, the tube wall thickened before ironing due to positive radial deviatoric stress σ_r' related to the tube shrinking in the hoop direction. The drawing load P was made large by reducing the thickened tube wall, and the tube elongated in the axial direction. In the expansion drawing, the deformation zone was divided into two regions, which are [A] the diameter expansion, and [B] the axial stretching. In region [A], the tube is mainly stretched in the hoop direction, leading to tube wall thinning due to the negative radial deviatoric stress σ_r'. The axial stress σ_z was small because the inner surface was fixed by the frictional force between the tube and the expansion plug. Therefore, the axial deviatoric stress σ_z' was negative, and the tube shrank in the axial direction. In region [B], the tube mainly stretched in the axial direction because the inner surface was not fixed by decreasing the frictional force with the expansion plug. Therefore, the thickness additionally decreased by axial elongation without changing the hoop length. Here, ε_z of the processed tube is determined by the amount of shrinkage and elongation in region [A] and [B]. In the conditions of Figure 13, the tube slightly shortened in the axial direction because the amount of shrinkage in region [A] was larger than the elongation in region [B].

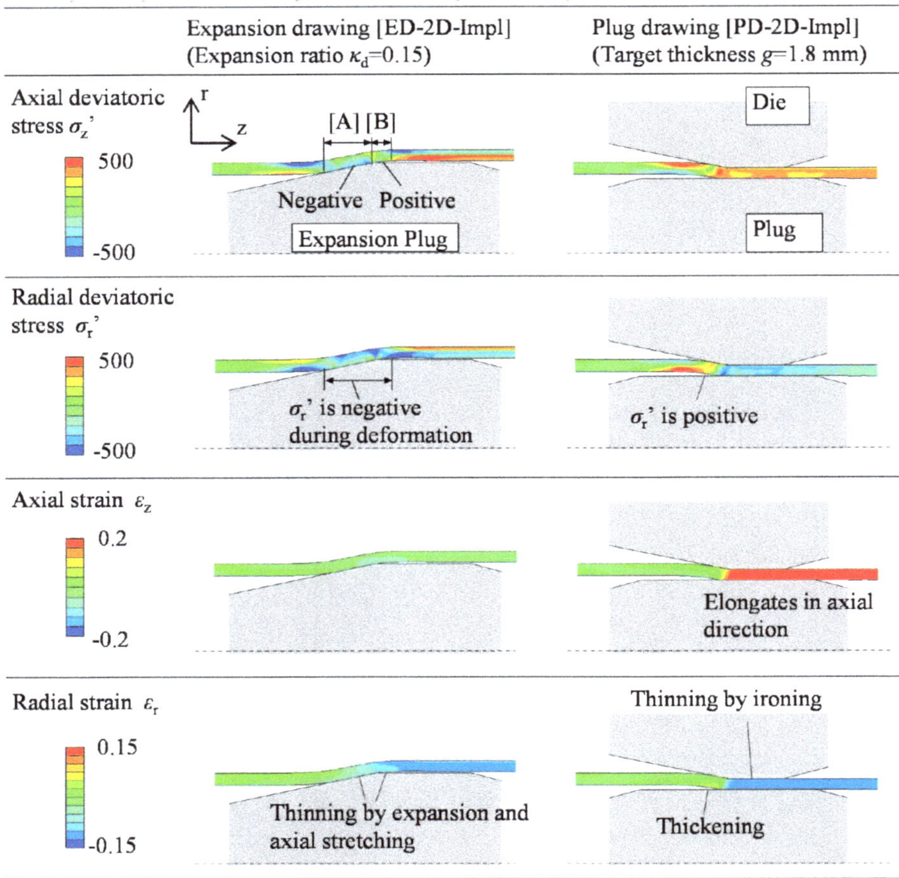

Figure 13. Distributions of deviatoric stress σ_z' and σ_r' and strain ε_z and ε_r for each drawing process (initial tube thickness, t_0 = 2 mm).

3.3. Experimental Investigation of Expansion Drawing

A series of experiments were conducted to investigate the formability on the expansion drawing, and the experimental results were compared to the FEM results. The FEM results were obtained using the models (ED-2D-Impl) and (ED-3D-Expl). STKM13C and AA1070 were used for the tube material in this investigation. Table 4 shows the feasibility of the drawing. The crack runs in the tube axial direction mainly due to the stretching in the hoop direction when the expansion ratio κ_d was too large, as shown in Figure 14. The maximum expansion ratio κ_{d_max} for producing a tube without a crack was 0.23 in the case of STKM13C and 0.31 in the case of AA1070. This is because STKM13C tube fractures easily due to a low elongation property, as shown in Figure 5. In addition, κ_{d_max} was lower than the maximum flaring ratio, κ_{f_max}, for both materials. Furthermore, since in the expansion drawing process, the axial tensile stress σ_z was larger than that of the flaring process, the tube wall easily cracked as a result of biaxial stretching.

Table 4. Feasibility of expansion drawing

Material	Expansion Ratio κ_d				
	0.08	0.15	0.23	0.31	0.38
STKM13C	s	s	s	c	-
AA1070	s	s	s	s	c

s: Success, c: Crack.

Free Chucked Free Chucked

(a) Success (κ_d=0.23) (b) Crack (κ_d=0.38)

Figure 14. Typical appearances of drawn tubes (AA1070).

Figure 15 shows the distribution of the thickness change ratio η in the tube axial direction. Here, η is the average value in the hoop direction for each axial position. The thickness t, reduced by drawing, is uniform, except for the portions, the lengths of which are l_{fr} and l_{ch}, near the tube ends. In the experiment, the portion near the chucked side (l_{ch}) was mainly expanded in the flaring process, in which the axial stress σ_z is lower, resulting in a small thickness reduction. The length l_{ch} becomes longer with the increase in the expansion ratio κ_d, because the taper length of the plug becomes longer with the increase in the plug diameter d_{ep}. The portion near the chucked side (l_{ch}) did not appear in the FEM analysis because the flaring process was not reproduced in the model (ED-2D-Impl). At the portion near the free end (l_{fr}), the thickness reduction was also low because σ_z was low during drawing.

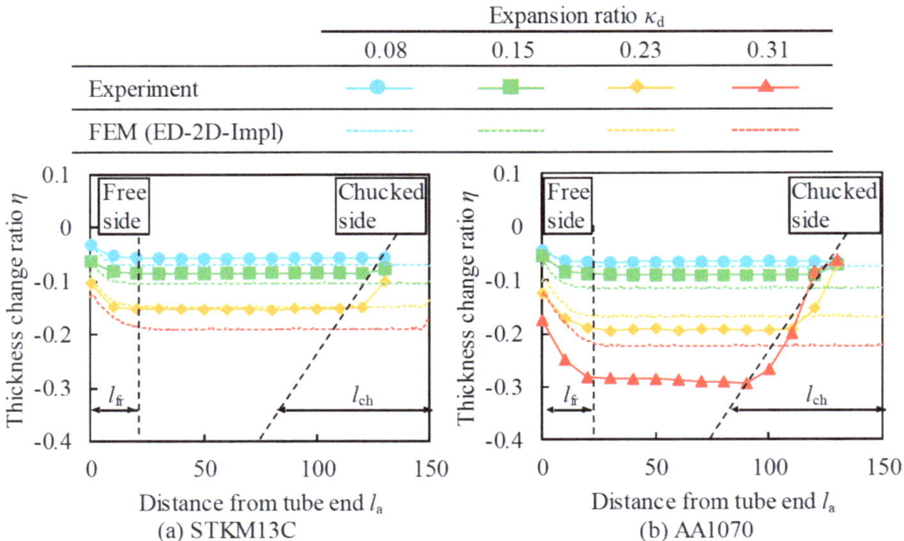

Figure 15. Distribution of thickness change ratio η in axial direction by drawing process.

Figure 16 shows the thickness reduction ratio γ at the axial middle portion of the drawn tube. Here, γ increased with the increase in the expansion ratio κ_d, reaching a maximum value of 0.29 when the κ_d of the AA1070 tube was 0.31 this experiment. In addition, γ of AA1070 was larger than that of STKM13C for each κ_d. This occurred because the AA1070 tube was strongly stretched in the axial direction by the large frictional force at region (B) in Figure 13. The experimental results almost fit the FEM results under the condition of κ_d = 0.1–0.23, although γ of the experiment was larger than that of the FEM analysis when the κ_d was 0.31. This is because local thinning occurred in the experiment.

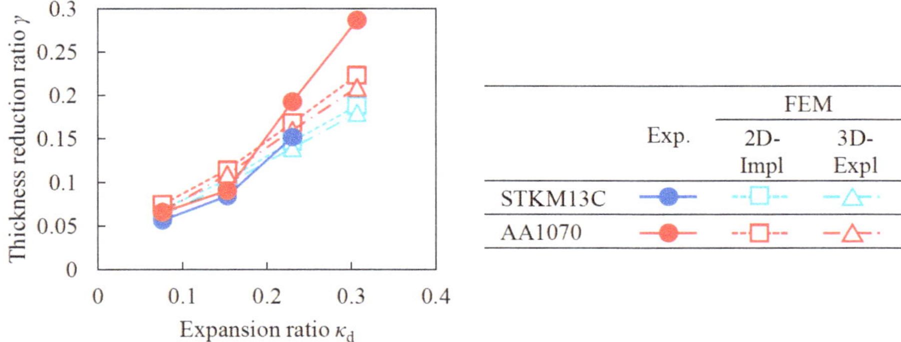

Figure 16. Effect of expansion ratio κ_d on thickness reduction ratio γ.

The dimensional accuracy of the cross section of the tube was evaluated. Figure 17 shows the ovality ratio ω of the tube. ω was small and less than 1%, although ω slightly increased by the expansion drawing. This result suggests that the ovality of the tube, which is processed by the expansion drawing, is good. However, the thickness deviation appeared for any forming conditions. Figure 18 shows the distribution of the thickness change ratio η in the tube hoop direction, where η is the average value in the axial direction for each angle θ. The thickness variation in the initial tube was observed, and initial thickness deviation ratios λ were 0.03 and 0.05 for the STKM13C and AA1070 tubes, respectively, in this experiment. Therefore, these initial thickness deviation ratios were set to that of the model (ED-3D-Expl) in the FEM analysis. As shown in Figure 18, the thickness variation of the processed tube follows the same trend as that in the initial tube, and the experimental results approximately fit the FEM results at the expansion ratio κ_d of 0.08–0.15. On the other hand, the thickness variation showed a different tendency from the initial tube when the expansion ratio κ_d was 0.23–0.31 due to the occurrence of local thinning. In particular, large local thinning appeared in the cross section of the AA1070 tube under the condition of κ_d = 0.31, as shown in Figure 19b.

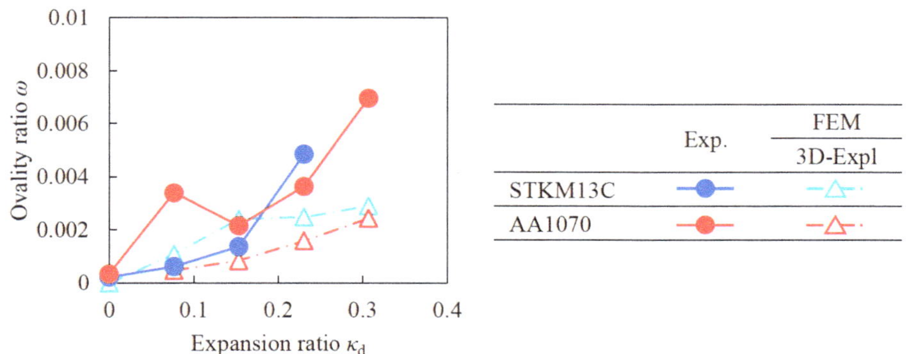

Figure 17. Effect of expansion ratio κ_d on ovality ratio ω.

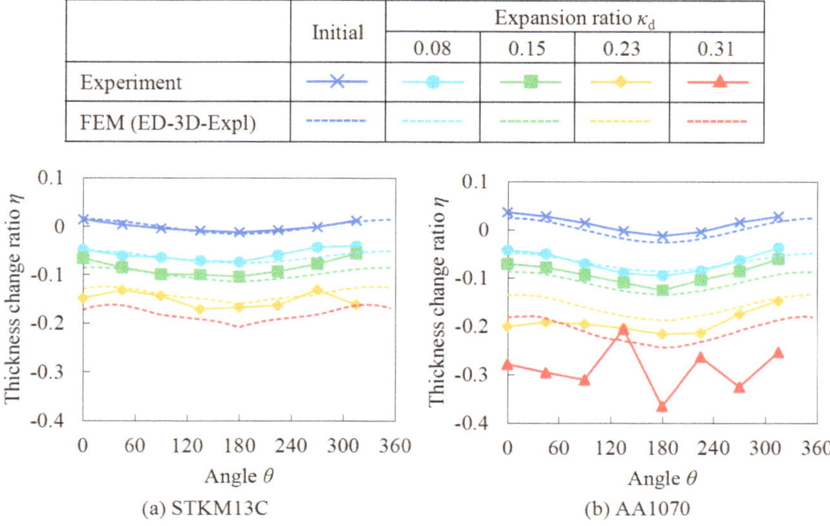

Figure 18. Distribution of thickness change ratio η in hoop direction by drawing process.

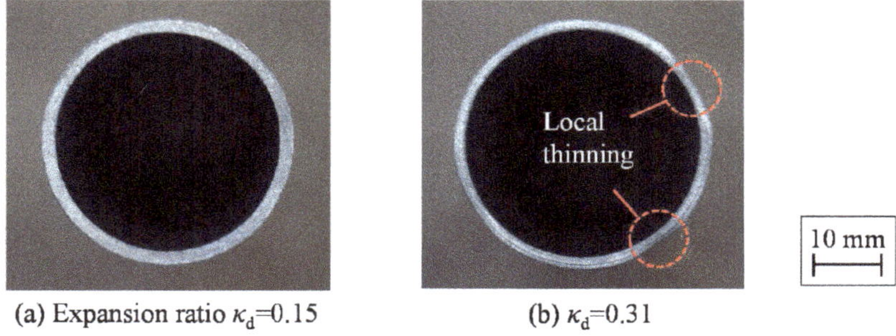

Figure 19. Typical cross-sectional view of drawn tube (AA1070).

Figure 20 shows the effect of the expansion ratio κ_d on thickness deviation ratio λ, which slightly increased with the increase in κ_d, except in cases in which large local thinning occurred, as shown in Figure 19b. This is because the local deformation occurred easily at the thin side with the increase in κ_d. Figure 21 shows the distribution of effective plastic strain ε_{eq} at the cross section in the FEM analysis. The distribution of ε_{eq} was comparatively uniform when κ_d was 0.15, whereas ε_{eq} locally increased at the thin side when κ_d was 0.31. Based on this result, the expansion drawing does not improve the thickness deviation. On the other hand, λ could be improved by slightly form the tube by hollow sinking or plug drawing, which reduce the thickness variation [2], after the expansion drawing.

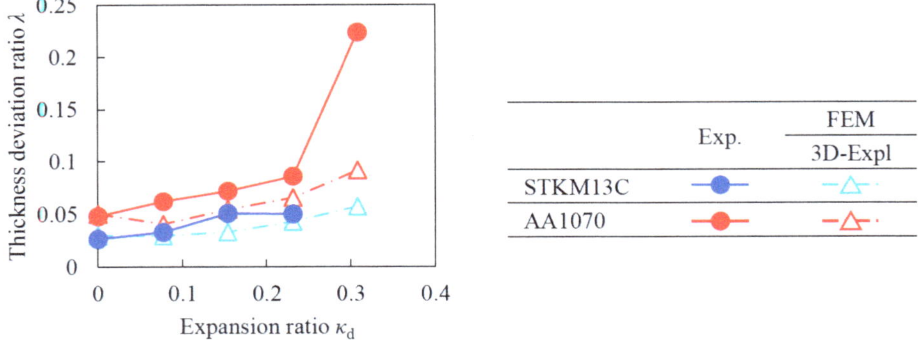

Figure 20. Effect of expansion ratio κ_d on thickness deviation ratio λ.

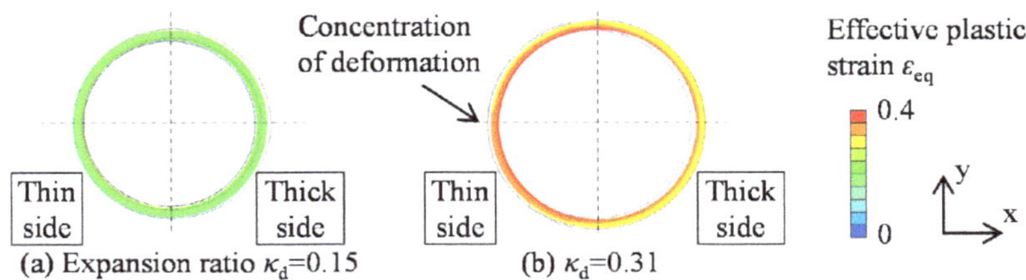

Figure 21. Distribution of effective plastic strain ε_{eq} at cross section of tube (AA1070).

4. Conclusions

The present paper proposed a tube drawing method with diameter expansion, which is referred to herein as 'expansion drawing', for producing thin-walled tubes effectively. The forming characteristics and effectiveness of the proposed method were investigated through a series of finite element method (FEM) analyses and experiments. Carbon steel (STKM13C) and aluminum alloy (AA1070) tubes were used as the test materials. According to the FEM analysis results, the thickness decreased with an increase in the expansion ratio. In addition, the expansion drawing reduced the tube thickness by a small load, when compared to the case of plug drawing with diameter shrinkage, which is the conventional method. In the expansion drawing experiment, it was possible to successfully produce thin-walled tubes, although for large expansion ratios the tube cracked due to biaxial stretching in the axial and hoop directions. The thickness reduction of the tube increased with the increase in the expansion ratio in a similar manner for both the experimental and FEM results. The maximum thickness reduction ratio was 0.15 when the expansion ratio was 0.23 for STKM13C, and the maximum thickness reduction was 0.29 when the expansion ratio was 0.31 for AA1070. The ovality ratio was small and less than 1% for each condition. The increase in the thickness deviation by the expansion drawing was small, but the thickness deviation drastically increased when local thinning occurred at the tube wall due to the use of expansion ratios that were too large. Therefore, this ratio should be set appropriately to produce tubes with small thickness deviations.

Author Contributions: Conceptualization, S.K., T.K. and M.A.; methodology, S.K., H.K., T.K. and M.A.; software, S.K. and T.K.; validation, S.K. and H.K.; formal analysis, S.K.; investigation, S.K. and H.K.; resources, H.K., I.A. and Y.T.; data curation, S.K. and H.K.; writing—original draft preparation, S.K.; writing—review and editing, S.K., H.K., T.K., I.A., Y.T. and M.A.; visualization, S.K. and H.K.; supervision, S.K.; project administration, S.K.; funding acquisition, S.K., I.A. and Y.T. All authors have read and agreed to the published version of the manuscript.

Funding: This research was funded by Amada foundation, grant number AF-2018034.

Conflicts of Interest: The authors declare no conflict of interest.

References

1. Kuboki, T.; Nishida, K.; Sakaki, T.; Murata, M. Effect of plug on levelling of residual stress in tube drawing. *J. Mater. Process. Technol.* **2008**, *204*, 162–168. [CrossRef]
2. Kuboki, T.; Tasaka, S.; Kajikawa, S. Examination of working condition for reducing thickness variation in tube drawing with plug. In *COMPLAS XIV, Proceedings of the XIV International Conference on Computational Plasticity: Fundamentals and Applications, Barcelona, Spain, 5–7 September 2017*; CIMNE: Barcelona, Spain, 2017; pp. 63–71.
3. Foadian, F.; Carradó, A.; Palkowski, H. Precision tube production: Influencing the eccentricity and residual stresses by tilting and shifting. *J. Mater. Process. Technol.* **2015**, *222*, 155–162. [CrossRef]
4. Zhang, L.; Xu, W.; Long, J.; Lei, Z. Surface roughening analysis of cold drawn tube based on macro–micro coupling finite element method. *J. Mater. Process. Technol.* **2015**, *224*, 189–199. [CrossRef]
5. Łuksza, J.; Burdek, M. The influence of the deformation mode on the final mechanical properties of products in multi-pass drawing and flat rolling. *J. Mater. Process. Technol.* **2002**, *125–126*, 725–730. [CrossRef]
6. Prakash, R.; Singhal, R.P. Shear spinning technology for manufacture of long thin wall tubes of small bore. *J. Mater. Process. Technol.* **1995**, *54*, 186–192. [CrossRef]
7. Chi, J.; Cai, Z.; Li, L. Optimization of spinning process parameters for long thin-walled cylinder of TC11 alloy based on processing map. *Int. J. Adv. Manuf. Technol.* **2018**, *97*, 1961–1969. [CrossRef]
8. Abe, H.; Iwamoto, T.; Yamamoto, Y.; Nishida, S.; Komatsu, R. Dimensional accuracy of tubes in cold pilgering. *J. Mater. Process. Technol.* **2016**, *231*, 277–287. [CrossRef]
9. Yasui, H.; Miyagawa, T.; Yoshihara, S.; Furushima, T.; Yamada, R.; Ito, Y. Influence of Internal Pressure and Axial Compressive Displacement on the Formability of Small-Diameter ZM21 Magnesium Alloy Tubes in Warm Tube Hydroforming. *Metals* **2020**, *10*, 674. [CrossRef]
10. Tamura, S.; Iguchi, K.; Mizumura, M. Effect of Punch Shape on Multiprocess Tube Flaring for Eccentric Parts. *J. Jpn Soc. Technol. Plast.* **2019**, *60*, 182–186. (In Japanese) [CrossRef]
11. Avalle, M.; Priarone, P.C.; Scattina, A. Experimental and numerical characterization of a mechanical expansion process for thin-walled tubes. *J. Mater. Process. Technol.* **2014**, *214*, 1143–1152. [CrossRef]
12. Al-Abri, O.S.; Pervez, T. Structural behavior of solid expandable tubular undergoes radial expansion process–Analytical, numerical, and experimental approaches. *Int. J. Solids Struct.* **2013**, *50*, 2980–2994. [CrossRef]
13. Seibi, A.C.; Barsoum, I.; Molki, A. Experimental and Numerical Study of Expanded Aluminum and Steel Tubes. *Procedia Eng.* **2011**, *10*, 3049–3055. [CrossRef]
14. Karrech, A.; Seibi, A. Analytical model for the expansion of tubes under tension. *J. Mater. Process. Technol.* **2010**, *210*, 356–362. [CrossRef]
15. Seibi, A.C.; Al-Hiddabi, S.; Pervez, T. Structural Behavior of a Solid Tubular Under Large Radial Plastic Expansion. *J. Energy Resour. Technol.* **2005**, *127*, 323–327. [CrossRef]
16. Yang, J.; Luo, M.; Hua, Y.; Lu, G. Energy absorption of expansion tubes using a conical–cylindrical die: Experiments and numerical simulation. *Int. J. Mech. Sci.* **2010**, *52*, 716–725. [CrossRef]
17. De Souza Neto, E.A.; Peric, D.; Dutko, M.; Owen, D.R.J. Design of simple low order finite elements for large strain analysis of nearly incompressible solids. *Int. J. Solids Struct.* **1996**, *33*, 3277–3296. [CrossRef]

Article

Pneumatic Experimental Design for Strain Rate Sensitive Forming Limit Evaluation of 7075 Aluminum Alloy Sheets under Biaxial Stretching Modes at Elevated Temperature

Jong-Hwa Hong [1], Donghoon Yoo [2], Yong Nam Kwon [3] and Daeyong Kim [2,*]

1. Department of Materials Science and Engineering, Seoul National University, Daehak-dong, Gwanak-gu, Seoul 151-744, Korea; jhong@snu.ac.kr
2. Materials Deformation Department, Korea Institute of Materials Science, 797 Changwon-daero, Changwon, Gyeongnam 642-831, Korea; dhyoo34@kims.re.kr
3. Aerospace Materials Center, Korea Institute of Materials Science, 797 Changwon-daero, Changwon, Gyeongnam 642-831, Korea; kyn1740@kims.re.kr

* Correspondence: daeyong@kims.re.kr; Tel.: +82-55-280-3509

Received: 30 October 2020; Accepted: 2 December 2020; Published: 5 December 2020

Abstract: A pneumatic experimental design to evaluate strain rate sensitive biaxial stretching forming limits for 7075 aluminum alloy sheets was attempted with the finite element method. It was composed of apparatus geometric design with pressure optimization as the process design. The 7075 aluminum alloy material was characterized by conventional Voce-type hardening law with power law strain rate sensitivity relationship. For optimization of the die shape design, the ratio of minor to major die radius (k) and profile radius (R) were parametrically studied. The final shape of die was determined by how the history of targeted deformation mode was well maintained and whether the fracture was induced at the pole (specimen center), thereby preventing unexpected failure at other locations. As a result, a circular die with $k = 1.0$ and an elliptic die with $k = 0.25$ were selected for the balanced biaxial mode and near plane strain mode, respectively. Lastly, the pressure inducing fracture at the targeted strain rate was studied as the process design. An analytical solution that had been previously studied to maintain constant strain rate was properly modified for the designed model. The results of the integrated design were compared with real experimental results. The shape and thickness distribution of numerical simulation showed good agreement with those of the experiment.

Keywords: forming limits; biaxial stretching; forming limit measurement; experimental design; strain rate sensitivity; elevated temperatures; pneumatic forming

1. Introduction

Pneumatic forming or superplastic forming (SPF) has been widely utilized in various different areas, such as the aerospace and automobile industries [1,2]. Pneumatic forming is a forming method that deforms sheets by gas pressure at elevated temperatures. Some researchers have developed the method to construct a forming limit diagram (FLD) by pneumatic forming [3,4]. Although conventional draw forming is widely utilized for constructing FLD, the apparatus for draw forming is vulnerable to elevated and high temperatures because of contact problems. Contact between specimen and tools at elevated and high temperatures causes degradation of tool life and less durability due to increased friction, heat expansion, and weakening tool strength. Moreover, the friction effect by contact can lead to inaccurate measurement of FLD, thereby triggering early failure. When it comes to constant deformation rate, it is quite difficult to maintain the targeted strain rate due to the instability coming from the contact. Pneumatic forming has no contact or only a little contact on the edge of the die during deformation. Only a little or no contact area can reduce abrupt material failure by friction without the use of lubricants, thus enduring high temperatures. Minimization of contact also enables pneumatic forming to maintain a constant strain rate more precisely than draw forming for construction of a rate-dependent forming limit diagram. In addition, pneumatic forming can be more accurately halted at the fracture point. The punching movement in draw forming is suspended after ample force reduction following a crack. However, the pneumatic test is halted right after a crack occurs. Blank deformation is stopped when there is not enough pressure to continue blowing because the air on one side of the blank is deflated to the other side as a crack occurs. Therefore, the pneumatic test has a big advantage as it does not bring disparity between fracture occurrence and halts at dome height. For all these reasons, pneumatic forming can be a good alternative to the draw forming method as it does not have the limitations of a mechanical contact test. There have been many studies to improve the pneumatic forming method to construct forming limit diagrams.

Some researchers have studied building biaxial tension–tension modes by changing the die cavity on the same specimen design [3,5]. Others have tried to construct axial tension–compression modes by changing the blank shape on the same die cavity design [4]. It is also challenging to maintain constant strain rate deformation to construct a rate-dependent forming limit diagram. An analytical solution formulated in a circular die cavity with various strain rates based on the Levy–Mises relations and linearly strain hardening material was studied [6]. The solution was modified for general die cavity shape with transversely isotropic yield function applied to Hollomon hardening material and strain rate-dependent superplastic material [5,7]. Some works of literature have researched strain rate- and temperature-dependent FLD using the Nakajima test and compared it to theoretical predictions with a Marciniak–Kaczynski (M–K) model and continuum damage mechanics (CDM)-based models [8,9]. However, such efforts to construct a FLD in an experiment are limited with the conventional draw forming method. Therefore, a FLD with rate sensitivity has to be constructed by a pneumatic forming method to eliminate failure variables, such as friction, by stopping at the exact moment with uniform deformation and constant strain rate history.

The aim of this research was to establish a design for the major factors of rate-dependent FLD of 7075 aluminum alloy sheets, such as proper die shape and pressure condition, to build fracture properties with strain rate sensitivity. In this study, an integral design for strain rate sensitive forming limit diagram by pneumatic forming was attempted with numerical simulation. The numerical procedures of the integral design consisted of two parts: apparatus geometric design and process condition design. First, two main geometric parameters were used to design a pneumatic forming die, namely, cavity radius ratio (k) and die profile radius (R), as shown in Figure 1. The cavity radius ratio (k) is the ratio of the minor to the major radius of the die. The goal of optimized k is to find the proper k value leading to proper fracture modes with a stable deformation path history. The aim of the optimization of R is to induce fracture occurrence at the pole by concentrating deformation on the center of the blank. This prevents fracture at an unexpected position that is not the center of the specimen and assists stable deformation at the pole, thereby leading to proportional load history in the domain of major and minor stress or triaxiality and effective strain. To trace the deformation mode history and compare the deformation quantity between the center of the blank and the next critical location (the largest deformed location except for the pole of the blank during deformation), numerical finite element simulation was utilized. Lastly, the pressure and time relationship for proper strain rates were studied as process conditions. An analytical solution to the pressure–time relationship was modified and compared with the results of a preexisting formulation. The effect of profile radius (R) was considered in the newly modified analytical model. For validation, the pneumatic forming die was fabricated based on the effect of the integral design process. Simulation results were finally compared with the experiments at optimized conditions.

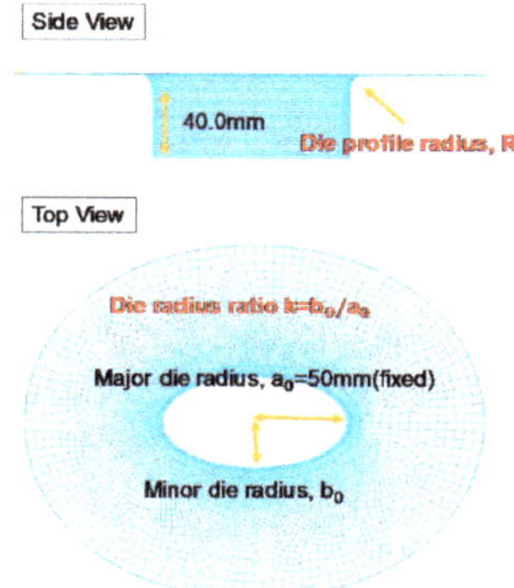

Figure 1. Description of the geometric parameters (cavity radius ratio (k) and die profile radius (R)) of the die used in the pneumatic test for sheets.

2. Finite Element Modeling

2.1. Materials

In this study, 7075 aluminum alloy sheets (Al–Zn–5.5 Mg–Cu) of 1.27 mm (0.05 inch) thickness was considered at 400 °C. Uniaxial tensile tests were performed for material plastic properties using a universal tensile test machine (Shimadzu, Kyoto, Japan). The tensile tests were examined along with the crosshead speeds to check the dependency of rate sensitivity. Every single curve representing engineering stress–engineering strain for each different crosshead speed is shown in Figure 2a. The mechanical behavior of 7075 aluminum alloy sheets at elevated temperatures shows an early, brief increase and large softening behavior in the engineering domain [10]. The engineering stress–engineering strain curves were also strongly dependent on the crosshead speed. The stress grew larger as the strain rate increased at the same temperature condition. Grip velocities of 0.02, 0.2, and 2.0 mm/s approximately corresponded to effective strain rates of 0.001, 0.01, and 0.1 s^{-1}, respectively, based on numerical simulation. When the engineering stress–engineering strains were converted to true stress–true strains, the curves displayed virtually perfect plasticity behavior, as shown in Figure 2b [8,9]. The ASTM E 8M subsize specimen was modified for the uniaxial tensile test at elevated temperatures, as shown in Figure 3 [11]. The mechanical properties are given in Table 1. Elastic modulus and Poisson ratio were measured from 7075 aluminum alloy sheets of 2.0 mm thickness and assumed to be applied to this material. The usual 7075 aluminum alloy sheets at elevated temperatures is assumed to show an isotropic plastic behavior [8,9]. Hence, von Mises yield function for stress potential was applied throughout the study.

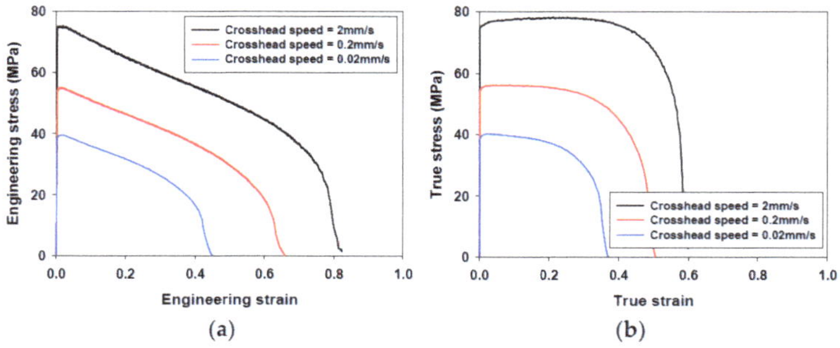

Figure 2. The engineering stress–engineering strain curves (**a**) and the true stress–true strain curves (**b**) along with crosshead speeds for strain rate sensitivity measurement.

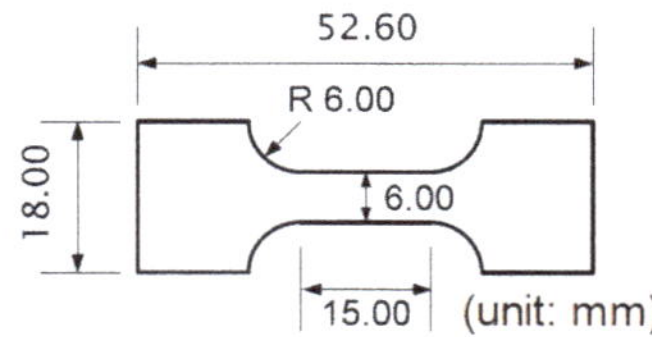

Figure 3. The dimension of the specimen utilized for uniaxial tensile test at elevated temperatures.

Table 1. The mechanical (engineering) properties of 7075 aluminum alloy sheets at 400 °C.

Temperature (°C)	Crosshead Speed (mm/s)	E (GPa)	Poisson's Ratio	UTS (MPa)	Elongation (%)	
					UTS	Total
400	2.0	51.82	0.35	76.15	1.48	82.36
	0.2			55.66	0.98	65.85
	0.02			40.14	1.25	44.85

The effective stress–effective strain was fitted by Voce-type hardening law with power law strain rate sensitivity relationship using Equation (1). The material coefficients of Voce-type hardening law, m-value of power law, and reference strain rate are given in Table 2. The curves based on power law rate sensitivity extrapolation were well matched with the hardening curve directly obtained from the uniaxial tensile test until ultimate tensile strength (UTS) at 0.001, 0.01, and 0.1 s^{-1}, as shown in Figure 4.

$$\overline{\sigma}_0 = \left\{K + C(1 - e^{-p\overline{\varepsilon}})\right\}\left(\frac{\dot{\overline{\varepsilon}}}{\dot{\varepsilon}_0}\right)^m \quad (1)$$

where K, C, and p are coefficients of Voce-type hardening law; $\dot{\varepsilon}_0$ is the reference strain rate; and m is the rate sensitivity coefficients.

Table 2. Parameters of Voce-type hardening law and m-value.

Temperature (°C)	K (MPa)	C (MPa)	p	m	$\dot{\varepsilon}_0$ (s^{-1})
400	38.63	1.68	171	0.14	0.001

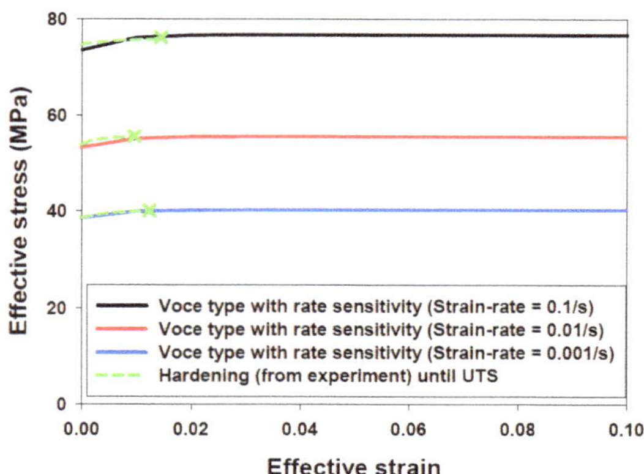

Figure 4. Comparison between hardening curve from Voce-type hardening law with power law rate sensitivity and hardening curve until ultimate tensile strength (UTS) directly obtained from the uniaxial tensile tests.

2.2. Numerical Simulation Procedure

Numerical simulation was performed to study the effects of the parameters. Isotropic elastic and Voce-type material with power law interpolation for strain rate sensitivity, as characterized in the previous section, were utilized for the numerical simulation. The ABAQUS/Implicit code (6.12-1, Dassault Systèmes, Vélizy-Villacoublay, France) for a hexahedral solid element with eight nodes and reduced integration (C3D8R) was used. Eight-node solid elements of 0.8 mm × 0.8 mm × 0.254 mm dimension and 8 mm × 8 mm × 0.254 mm dimension were used for the fine mesh region and the coarse mesh region, respectively, with five layers of thickness (1.27 mm), as shown in Figure 5. A fixed boundary condition was assumed without a blank holding force because no slip was expected with the double beads at the blank clamping area. As for the process conditions, a fixed boundary condition was utilized, as shown in Figure 6, instead of applying a blank holding force. Other process conditions, such as friction coefficient and pressure, were assumed as specific values. The friction coefficient is relatively unimportant due to minimal contact in the pneumatic forming process. However, contact of the blank with the die occurs in some parts, such as the edge of the die (or die profile). Therefore, the friction coefficient was set as 0.1, assuming that the apparatus was well lubricated and had no lubrication issue. The pressure was applied 5 mm from the finish line of the die profile covering the whole cavity and corner of the hole, as shown in Figure 6. Throughout the study, the conditions mentioned above were applied to the numerical model.

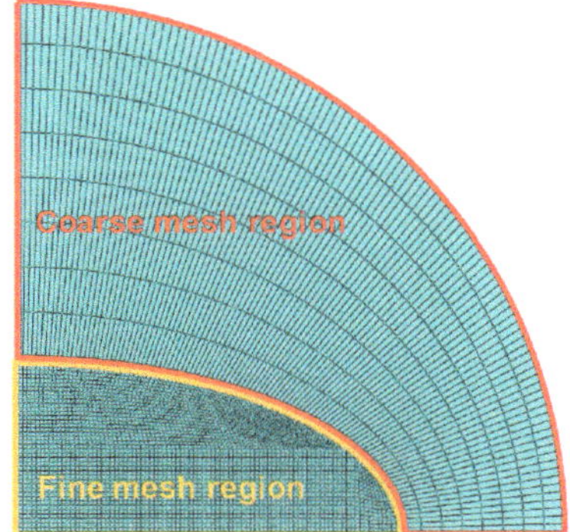

Figure 5. The fine mesh region and the coarse mesh region in the blank.

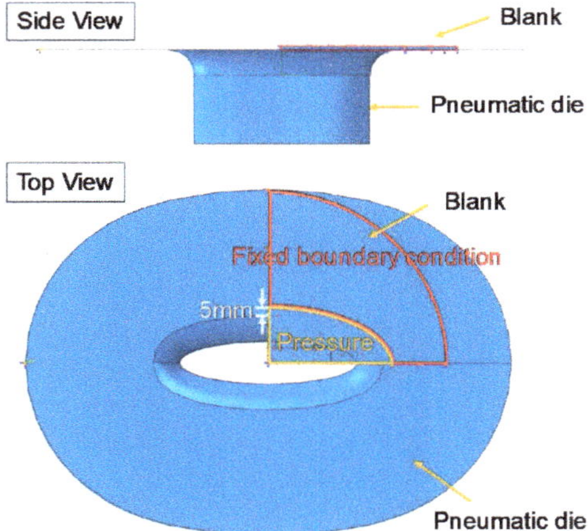

Figure 6. The boundary condition of the pneumatic test.

3. Design of Geometric and Process Conditions

The design of the geometric and process conditions in pneumatic forming are discussed in this section. For proper deformation modes and fracture occurrence at the proper location, the ratio of minor to major radius, k, and the profile radius, R, were studied as the geometric design. In pneumatic forming, the pole of the blank can be deformed with approximately constant strain rate by adjusting the pressure history. The pressure history, which determines approximate constant strain rate, was found as the process condition design.

3.1. Geometric Design: The Ratio of Minor to Major Radius (k)

To determine deformation modes, a parametric study on the ratio of the minor to the major die radius, k, was performed. As shown in Figure 1, other geometric factors, such as major radius and die profile radius (R), were fixed to determine the effect of k. The major radius and die profile radius (R) have to be carefully determined according to the circumstances. As the major radius becomes small, it secures enough blank holding area to reduce the possibility of a slip in the size limit of the machine, but a small cavity requires more pressure to deform the blank in the pressure capacity limit of the machine. In this study, the major radius and the die profile radius (R) were fixed as 100.0 and 15.0 mm, respectively. The constant pressure was arbitrarily decided for each k die for simplicity. The conditions for parametric study are shown in Table 3. As can be seen, higher pressure was required as k decreased.

Table 3. The conditions for parametric study of the die insert radius ratio (k).

Die Insert Radius Ratios (k)	Major Die Radius (a_0)	Minor Die Radius (b_0)	Die Profile Radius (R)	Pressure Values
1	50.0 mm	50.0 mm	15.0 mm	1 MPa
0.75		37.5 mm		2 MPa
0.5		25.0 mm		2 MPa
0.25		12.5 mm		4 MPa
0.1		5.0 mm		6 MPa

As k decreased, the deformation modes approached the plane strain mode, as shown in Figure 7. However, when k was too small, the deformation history became unstable and not proportional. For instance, although the deformation mode of the elliptic die with $k = 0.1$ seemed closer to the plane strain mode at the initial stage, the deformation history on the domain of triaxiality with respect to effective plastic strain and of the major/minor differed a lot without any proportional increase. This is because the 3D effect increases as k diminishes, meaning that the mode of the critical element is no longer plane stress, which is important for sheet metal forming. With smaller k, not only is more pressure needed to complete the same deformation, but the effective strain is also more likely to be concentrated on the edge, which will be discussed in the profile radius (R) design. In conclusion, the optimized k value was 0.25 for near plane strain mode (PSA) die with the already selected die ratio of $k = 1$ for balance biaxial mode.

3.2. Geometric Design: The Profile Radius (R)

Another significant geometric design factor that has been negligently treated before is the profile radius (R), as shown in Figure 1. As can be seen, as k diminished, the effective strain was more likely to be concentrated on the edge. This is because the profile radius (R) is crucial to induce a fracture on the proper location during the experiment, preventing an unfavorable fracture on the edge. The parametric study conditions for optimized circular and elliptic die are given in Table 4. It can be seen that more pressure was required for the same amount of deformation as k became smaller. The amount of deformation of the center and edge of the blank was compared to each other during the deformation to determine the minimum profile radius (R) of the die to concentrate fracture on the center of the blank. The minimum profile radius (R) of the die has some advantages. A smaller R secures enough blank holding area to reduce the possibility of a slip. Aside from area cavity, larger R means less blank holding area or increased size of the machine with the same blank holding domain. Therefore, smaller but safe R can be used depending on the preference of the user and machine.

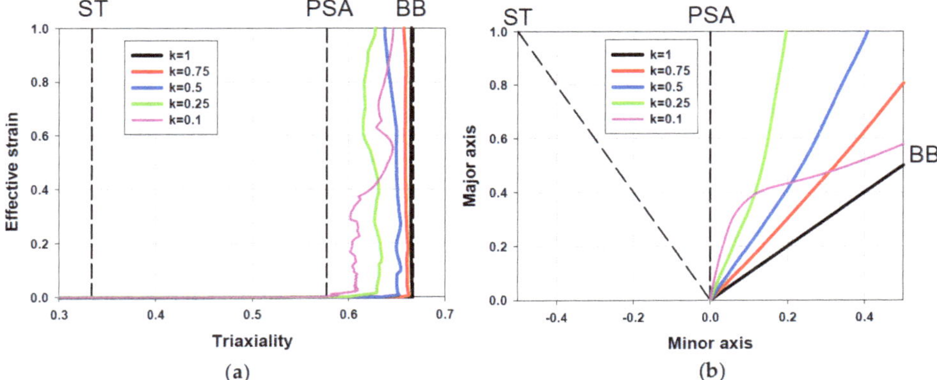

Figure 7. History of deformation mode along the die insert radius ratio (k) with regard to effective strain and triaxiality (**a**) and major strain and minor strain (**b**).

Table 4. The parametric study conditions for the circular and elliptic die profile radius.

Die Insert Radius Ratio (k)	Die Profile Radius (R)	Major Die Radius (a_0)	Pressure Value
1 (Circular die)	5.0 mm	50.0 mm	1 MPa
	10.0 mm		
	15.0 mm		
	20.0 mm		
0.25 (Elliptic die)	5.0 mm		4 MPa
	10.0 mm		
	15.0 mm		
	20.0 mm		

Even though R was changed, it did not occupy cavity size and did not influence the dimension of the major radius and die insert ratios (k). To reduce the possibility of fracture at an unfavorable area like the die edge, 0.4 of effective strain was set as a quantitative standard to distinguish whether the strain was largely concentrated on the center or not. Figure 8 shows a comparison of effective strain at the center element with those at the edge element as the deformation progressed with various die profile radii (R) for the circular die. The solid and dashed lines were obtained from a single center element and the critical element at the edge, respectively. The two black lines were obtained from the element of the circular die with 5 mm die profile radius (R) and the red, blue, and green were obtained from 10, 15, and 20 mm die profile radii (R), respectively. Although the deformation tended to focus on the edge as k decreased, all parameters were within the standard and did not exceed 0.4 of effective strain on the edge. The deformation history of the critical element was uniform for all the profile radii (R) for the circular shape die, as shown in Figure 9. A comparison of effective strain at the center element with those at the edge element as deformation progressed with various die profile radii (R) for elliptic die is also shown in Figure 10. The solid and dashed lines were obtained from a single center element and the critical element at the edge, respectively. The two black lines were obtained from the element of the circular die with 5 mm die profile radius (*R*) and the red, blue, and green were obtained from 10, 15, and 20 mm die profile radii (*R*), respectively. For the elliptic die, deformation also tended to focus on the edge as *k* decreased. However, the deformation was so severe that the effective strain on the edge exceeded 0.4 for *R* below 15 mm (blank and red lines). It was also further from the targeted deformation history of the critical element for *R* below 15 mm, as shown in Figure 11. To meet the safety condition for both circular and elliptic cases, *R* = 15 mm was finally determined for die profile radius (*R*) based on the numerical simulation results.

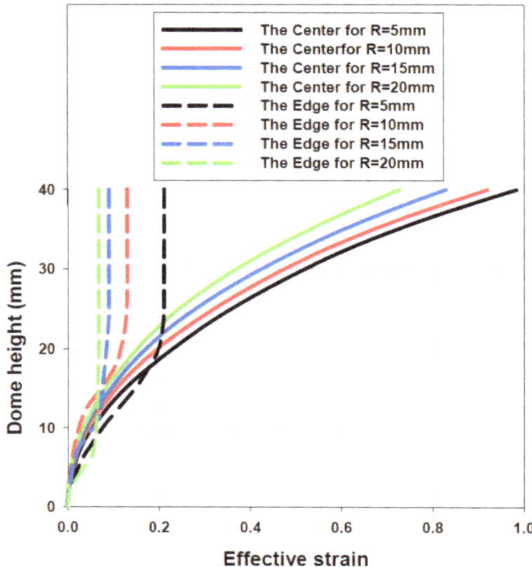

Figure 8. Comparison of deformation at the center element with those at the edge element with various die profile radii (*R*) for the circular die (solid line is for the center elements, dash line is for the edge elements, black color is for *R* = 5 mm, red color is for *R* = 10 mm, blue color is for *R* = 15 mm, and green color is for *R* = 20 mm).

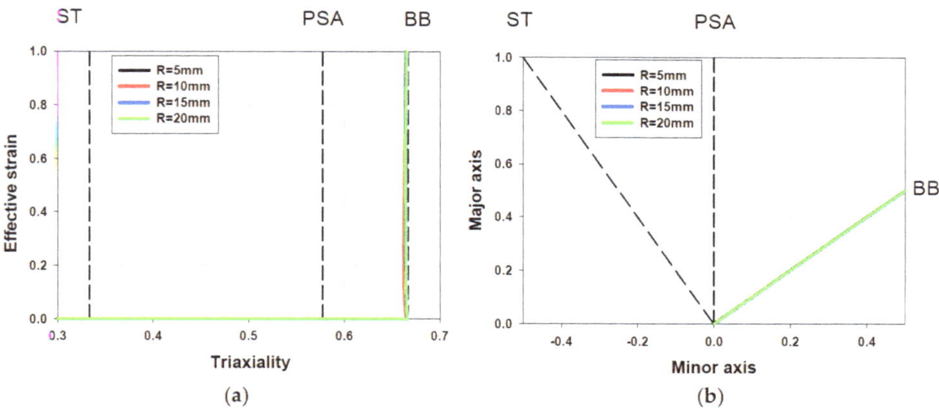

Figure 9. Deformation history with various die profile radii (*R*) for the circular die with regard to effective strain and triaxiality (**a**) and major strain and minor strain (**b**).

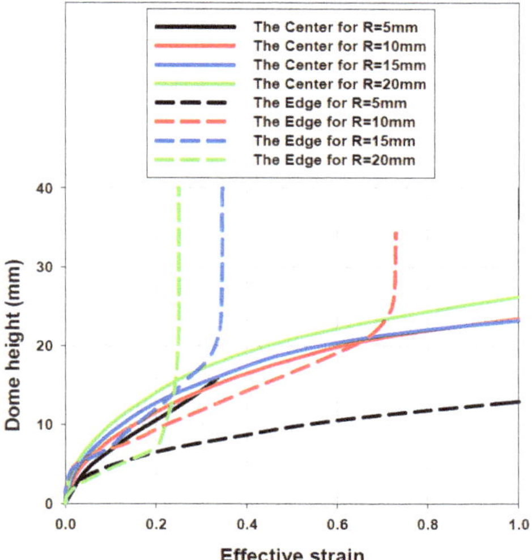

Figure 10. Comparison of deformation at the center element with those at the edge element with various die profile radii (R) for the elliptic die (solid line is for the center elements, dash line is for the edge elements, black color is for R = 5 mm, red color is for R = 10 mm, blue color is for R = 15 mm, and green color is for R = 20 mm).

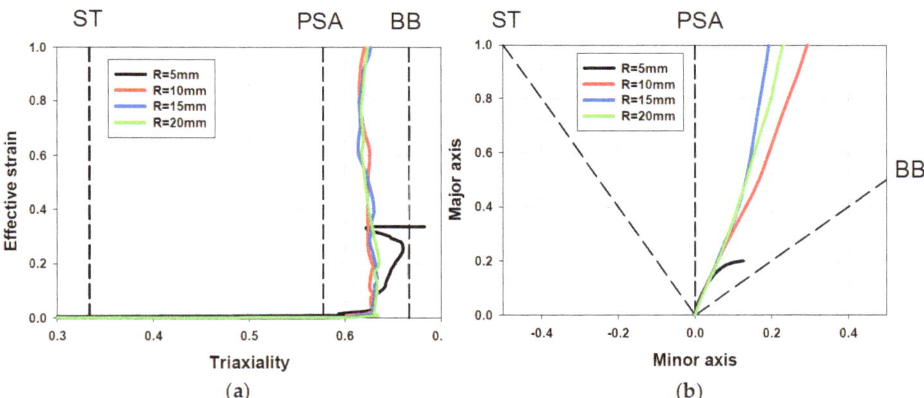

Figure 11. Deformation history with various die profile radii (R) for the elliptic die with regard to effective strain and triaxiality (**a**) and major strain and minor strain.

3.3. Process Condition Design

Due to its fast and easy application, an analytical solution has an advantage in the early stage of the test design and experimental setting. The relationship of the pressure and height of the blank in a circular bulge was derived with linear plasticity [6]. A closed-form analytical model with transversely isotropic material was extended to the elliptic bulge test, which is a general form of the circular bulge test. The extended model was applied to Hollomon-type hardening law and strain rate sensitive superplastic material law [7]. Pressure and time relationship for general k was obtained in full three-dimensional isotropic yield function with the general material hardening model [5]. In this study, a previous analytical model utilized for extended elliptic bulge test was modified by considering the profile radius (R) effect, as shown in Equation (2). The correlation among variables is also written in Equation (3).

In this model, ω or correction factor is newly introduced. a_0, b_0, and k are substituted with A, B, and K, respectively.

$$p(t) = \frac{2s_0\overline{\sigma}}{B_0}\frac{1+\alpha K^2}{\rho}\left(e^{\frac{(2-\alpha)\dot{\overline{\varepsilon}}t}{2\rho}} - 1\right)^{\frac{1}{2}} e^{-\frac{3\dot{\overline{\varepsilon}}t}{2\rho}} \tag{2}$$

$$K = \frac{B_0}{A_0} = \frac{b_0+\omega R}{a_0+\omega R},\ \omega = \omega_0 k_0 = \omega_0\frac{b_0}{a_0},\ \alpha = \frac{1}{2}(1+e^{1-(\frac{1}{K})}),\ \rho = \sqrt{1-\alpha+\alpha^2} \tag{3}$$

where p is the blow forming pressure; s_0 is the initial blank thickness; a_0 and b_0 are the major and minor radii of elliptic die, respectively; A and B are modified major and minor radii of elliptic die considering the effect of profile radius (R); t is the time; $\dot{\overline{\varepsilon}}$ is the targeted strain rate; $\overline{\sigma}$ is the effective stress; ω is the correction factor taking into account the effect of the profile radius (R) and is a function of k, which is the ratio of minor radius to major radius without considering the profile radius; and ω_0 is a coefficient of the correction factor. As ω_0 becomes zero, a_0, b_0, and k become A, B, and K, respectively. Then, the pressure and time relationship is identical with the model developed by D. Banabic and M. Vulcan. As ω_0 becomes unity, the profile radius (R) is fully considered as part of the die hole radius in the relationship. In general, ω_0 is maintained in between $0 \le \omega_0 \le 1$.

Voce-type law with power law rate sensitivity model was applied to the relationship for each targeted strain rate. Two extreme cases were dealt with to show the effect of the profile radius (R). To begin with, no consideration was taken into account of the effect of R, i.e., $\omega_0 = 0$, which is the same as the previous model. Then, full consideration of the effect of R was taken into, i.e., $\omega_0 = 1$, which is the other extreme case. For the case of $\omega_0 = 0$, the pressure and time relationship is shown in Figure 12.

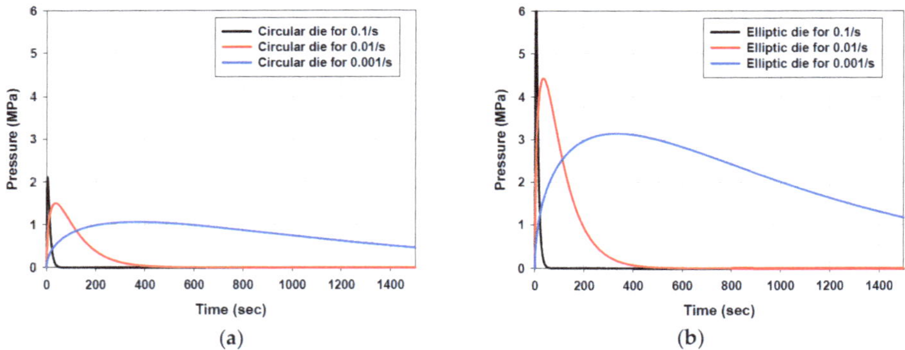

Figure 12. Pressure and time relationship of the circular die (**a**) and the elliptic die (**b**) for targeted strain rate at $\omega_0 = 0$.

There was a discrepancy between the input strain rate in the analytical solution and output strain rate in the numerical model, as shown in Figure 13. The results were faster than each targeted strain rate. The error could have occurred because of some limitations, such as the constraint of approximated geometric assumption on the bulge test and assumption that strain increases linearly in time, as mentioned in [7]. However, the major discrepancy of the strain rate results is due to the introduction of the profile radius (R) in the geometric design. As a comparison, the effect of profile radius (R) is fully considered in Figure 14. It can be seen that as the modified major radius, $a_0 + \omega R$ became 65 mm, pressure became approximately 0.77 smaller than the previous model for $\omega = 0$ for the circular die. The effective strain rate result in numerical simulation for the circular die and elliptic die is shown in Figure 15. However, the results did not have enough deformation and did not satisfy targeted values of strain rate history.

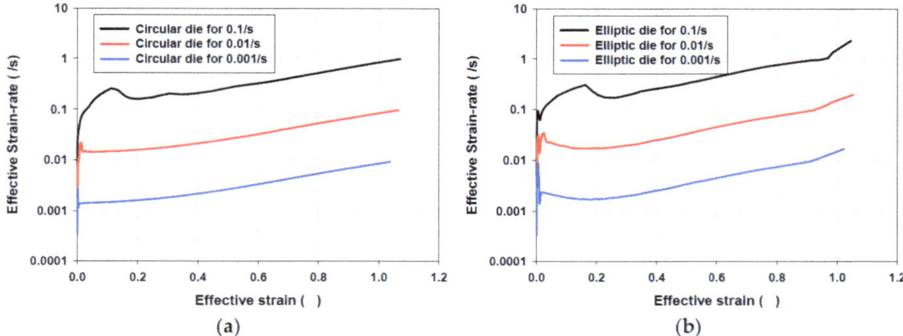

(a) (b)

Figure 13. Effective strain rate and effective strain result in numerical simulation for the circular die (**a**) and the elliptic die (**b**) at $\omega_0 = 0$.

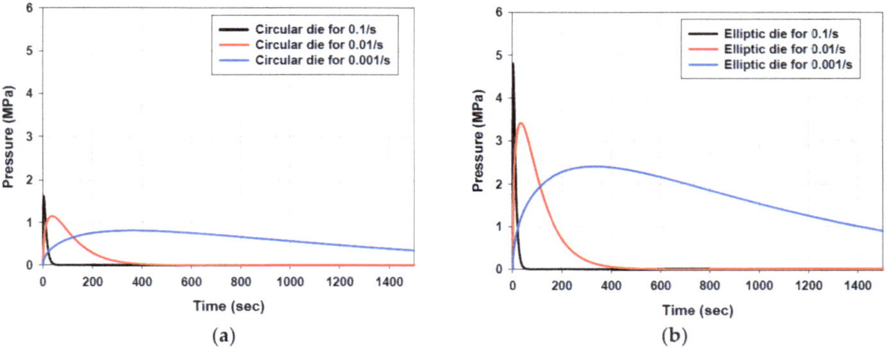

(a) (b)

Figure 14. Pressure and time relationship of the circular die (**a**) and the elliptic die (**b**) for targeted strain rate at $\omega_0 = 1$.

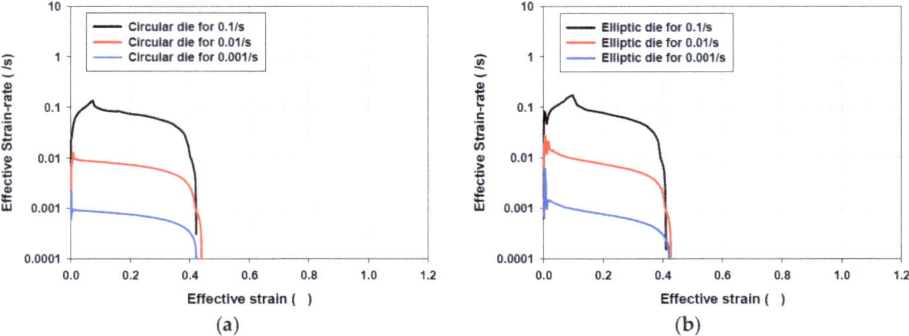

(a) (b)

Figure 15. Effective strain rate and effective strain result in numerical simulation for the circular die (**a**) and the elliptic die (**b**) at $\omega_0 = 1$.

A geometric relationship was assumed to obtain proper coefficient of correction factor, ω_0, in between the two extreme cases, as shown in Figure 16. The sphere (red short line) was in contact with the profile radius (R) of the die. The center of the geometrical figure had a fixed distance of 10 mm from the center of the blank. The radius of the new geometrical figure was calculated with the geometric constraint. Then, ω_0 was found to be approximately 0.31, which is between zero and unity, indicating that ω was equal to ω_0 in the circular case. For the general ellipsoid, the effect of profile radius (R) was compensated by the ratio of minor radius to major radius (k) being ω. The pressure and time relationship for each circular and elliptic die is shown in Figure 17. The effective strain rate results in numerical simulation for the circular die and elliptic die are shown in Figure 18. The results were improved than the previous analytical models in maintaining approximately constant targeted strain rate.

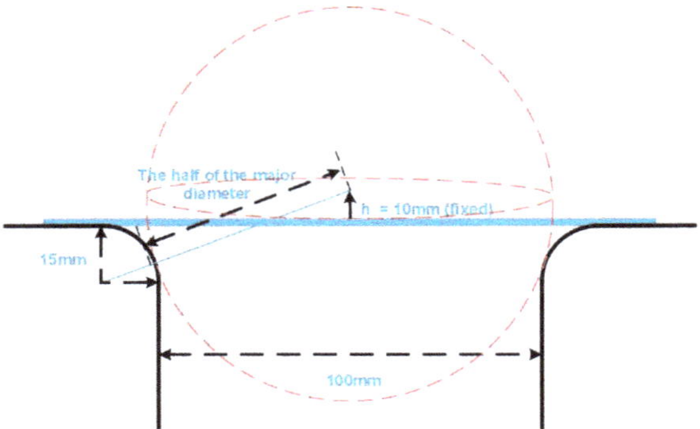

Figure 16. A geometric assumption to obtain coefficient of correction factor considering modified radius of the sphere in the circular die.

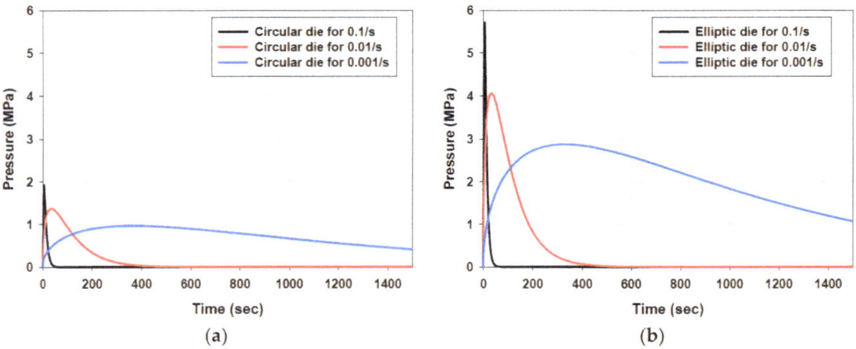

Figure 17. Pressure and time relationship of the circular die (**a**) and the elliptic die (**b**) for targeted strain rate at $\omega_0 = 0.31$.

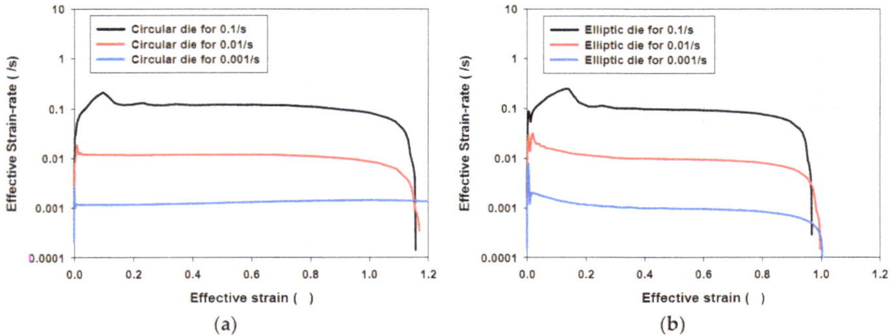

Figure 18. Effective strain rate and effective strain result in numerical simulation for the circular die (**a**) and the elliptic die (**b**) at $\omega_0 = 0.31$.

4. Validation

To validate the integral experimental design, tests for pneumatic forming for $k = 1$ and 0.25 with the pressure condition maintained at approximately constant strain rate of 0.01 s^{-1} were performed for 7075 aluminum alloy sheets. The circular and elliptic pneumatic forming dies were fabricated, as shown in Figure 19, based on the parametric study for k and R. A double bead was also introduced to prevent slip of the blank and gas leakage. The blank dimensions for circular and elliptic pneumatic forming die used in this study are also shown in Figure 20. The pneumatic process was performed using a pneumatic forming machine. The photograph and schematic view of the machine are also shown in Figure 21.

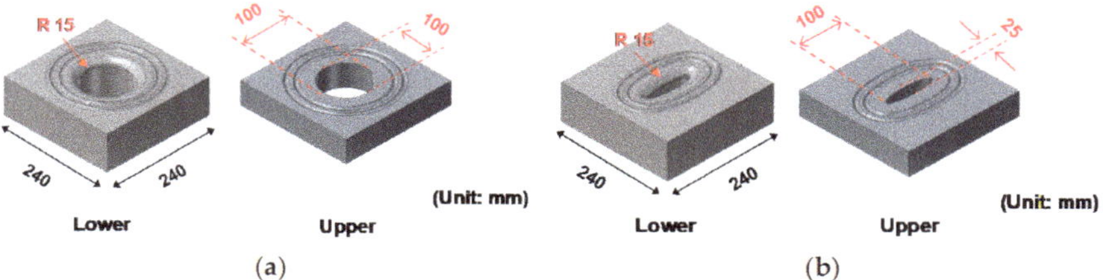

Figure 19. The pneumatic forming dies for the circular die (**a**) and the elliptic die (**b**).

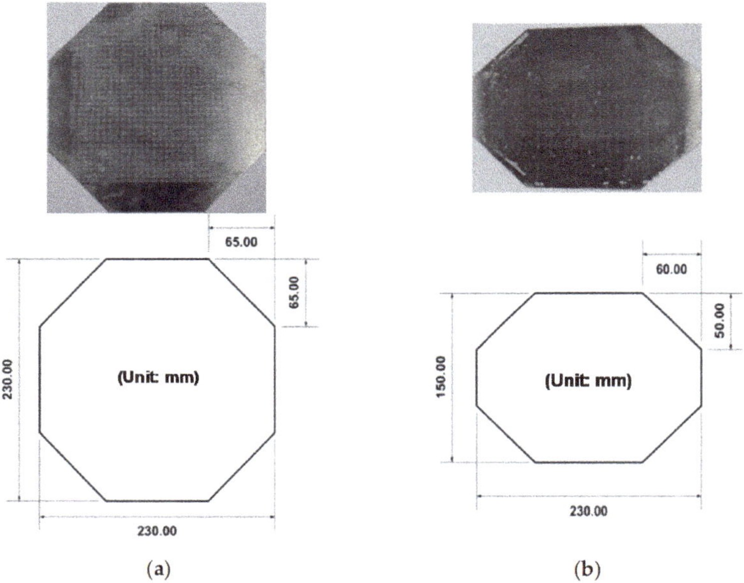

Figure 20. The blank dimensions for the circular die (**a**) and the elliptic die (**b**).

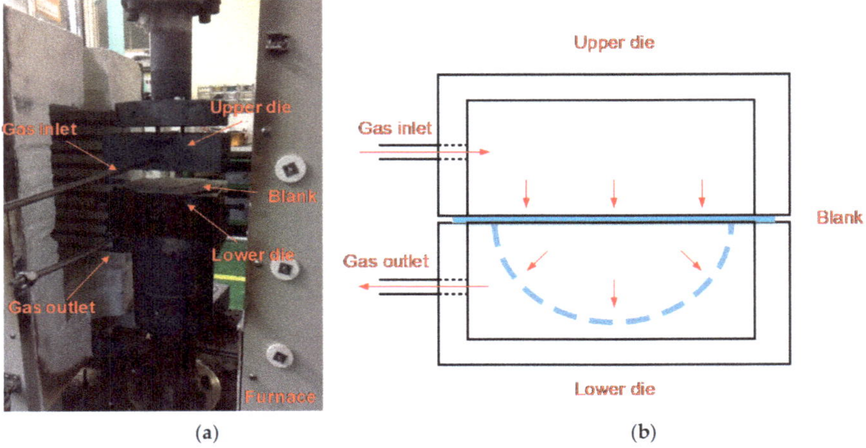

(a) (b)

Figure 21. Photograph (**a**) and schematic view (**b**) of the pneumatic forming machine.

The final blanks after deformation for circular and elliptic dies are shown in Figure 22. The pneumatic forming tests were completed due to crack formation on the pole of each blank. The shape (z coordinate) and thickness distribution results of the experiment and simulation are compared in Figures 23 and 24. The shape and thickness evolution obtained from pneumatic forming numerical simulation of 7075 aluminum alloy sheets were superimposed. The black solid line was measured from the final experimental results. Dashed red lines were obtained from numerical tests at each shape height with 10 mm difference (for example, 10, 20 mm, etc.) to show the evolution of pneumatic forming. However, as no fracture properties were applied on the finite element simulation, numerical deformation continued until the onset of localized necking coming from instability of the mathematical effect. Therefore, the simulations were stopped when the height of each blown blank reached the same height of each experimental result. The simulation results at the height are plotted with a red solid line. The results showed good agreement with the experimental results. When it came to improvement of the small thickness discrepancy at the pole, a material model considering material deterioration with hardening and rate sensitivity is needed to represent the material softening behavior [12,13]. This is because of the material properties of 7075 aluminum alloy sheets at elevated temperatures, which display material softening behavior after short arrival of UTS, as shown in the Figure 2. However, sophisticated material characterization is beyond the scope of our research. Hence, the conventional Voce-type hardening law, which has been widely used for aluminum alloy, was utilized in this study to highlight and focus on the integral pneumatic die and pressure optimization design process.

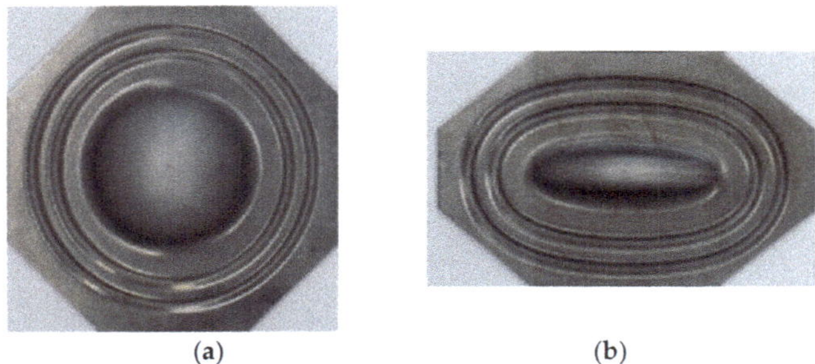

(a) (b)

Figure 22. The blanks after pneumatic forming test for the circular die (**a**) and the elliptic die (**b**).

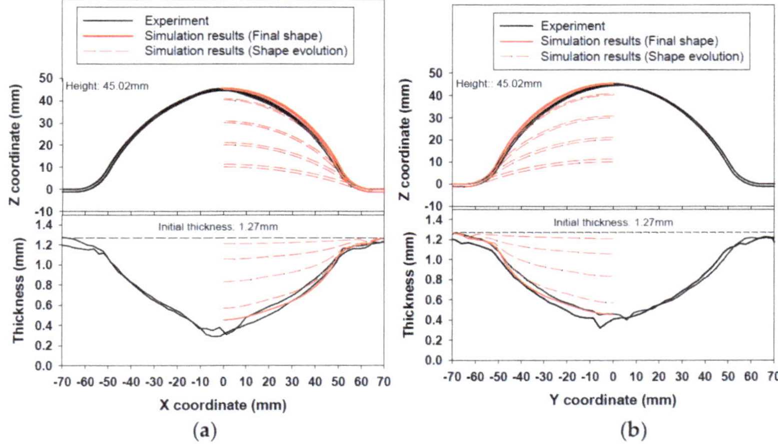

Figure 23. The shape and thickness distribution of the experimental and simulation results for the circular die at $k = 1$ in the x-direction (**a**) and y-direction (**b**).

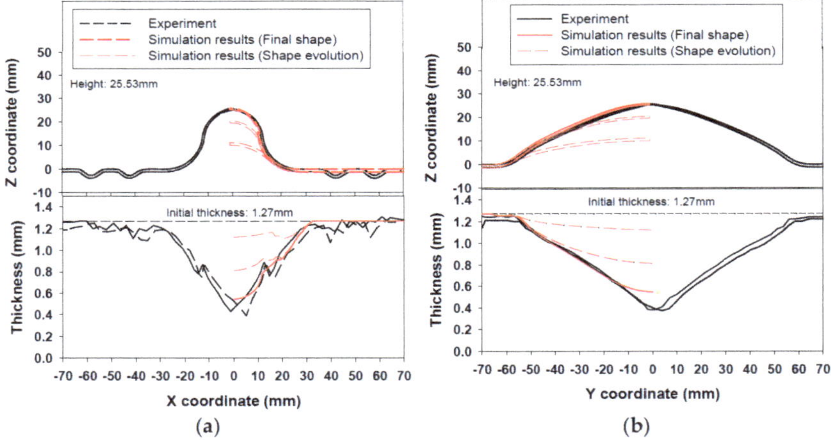

Figure 24. The shape and thickness distribution of the experimental and simulation results for the elliptic die at $k = 0.25$ in the x-direction (**a**) and the y-direction (**b**).

5. Conclusions

An integral experimental design of strain rate sensitive forming limits in the biaxial stretching modes for 7075 aluminum alloy sheets by pneumatic forming was attempted with the finite element method. The integral experimental design work consisted of apparatus geometric design and pressure optimization as the process design. The material was characterized by conventional Voce-type hardening law with power law strain rate sensitivity relationship. For the design of the die shape, the ratio of minor to major radius (k) and profile radius (R) were parametrically studied for best fit to 7075 aluminum alloy sheets. For the design of process condition, pressure and time relationship that maintained targeted strain rate at the pole was obtained. The following conclusions were established:

- As the ratio of minor to major radius (k) became smaller, more effective strain was likely concentrated on the edge and more pressure was needed to complete the forming process. As the ratio of minor to major radius (k) became smaller, the history of the deformation mode in triaxiality (η) and major/minor domain approached the near plane strain (PSA) mode; however, when the k was too small, it did not show uniform deformation history because of the 3D effect (no more plane strain). As for the ratio of minor to major radius (k) of the die, a circular die with $k = 1.0$ and elliptic die with $k = 0.25$ were selected for balanced biaxial mode and near plane strain mode, respectively.

- As the profile radius (R) of the die got larger, deformation was more likely concentrated on the center for both circular and elliptic dies. $R = 15$, which is safe and small enough for favorable fracture occurrence at the center of the blank, was also chosen as profile radius (R) for both circular and elliptic dies.
- For the process design, a preexisting analytical model was modified with geometrical consideration of the profile radius (R). The modified analytical model induced fracture at the pole with approximately constant targeted strain rate.
- Finally, the simulation results of the designed geometric and process conditions were compared with the experimental results, and they showed good agreement with regard to shape and thickness distribution.

Author Contributions: Conceptualization, D.K.; data curation, J.-H.H.; investigation, J.-H.H. and D.Y.; supervision, D.K. and Y.N.K.; visualization, J.-H.H.; writing–original draft, J.-H.H.; writing–review & editing, D.K. and J.-H.H. All authors have read and agreed to the published version of the manuscript.

Funding: This study was financially supported by the Industrial Technology Innovation Program (No. 10077492) funded by the Ministry of Trade, Industry, and Energy (MOTIE) and the Fundamental Research Program of the Korea Institute of Materials Science (PNK6850) funded by the Ministry of Science and ICT (MSIT), Republic of Korea.

Conflicts of Interest: The authors declare no conflict of interest.

References

1. Barnes, A. Superplastic forming 40 years and still growing. *J. Mater. Eng. Perform.* **2007**, *16*, 440–454. [CrossRef]
2. Friedman, P.A.; Luckey, S.G.; Copple, W.B.; Allor, R.; Miller, C.E.; Young, C. Overview of superplastic forming research at ford motor company. *J. Mater. Eng. Perform.* **2004**, *13*, 670–677. [CrossRef]
3. Abu-Farha, F.; Verma, R.; Hector, L.G. High temperature composite forming limit diagrams of four magnesium AZ31B sheets obtained by pneumatic stretching. *J. Mater. Proc. Technol.* **2012**, *212*, 1414–1429. [CrossRef]
4. Mitukiewicz, G.; Antheshwara, K.; Zhou, G.; Mishra, R.K.; Jain, M.K. A new method of determining forming limit diagram for sheet materials by gas blow forming. *J. Mater. Proc. Technol.* **2014**, *214*, 2960–2970. [CrossRef]
5. Banabic, D.; Vulcan, M.; Siegert, K. Bulge testing under constant and variable strain rates of superplastic aluminium alloys. *CIRP Annals* **2005**, *54*, 205–208. [CrossRef]
6. Hill, R.C. A theory of the plastic bulging of a metal diaphragm by lateral pressure. *Lond. Edinb. Dublin Philos. Mag. J. Sci.* **1950**, *41*, 1133–1142. [CrossRef]
7. Banabic, D.; Bălan, T.; Comşa, D.-S. Closed-form solution for bulging through elliptical dies. *J. Mater. Proc. Technol.* **2001**, *115*, 83–86. [CrossRef]
8. Rong, H.; Hu, P.; Ying, L.; Hou, W.; Zhang, J. Thermal forming limit diagram (TFLD) of AA7075 aluminum alloy based on a modified continuum damage model: Experimental and theoretical investigations. *Int. J. Mech. Sci.* **2019**, *156*, 59–73. [CrossRef]
9. Wang, N.; Ilinich, A.; Chen, M.; Luckey, G.; D'Amours, G. A comparison study on forming limit prediction methods for hot stamping of 7075 aluminum sheet. *Int. J. Mech. Sci.* **2019**, *151*, 444–460. [CrossRef]
10. Shojaei, K.; Sajadifar, S.V.; Yapici, G.G. On the mechanical behavior of cold deformed aluminum 7075 alloy at elevated temperatures. *Mater. Sci. Eng. A* **2016**, *670*, 81–89. [CrossRef]
11. Kim, H.; Kim, D.; Ahn, K.; Yoo, D.; Son, H.; Kim, G.; Chung, K. Inverse characterization method for mechanical properties of strain/strain-rate/temperature/temperature-history dependent steel sheets and its application for hot press forming. *Metals Mater. Int.* **2015**, *21*, 874–890. [CrossRef]
12. Kim, H.; Yoon, J.W.; Chung, K.; Lee, M. A multiplicative plastic hardening model in consideration of strain softening and strain rate: Theoretical derivation and characterization of model parameters with simple tension and creep test. *Int. J. Mech. Sci.* **2020**, *187*, 105913. [CrossRef]

13. Kim, H. Characterization Method for Hardening and Strain-Rate Sensitivity at Elevated Temperature Based on Uniaxial Tension and Creep Tests. Ph.D. Thesis, Seoul National University, Seoul, Korea, 2015.

 metals

Article

Ball Spin Forming for Flexible and Partial Diameter Reduction in Tubes

Shota Hirama [1,*], Takayuki Ikeda [1], Shiori Gondo [2], Shohei Kajikawa [1] and Takashi Kuboki [1]

1 Department of Mechanical and Intelligent Systems Engineering, The University of Electro-Communications, 1-5-1 Chofugaoka, Chofu, Tokyo 182-8585, Japan; ikeda@mt.mce.uec.ac.jp (T.I.); s.kajikawa@uec.ac.jp (S.K.); kuboki@mce.uec.ac.jp (T.K.)

2 Advanced Manufacturing Research Institute, Department of Electronics and Manufacturing, National Institute of Advanced Industrial Science and Technology (AIST), 1-2-1 Namiki, Tsukuba, Ibaraki 305-8564, Japan; shiori-gondo@aist.go.jp

* Correspondence: hirama@mt.mce.uec.ac.jp

Received: 30 October 2020; Accepted: 2 December 2020; Published: 4 December 2020

Abstract: This paper proposed a new ball spin forming equipment, which can form a reduced diameter section on the halfway point of a tube. The effects of forming process parameters on the surface integrity and deformation characteristics of the product were investigated. The proposed method can reduce the diameter in the middle portion of the tube, and the maximum diameter reduction ratio was over 10% in one pass. When the feed pitch of the ball die was more than 2.0 mm/rev, spiral marks remained on the surface of tube. Torsional deformation, axial elongation and an increase in thickness appeared in the tube during the forming process. All of them were affected by the feed pitch and feed direction of the ball die, while they were not affected by the rotation speed of the tube. By sending the ball die towards the fixed part of the tube or by increasing the feed pitch, torsional deformation and elongation decreased, and the amount of thickening increased. When the tube was pressed perpendicularly to the axis without axial feed, a diameter reduction ratio of 21.1% was achieved without defects using a ball diameter of 15.9 mm. The polygonization of the tube was suppressed by reducing the pushing pitch. The ball spin forming has a high advantage in flexible diameter reduction processing on the halfway point of the tube for producing different diameter tubes.

Keywords: tube forming; metal tube; planetary ball dies; diameter reduction process

1. Introduction

Tubes with variable diameters in the axial direction are in demand but it is costly to manufacture them. For instance, changing the tube diameter is achieved by connecting different diameter tubes using joints. It takes time and effort to connect, and in some cases, the connection often causes low airtightness. Therefore, the demand for different diameter continuous tubes (DDC tubes) and the process for DDC tubes without joining processes is high. There are some methods for manufacturing DDC tubes, such as U-O bending [1–3], and hydro forming [4,5]. U-O Bending is a processing method to form a tube from plate. Since these methods require tools whose shapes are uniquely related to the final shapes, they lack the flexibility of the product shape. Incremental forming is a flexible process. Incremental forming does not use die and can be used for a wide variety of product shapes [6–8]. Spinning is a typical incremental forming method [9,10]. Spinning with rollers does not require tools with unique shapes but needs repetitive processing because the diameter reduction rate per one processing is small due to the small area between the roller and tube, resulting in complex control of the roller position. In addition, the rigidity of the device is low due to the deformation of the roller struts when high stress is applied [11–13]. Therefore, Kanayama et al. developed a new method—"Ball spin forming" [14]. Ball spin forming is an efficient processing method that can solve the above problems. The essence of ball spin forming is the usage of "ball die", which is a processing unit with three balls inside. The ball die can apply a large deformation to the tube at one pass as the balls are set inside a rigid and compact box. Kanayama et al. developed a planetary ball die for processing tube end and obtained tubes with smooth surface integrity. Kanayama at el. also reported that when the diameter reduction is performed using ball die, torsional deformation, axial elongation, and thickness increase occurred in tubes [15]. This type of die is applied in the manufacture of internally grooved tubes [16]. However, a ball die that cannot change the ball position in the radial direction can only process a simple shape. In this study, we developed a new ball forming ball die that can control the ball position in the radial direction of the tube for forming tubes with variable diameters in the axial direction. The new ball die combines the advantages of both spinning and planetary ball dies. The kinematics are the same as for roller spinning, but the rigidity and productivity of the equipment are improved, which solves the problems of spinning.

The structures of the developed ball die and deformed tube on the halfway portion are shown in Figure 1. Iron balls, which are constrained in the circumferential direction of the tube, are pushed in the radial direction of the tube by the rotation of the ball guide. The ball die travels along the tube axis while the tube rotates. As the result, the diameter reduction forming is performed by expanding the local concavity formed by the balls.

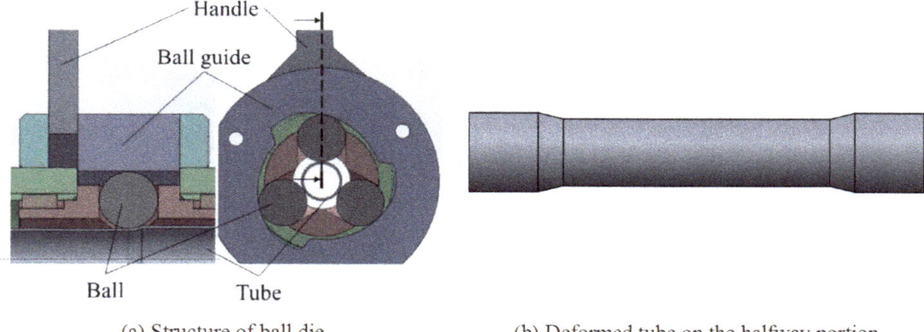

Figure 1. Schematics of ball die and deformed tube.

The remarkable feature of the new ball die, which is developed in this study, is that the balls move in the radial direction of the tube. The inner surface of the ball guide has the shape of a cam, and the shape realizes the control of the ball position in the tube-radial direction just by rotating the ball guide, to which a handle is connected. This feature enables flexible forming at the halfway point of the tube. This paper investigates the effect of parameters such as diameter reduction ratio, feed pitch, and rotational speed on the product shape and surface integrity. The experiments were carried out to the halfway point of the tube using the developed ball die, and the appropriate conditions were clarified for reducing the tube diameter without defects.

2. Experiment with a Small Ball Die for Lathe

2.1. Materials and Methods

The experimental material was an extruded tube of aluminum alloy 1050 (AA1050). The ball material was high carbon chrome bearing steel of SUJ2 (Japan Industrial Standard, JIS). The outer diameter and thickness of tubes were 19 and 0.9 mm, respectively. The outer diameter of the tool ball was 20 mm. Grease with molybdenum disulfide was applied to the surface of the tube as lubricant.

A schematic diagram of the experimental ball die is shown in Figure 2. The tube end was fixed to the chuck of the lathe. The ball die was mounted on the tool stand of the lathe, which moves in the axial direction automatically at a constant feed pitch f. The working conditions are shown in Table 1, including the diameter reduction ratio κ_D, the tube rotation speed n, and the feed pitch f. The procedure is as follows:

(1) Push the balls into the tube surface without giving the feed pitch f.
(2) Stop pushing the balls and feed the ball die at the constant feed pitch f.
(3) Set the feed pitch to zero and release the ball from the tube.

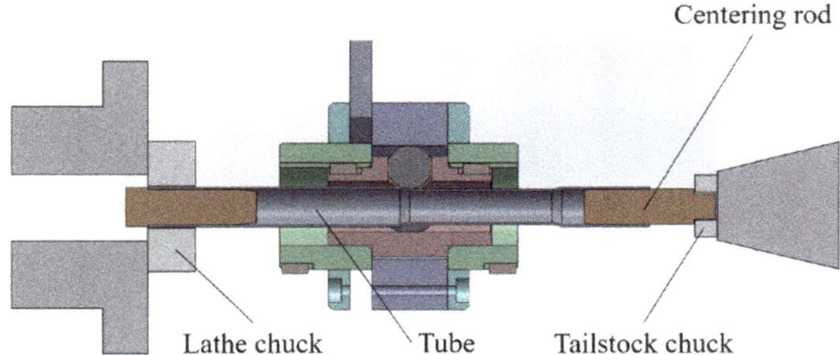

Figure 2. Schematic diagram of the experimental ball die.

Table 1. Conditions of the experiment.

Tube	Outer diameter D_P/[mm]	19.0		
	Thickness t/[mm]	0.9		
Experimental conditions	Diameter reduction ratio κ_D/[%]	5.3	10.5	15.8
	Target diameter D_T/[mm]	18	17	16
	Rotational speed n/[rpm]	50, 85, 140		
	Feed pitch f/[mm/rev]	0.5, 1.0, 1.5, 2.0, 2.5, 3.0		
	Formed length l/[mm]	50, 120		

Torsional deformation and elongation were evaluated by scribe lines. Scribe lines were marked on the tube surface in the longitudinal direction before forming as shown in Figure 3. The torsional deformation and elongation of the tube were measured by comparing the shape of scribe lines before and after forming. One end of the tube was chucked, while the other end was supported by a centering rod without axial constraint. That is to say, the constraint condition is different between the two tube ends. Therefore, the effect of moving direction of the ball die on the formability was investigated. The longitudinal section of the tube was observed using an optical microscope for evaluating the thickness distribution.

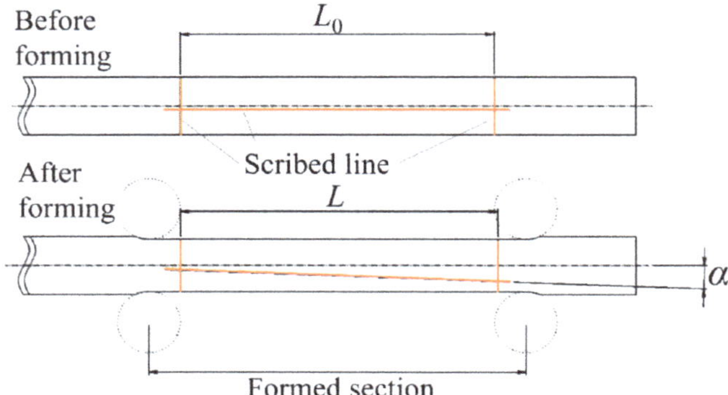

Figure 3. Schematic diagram of scribe lines on the tube.

2.2. Results and Discussion

2.2.1. Effect of Feed Pitch on Surface Integrity

Figure 4 shows the tube surface after forming. When the feed pitch f exceeded 2.5 mm/rev, spiral marks appeared on the tube surface after the ball die passed, as shown in Figure 4a. When the feed pitch f was small, a smooth surface was obtained, as shown in Figure 4b. The effect of the rotational speed n on the surface integrity was small. Therefore, it is concluded that the formed surface integrity is determined by the feed pitch of the ball.

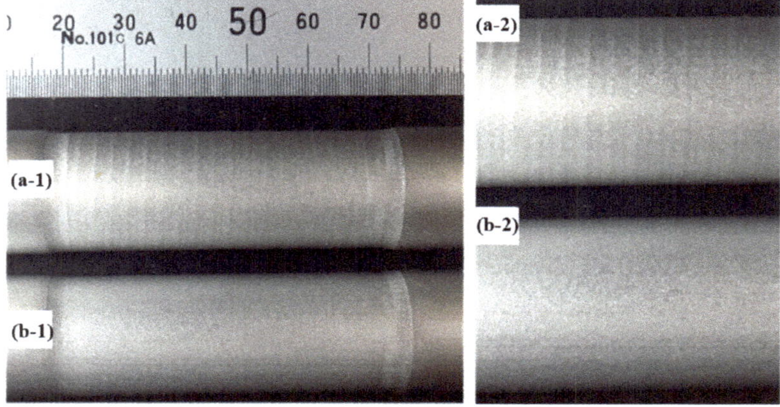

Figure 4. Appearance of the formed tubes (**a-1**) spiral mark: the diameter reduction ratio $\kappa_D = 15\%$, $n = 85$ rpm, $f = 3.0$ mm/rev, (**b-1**) smooth surface: $\kappa_D = 15\%$, $n = 85$ rpm, $f = 0.5$ mm/rev. The photographs (**a-2**) and (**b-2**) are high magnification version of (**a-1**) and (**b-1**), respectively.

2.2.2. Forming Limit in One and Two-Step Processing

When processing was performed under a diameter reduction ratio κ_D of 15.8%, the tubes could not be formed into the target shape due to polygonization. In this case, the axis of the tube vibrated when the balls were excessively pushed to the tube. Then, the cross-section of the tube was deformed into a pentagonal shape as shown in Figure 5. In the case that κ_D was 10.5%, it was possible to push the ball to the tube at the beginning of the forming, but the pentagonal deformation occurred when the ball die proceeded to the middle portion of the tube.

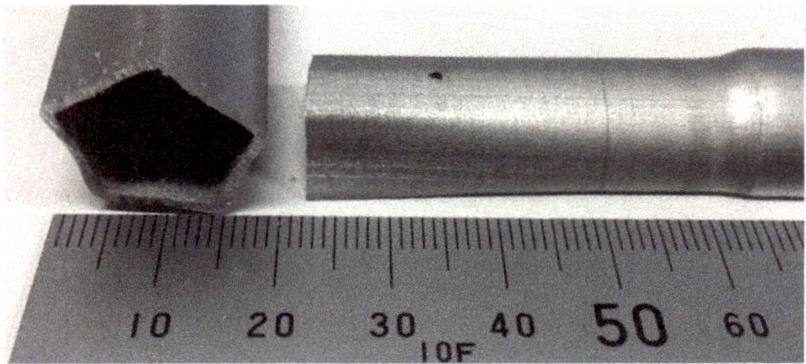

Figure 5. Pentagonal deformation.

On the other hand, two-step processing prevented the pentagonal deformation. In the first step, the tube was processed under a diameter reduction ratio κ_D of 5.3%. In the second step, the tube was processed again at the target diameter D_T such as 17 or 16 mm. It is considered that the pentagonal deformation occurred because the tube axis was not held tightly against the large pushing force of the ball.

2.2.3. Dimensional Accuracy of Formed Tube

Figure 6 shows the distribution of the outer diameter of the formed tube in the processed section. The variation of the outer diameter in the processing section of the tube for 50 mm length was 0.04 mm or less in each processing condition. The deviation between the outer diameter of the formed tube and target was approximately 0.2 mm.

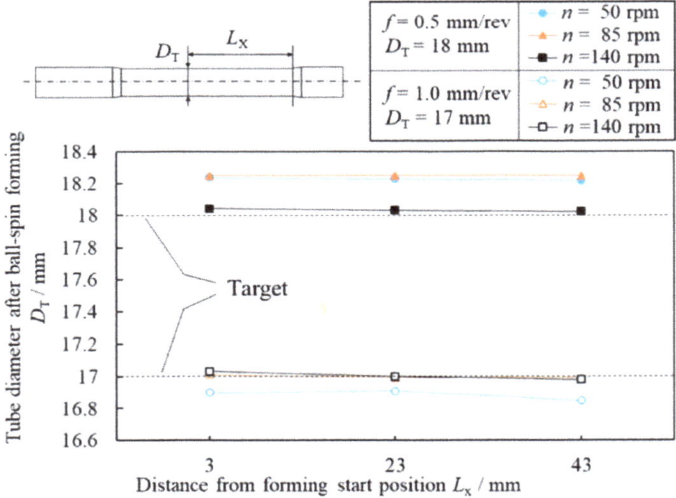

Figure 6. Distribution of the outer diameter of the formed tubes.

2.2.4. Torsional Deformation of Tube

When the ball die moved towards the free end in the processing, the formed portion of the tube was twisted. The effect of the feed pitch f on the twist angle α of the tube is shown in Figure 7. The twist angle α decreased with the increase in the feed pitch f as the increase in f lead to the decrease in tube-rotation number L/f. On the other hand, no twisting occurred when the ball die was moved towards the chucked end. Therefore, the effect of the feed direction of the ball die on the torsional deformation of the tube was significant.

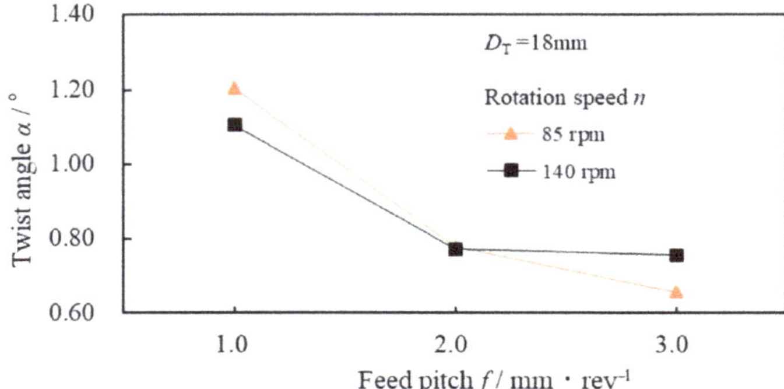

Figure 7. Relationship between the feed pitch and twist angle.

2.2.5. Thickness Increase in the Tube

Figure 8 shows the longitudinal section of the portion at the beginning of processing. The thickness increasing ratio η was defined as the following equation.

$$\eta = (t - t_0)/t_0 \tag{1}$$

where t_0 and t were the tube thickness before and after forming, respectively. The thickness increasing ratio η showed the opposite tendency to the tube elongation rate. As mentioned in the previous section, the tube elongation was small in the case that the ball die moved from the free to chuck side. In Figure 8a, the thickness increasing ratio η was 11.1% when the ball die moved towards the free end, that is, when elongation deformation is likely to occur. On the other hand, as shown in Figure 8b, the thickness increasing ratio η was 13.0% when the ball die moved towards the chuck end. This is a reasonable result in the view of the constant volume law.

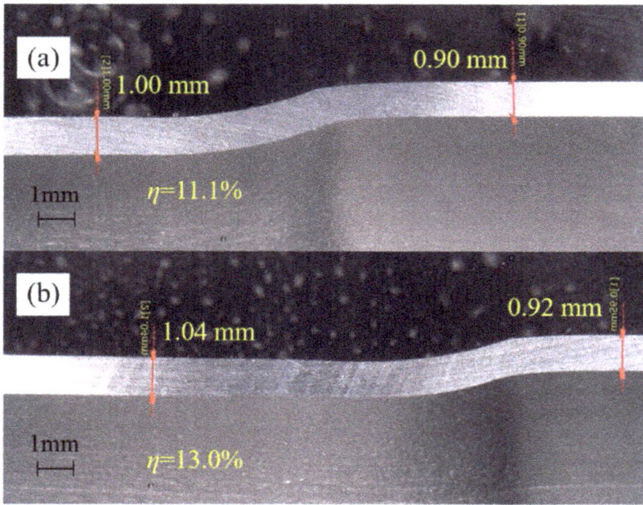

Figure 8. Cross-section of the tube (**a**) κ_D = 10.5 %, n = 85 rpm, f = 1.5 mm/rev, feed direction: chuck to free side (**b**) κ_D = 10.5 %, n = 85 rpm, f = 1.5 mm/rev, feed direction: free side to chuck.

3. Experiment with a Cam-Implemented Ball Die

3.1. Development of a Cam-Implemented Ball Die

A cam-implemented ball die was developed to investigate the effects of the radial feeding ratio and the diameter of the steel balls on the formability. In the above-mentioned experiments, the ball die was mounted on a lathe and the radial position of the balls was manually controlled by the handle, and then it was not possible to adjust the radial feeding ratio of the balls or to change the ball size without the major modification of the ball die. Therefore, a new ball die was developed for the stable control of the radial position of the balls and the concise exchange of the balls.

The structure of the cam-implemented ball die is shown in Figure 9. The ball diameter can be changed by using a "free bear" in place of steel balls. The free bear is composed of a steel ball, a cylinder, and dozens of supporting small balls. The small balls help the steel ball to rotate in a frictionless manner. The servo motor applies rotational displacement to the ball guide. The inner surface of the ball guide is designed so that the rotational displacement of the ball guide is proportional to the pushing depth. For each 8600° rotation of the servo motor, the ball is pushed 1 mm. The pushing pitch and pushing depth were controlled by controlling the number of revolutions and rotation speed of the servo motor. The steel ball in the free bear is pushed in the radial direction of the tube. The curved surface of the inner surface of the ball guide was designed so that the rotational displacement of the ball guide is proportional to the pushing amount of the free bear. The newly developed ball die was mounted on a LM guide instead of a lathe. LM Guide is a machine element part that uses "rolling" to guide the linear motion part of a machine.

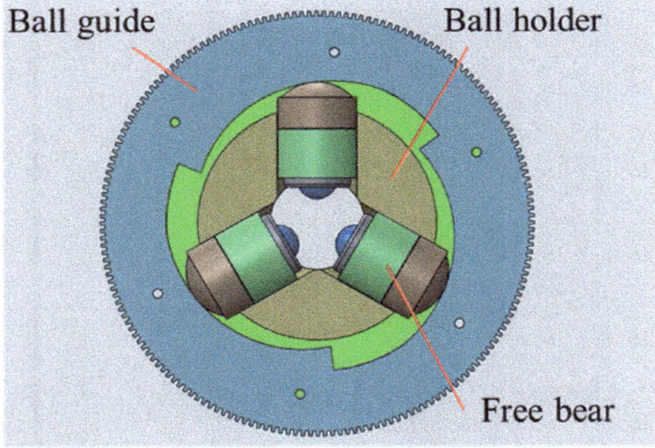

Figure 9. Schematic diagram of cam-implemented ball die.

An overall view of the experimental machine is shown in Figure 10. A characteristic of the newly developed experimental machine is that the axial feed and the radial push are controlled by separate servo motors. This makes it possible to change the rate of reduction in diameter while the ball die travels in the axial direction, making it possible to deform tubes into a greater variety of shapes.

The effects of pushing pitch of the ball and ball diameter on the formability were investigated using this machine.

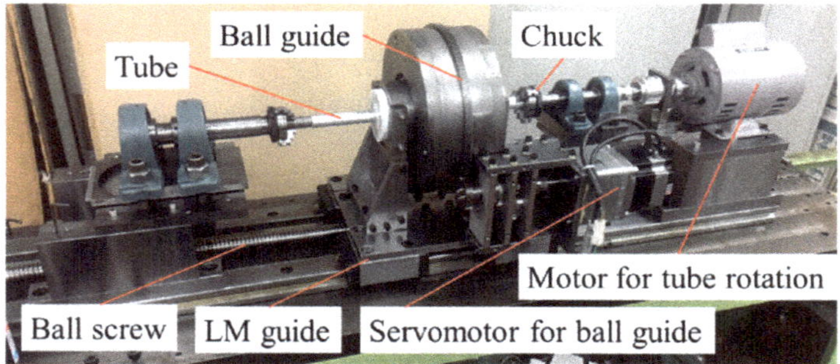

Figure 10. Overall view of the experimental equipment.

3.2. Materials and Methods

AA1050 tubes with a wall thickness of 1 mm and an outer diameter of 19 mm were used in the experiments. The steel balls in the free bear were made of high carbon-chromium bearing steel SUJ2 (Japan Industrial Standard, JIS). Machine oil was applied to the surface of the tube during processing.

A schematic diagram of the experimental machine is shown in Figure 11. One end of the tube was fixed to a chuck connected to the spindle motor, and the other end was held by a centering rod inside. In addition, a centering cap was set near the processing portion for preventing axial and radial vibration during processing. A schematic diagram of the experimental procedure is shown in Figure 12. The experiment was carried out by changing the amount of pushing of the steel balls into several locations of a single tube. Three free bears with steel balls with diameter Φ_b were pushed into the tube in the radial direction until the outer diameter of the part became the target outer diameter D. The pushing pitch of the ball f_R was set to be constant. After holding the ball position for 3 seconds at the target diameter, the balls were released. After that, the ball die was moved in the axial direction of the tube, and the steel balls were pushed in again. This process was performed for different ball diameters Φ_b and pushing pitches of the ball f_R. The parameters of the experimental conditions are shown in Table 2.

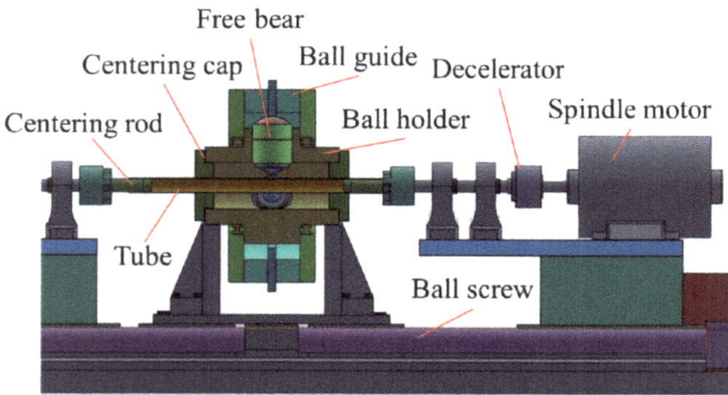

Figure 11. Schematic diagram of the experimental machine.

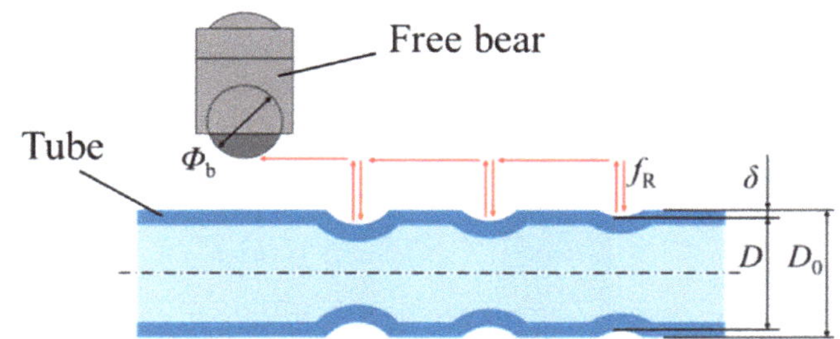

Figure 12. Schematic diagram of processing.

Table 2. Conditions of experiment.

Tube diameter D_0/mm	19						
Wall thickness t_0/mm	1.0						
Ball diameter Φ_b/mm	15.9, 19.1, 25.4						
Pushing pitch f_R/mm·rev^{-1}	0.042, 0.083, 0.125, 0.167						
Pushing depth δ/mm	0.5	1.0	1.5	2.0	2.5	3.0	3.5
Target diameter D/mm	18	17	16	15	14	13	12
Diameter reduction ratio κ/%	5.3	10.5	15.8	21.1	26.3	31.6	36.8

3.3. Results and Discussion

3.3.1. Effect of Steel Ball Diameter

Figure 13 shows the effect of ball diameter Φ_b on tube appearance and formability at pushing pitch f_R = 0.042 mm/rev. The cross-section of the formed tubes is shown in Table 3. The maximum diameter reduction ratio without the defects of polygonization increased with decreasing the steel ball diameter. When the ball diameter Φ_b = 15.9 mm, the tube could be formed under reduction ratio κ = 21.1% without defects. For steel ball diameters Φ_b = 19.1 mm and 25.4 mm, the tube was polygonalized at the target diameters of D = 16 mm and 15 mm. The polygonization of the tube is thought to occur when the central axis of the rotating tube deviates from the central axis of the tool.

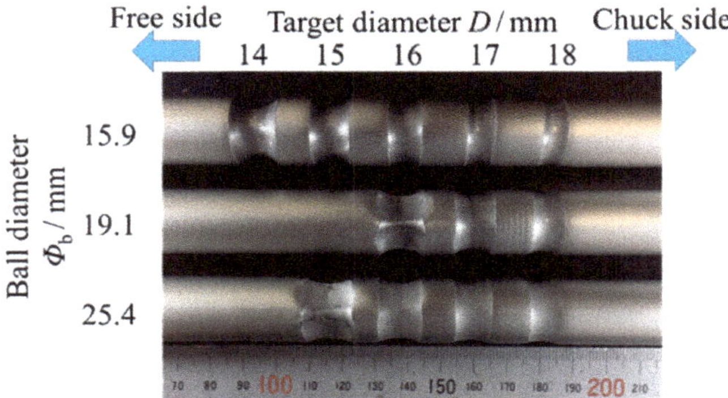

Figure 13. Appearance of formed tube when f_R = 0.042 mm/rev.

Table 3. Cross-section of formed tube when f_R = 0.042 mm/rev.

	Target Diameter *D*/mm				
	14	15	16	17	18
15.9					
19.1					
25.4					

Figure 14 shows the longitudinal sections of the formed tubes for the examination of wall thickness. Table 4 shows the thickness increasing ratio η, which has been introduced by Equation (2). The thickness of the necked area decreased with decreasing target diameter D. In the case of $D \leq 15$ mm, the surface integrity deteriorated at the boundary between the formed and un-formed areas. The axial distance from the neck point with the smallest diameter [A] to the boundary of the formed area on the free end [B] is larger than from [A] to the boundary on the chuck side [C]. From these facts, it is considered that the force to stretch the material around the center of the workpiece is acting in the axial direction by pushing the steel ball into the workpiece. In particular, this stretching effect is strong on the free side.

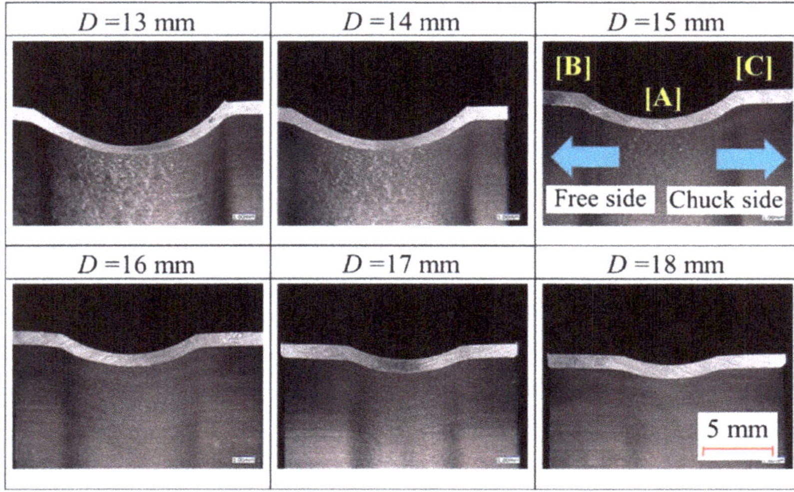

Figure 14. Cross-section of the tube (ball diameter Φ_b = 15.9 mm, pushing pitch f_R = 0.042 mm/rev).

Table 4. The thickness increasing ratio η.

D/mm	η/%
18	0.0
17	1.1
16	−6.4
15	−20.4
14	−44.7
13	−40.2

3.3.2. Effect of Pushing Pitch

Figure 15 shows the effect of pushing pitch on the appearance and formability with a steel ball diameter Φ_b = 15.9 mm. The cross-sectional states of the formed tubes are shown in Table 5. The results indicate that the tube was polygonalized with increasing pushing pitch. It is considered that as the pushing pitch increases, the contact area between the tube and the ball increases per rotation, and thus the tube perimeter did not shrink sufficiently and the surplus perimeter caused the polygonization force to disperse.

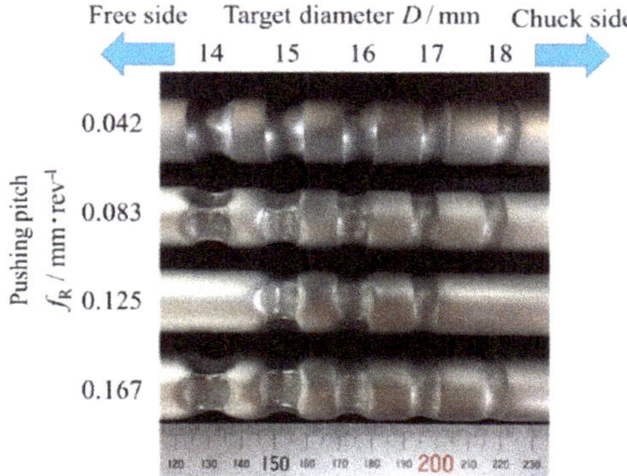

Figure 15. Appearance of formed tube when Φ_b = 15.9 mm.

Table 5. Cross-sectional state of the formed tubes.

Ball Diameter **Φ_b = 15.9 mm**	**Target Diameter D/mm**				
Pushing Pitch **f_R/mm·rev^{-1}**	**14**	**15**	**16**	17	18
0.042	△	○	○	○	○
0.083	×	×	△	○	○
0.125	×	×	×	○	○
0.167	×	×	△	○	○

4. Conclusions

The following conclusions can be drawn from the present research:

1 A ball die, which can push the balls flexibly in the radial direction, was developed.

2 It was possible to reduce the tube diameter at the halfway point of an AA1050 tube whose diameter and thickness were 19 and 1 mm, respectively. The maximum diameter reduction ratio for forming the tube without any defects was 10.5% in the case of one-pass forming.

3 When processing with a large diameter reduction rate was performed, the cross-section deformed to a pentagonal shape. This defect was improved by two-pass forming.

4 A smooth formed surface was obtained under the condition of the appropriate feed pitch. If the feed pitch was too large, spiral marks appeared on the surface of the tube.
5 Elongation and twisting of the tube could be suppressed by increasing the feed pitch or feeding the ball die towards the chucked end.
6 The smaller diameter of the steel ball made it possible to process a larger diameter reduction ratio.
7 The maximum diameter reduction ratio got bigger with decreasing the pushing pitch of the steel balls.
8 When a steel ball was pressed into the tube, the wall thickness of the tube decreased with an increase in the pushing depth.

Author Contributions: Conceptualization, S.H., T.I., S.K. and T.K.; methodology, T.I., S.K. and T.K.; validation, S.H. and T.I.; investigation, S.H. and T.I. and T.K.; data curation, S.H. and T.I.; formal analysis, S.H. and T.I.; resources, T.K.; writing—original draft preparation, S.H. and T.I.; writing—review and editing, S.H., T.I., S.G., S.K. and T.K.; visualization, S.H. and T.I.; supervision, T.K.; project administration, S.H., T.I., S.G., S.K. and T.K.; funding acquisition, T.K. All authors have read and agreed to the published version of the manuscript.

Funding: This research received no external funding.

Conflicts of Interest: The authors declare no conflict of interest.

References

1. Sato, M.; Mizumura, M.; Yoshida, T.; Kuriyama, Y.; Suzuki, K.; Tomizawa, A. Deformation type in forming of horn tubes—Fundamental research for forming of closed-section parts from sheet metal I. *J. JSTP* **2017**, *59*, 27–31. [CrossRef]
2. Sato, M.; Mizumura, M.; Kuriyama, Y.; Suzuki, K.; Tomizawa, A. Deformation type in forming of curved conical tubes—Fundamental research for forming of closed-section parts from sheet metal II. *J. JSTP* **2018**, *59*, 229–233. [CrossRef]
3. Sato, M.; Mizumura, M.; Kuriyama, Y.; Suzuki, K. Deformation paths of horn tube, curved circular tube and curved conical tube-Fundamental research for forming of closed-section parts from sheet metal III. *J. JSTP* **2020**, *61*, 131–135. [CrossRef]
4. Wang, X.; Li, P.; Wang, R. Study on hydro-forming technology of manufacturing bimetallic CRA-lined pipe. *Int. J. Mach. Tools Manuf.* **2005**, *45*, 373–378. [CrossRef]
5. Kumar, R.U.; Reddy, P.R.; Sitaramaraju, A.V. Role of Viscosity in Hydro-forming Process. *Mater. Today Proc.* **2017**, *4*, 790–798. [CrossRef]
6. Cristino, V.; Magrinho, J.; Centeno, G.; Silva, M.; Martins, P.A.F. Theory of single point incremental forming of tubes. *J. Mater. Process. Technol.* **2020**, *287*, 116659. [CrossRef]
7. Cui, X.; Mo, J.; Li, J.; Zhao, J.; Zhu, Y.; Huang, L.; Li, Z.; Zhong, K. Electromagnetic incremental forming (EMIF): A novel aluminum alloy sheet and tube forming technology. *J. Mater. Process. Technol.* **2014**, *214*, 409–427. [CrossRef]
8. Becker, C.; Tekkaya, E.; Kleiner, M. Fundamentals of the incremental tube forming process. *CIRP Ann.* **2014**, *63*, 253–256. [CrossRef]
9. Xu, W.; Wu, H.; Ma, H.; Shan, D. Damage evolution and ductile fracture prediction during tube spinning of titanium alloy. *Int. J. Mech. Sci.* **2018**, *135*, 226–239. [CrossRef]
10. Wang, X.; Hu, Z.; Yuan, S.; Hua, L. Influence of tube spinning on formability of friction stir welded aluminum alloy tubes for hydroforming application. *Mater. Sci. Eng. A* **2014**, *607*, 245–252. [CrossRef]
11. Ozer, A.; Sekiguchi, A.; Arai, H. Experimental implementation and analysis of robotic metal spinning with enhanced trajectory tracking algorithms. *Robot. Comput. Integr. Manuf.* **2012**, *28*, 539–550. [CrossRef]
12. Murata, M.; Kuboki, T.; Murai, T. Compression spinning of circular magnesium tube using heated roller tool. *J. Mater. Process. Technol.* **2005**, *163*, 540–545. [CrossRef]
13. Kuboki, T.; Takahashi, K.; Sanda, K.; Moriya, S.; Ishida, K. Development of a Tube-Spinning Machine for Thin Tubes with a Large Diameter. *Mater. Trans.* **2012**, *53*, 853–861. [CrossRef]
14. Kanayama, K.; Yoshioka, K.; Shigematsu, I.; Hirai, Y. Effect of friction on the processing force in tube reducing of aluminum alloy by using planetary ball die. *J. Jpn. Inst. Light Met.* **1992**, *42*, 61–66. [CrossRef]

15. Kanayama, K.; Yoshioka, K.; Shigematsu, I.; Kozuka, T. Tube reducing of welded titanium tube by using planetary ball die. *J. Jpn. Inst. Light Met.* **1992**, *42*, 657–662. [CrossRef]
16. Guang-Liang, Z.; Shi-Hong, Z.; Bing, L.; Hai-Qu, Z. Analysis on folding defects of inner grooved copper tubes during ball spin forming. *J. Mater. Process. Technol.* **2007**, *184*, 393–400.

Article

Movable Die and Loading Path Design in Tube Hydroforming of Irregular Bellows

Yeong-Maw Hwang * and Yau-Jiun Tsai

Department of Mechanical and Electro-Mechanical Engineering, National Sun Yat-sen University, Lien-Hai Rd., Kaohsiung 804, Taiwan; jim0727jim@gmail.com

* Correspondence: ymhwang@mail.nsysu.edu.tw

Received: 18 October 2020; Accepted: 14 November 2020; Published: 16 November 2020

Abstract: Manufacturing of irregular bellows with small corner radii and sharp angles is a challenge in tube hydroforming processes. Design of movable dies with an appropriate loading path is an alternative solution to obtain products with required geometrical and dimensional specifications. In this paper, a tube hydroforming process using a novel movable die design is developed to decrease the internal pressure and the maximal thinning ratio in the formed product. Two kinds of feeding types are proposed to make the maximal thinning ratio in the formed bellows as small as possible. A finite element simulation software "DEFORM 3D" is used to analyze the plastic deformation of the tube within the die cavity using the proposed movable die design. Forming windows for sound products using different feeding types are also investigated. Finally, tube hydroforming experiments of irregular bellows are conducted and experimental thickness distributions of the products are compared with the simulation results to validate the analytical modeling with the proposed movable die concept.

Keywords: tube hydroforming; movable die; loading path; finite element simulation; irregular bellows

1. Introduction

Tube hydroforming is a forming technique that utilizes a high internal pressure to bulge the tube material to fit the die shape and then obtain the desired product shape. Nowadays, tube hydroforming technologies have been widely applied in automotive and aerospace industries for manufacturing stronger and lighter products. However, higher pressures and complicated loading paths are required in manufacturing of complex shape products. If loading path or the relationship between the internal pressure and axial feeding is not controlled appropriately, various defects such as bursting and wrinkling, etc., would probably occur [1].

Metal bellows have been widely applied in various industries, such as chemical plants, power systems, heat exchangers, automotive vehicle parts, etc., for absorbing the irregular expansions of the pipes, damping vibration of the circumference and mechanical movements [2,3]. Conventional regular metal bellows with multiple convolutions are usually manufactured by hydraulic bulging and die folding. At first, a blank tube is bulged outward with an internal hydraulic pressure and then an open die set of multiple plate dies moves axially to fold the bulged tube into the desired shape [4,5]. However, for irregular bellows, in which the outer diameter of the bellows is not much larger than its inner diameter, the conventional manufacturing method combining hydraulic bilging and die folding cannot be applied to this kind of irregular bellows. This kind of irregular bellows can only be made by hydraulically bulging the tube into the desired shape under an irregular closed die set. Especially for an irregular bellows with small radii and sharp angles, it is difficult to make the tube material flow into the corner of the closed die completely using a hydraulic bulging process.

Furthermore, a quite large internal pressure is required and a non-uniform thickness distribution is probably obtained.

Recently, several die design concepts and optimizations of loading paths have been proposed to increase the formability and obtain a more uniform thickness distribution in the product. For example, Ngaile and Lowrie [6] conducted several kinds of punch shape designs in floating-based micro-tube hydroforming to manufacture Y, T, and bulged-shape tubes. The length of the punches was investigated to prevent occurrence of buckling during the hydroforming process. They found that the notched punch is stronger in withstanding the axial loads. Shinde et al. [7] used a high internal pressure to bulge a circular cold-rolled steel tube into a square shape. The effects of die corner radius, length of tube, tube thickness and internal pressure on thinning ratios were discussed. The results showed that the tube length and die corner radius affect tube thinning significantly. In addition, no matter how long the tube length is, a thicker initial tube thickness and a higher internal pressure can improve the product quality. Elyasi et al. [8] proposed a die design concept to obtain a sharp corner tube with a thickness thinning ratio of 20%. Zhang et al. [9] added two movable sleeves in the die set to enhance the formability in a hydroforming process of stainless steel sheets. Comparing with the conventional hydroforming without movable sleeves, the internal pressure was significantly reduced from 180 to 47 MPa, and the die corner was well filled with the blank material. Meanwhile, a more uniform thickness distribution was obtained. Tomizawa et al. [10] proposed an axial movable die concept in a hydroforming process to obtain a small corner radius and a uniform thickness distribution of the formed product with a lower internal pressure. Hwang and Chen [11] proposed a compound hydroforming concept to manufacture an asymmetric profiled tube. The loading paths are composed of a die crushing stage, a punch feeding stage and an internal pressurization stage. The loading path was well designed and controlled so that a corner radius of 6 mm was formed with a low internal pressure of 18 MPa. Hwang et al. [12] used finite element codes LS-DYNA and DYNAFORM to analyze the plastic flow pattern of a tube hydroformed into a product with large expansion ratio and eccentric axes. They proposed a movable die set concept to enhance the forming capacity of tube hydroforming technology. From the geometric analysis, the relative speed of the axial feeding to the movable die was determined and a more uniform thickness distribution was obtained.

In addition to the design of hydroforming die sets, loading paths also have a great influence on the thickness uniformity of the product. Yuan et al. [13] investigated the wrinkling behavior in tube hydroforming processes. They divided it into two types: useful wrinkles and dead wrinkles. If the axial feeding and the internal pressure matched up well, the so-called useful wrinkles could be an effective way to accumulate tube materials in the expansion zone and the formability could be increased. Xu et al. [14] carried out experiments to show the superiority of pulsating loading path over the monotonic loading path. A higher pressure and a larger axial stroke could be applied without tube bursting, which meant the forming limit could be extended. Imaninejad et al. [15] used the finite element simulations and an optimization software to optimize the loading paths in T-joint tube hydroforming processes. Experiments were carried out on aluminum tubes to validate the reliability of the simulation results. Hama et al. [16] used the finite element analysis to discuss the effects of three loading paths on the tube formability in a hydroforming process. If local wrinkling occurred, the tube formability became worse. They applied an axial feeding after the free bulging stage to obtain a product without defects.

The above literature is focused on hydroforming die sets and loading path design to solve problems on occurrence of defects, excessive thinning or nonuniform thickness distribution in T or Y-shape tube hydroforming processes. In this study, a novel movable die concept is proposed to manufacture irregular bellows with sharp angles. According to hydroforming machine categories, two kinds of movable die feeding types for the loading path are designed. The forming windows for sound products without defects using a lower forming pressure are discussed. Experiments of hydroforming are also conducted to validate the proposed novel movable die concept and loading paths.

2. Movable Die Design and Loading Path Definition

2.1. Shape and Geometry

Figure 1 shows the geometrical shape of an irregular bellows. The irregular bellows has a symmetric axis, but it is not symmetric at its both ends. It consists of two bulged regions (zones I and II) and a conical region (zone III) with an inclination angle θ. In the two bulged regions, the tube surfaces are two circular arcs with different curvature radii (r_1 and r_3) and are connected by a circular arc of radius r_2. The conical part (zone III) connects zone II with another circular arc of radius r_4. t is the bellows thickness. d_1, d_2 and d_3 are the diameters of the bellows at the left end, middle part, and the right end, respectively. *l* is the bellows length. Since the maximal thinning ratio usually occurs at points A, B, or C, the thickness distribution and thinning ratios at points A, B, and C are discussed in the following sections. Conventional metal bellows have multiple wrinkles or convolutions. As a result that the outer diameter of each convolution is much larger than its inner diameter, they can be manufactured by hydraulic bulging and die folding processes. The outer diameter d_3 in Figure 1 is not much larger than its original tube diameter d_1. This kind of irregular bellows can only be made by a hydraulic bulging process under a closed die set.

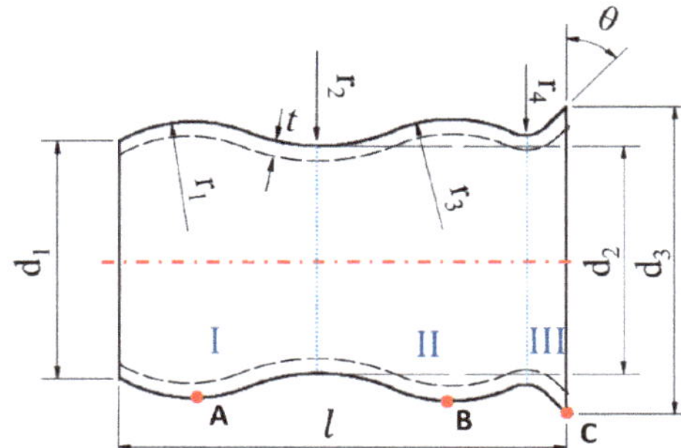

Figure 1. Geometrical configuration of an irregular bellows.

2.2. Movable Die Design

To increase the productivity and balance the axial forces from the left and right punches, a hydroforming die set is designed to manufacture two bellows in one forming pass. Another irregular bellows with inverse direction is allocated at the right hand side, as shown in Figure 2a. The central part (zone C) connecting the conical regions of the two bellows is designed to let the tube flow into the conical part easily. Part A at both ends of the die set is designed as a guiding zone. After forming, parts A and C of the formed product would be cut off and part B would be the desired irregular bellows products.

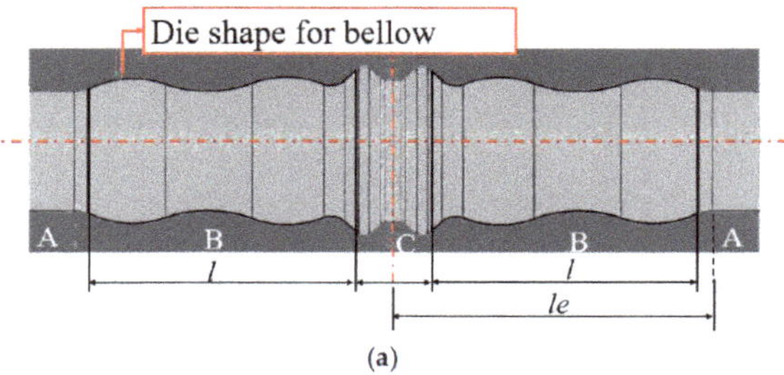

(**a**)

Figure 2. *Cont.*

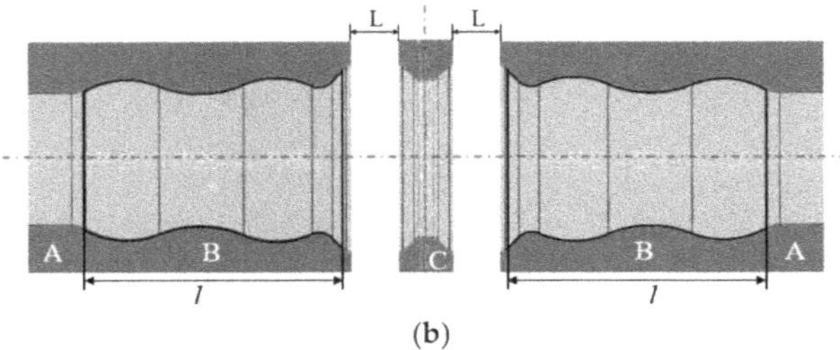

(b)

Figure 2. Cross-sectional configurations of designed die set. (**a**) Traditional die set; (**b**) Movable die set.

For a movable die design, the hydroforming die set is divided into three parts: a fixed central die (zone C) and two movable dies (zones A and B) as shown in Figure 2b. The movable dies are separated from the fixed die by a gap of width L at the beginning of the process. The hydroforming process is divided into three stages. The relationship between the internal pressure and the die movement at each stage is shown in Figure 3, where L is the die gap width, P_1 and P_2 are the pressures at the free bulging stage (stage I) and die feeding stage (stage II), respectively, and P_f is the final pressure at the calibration stage.

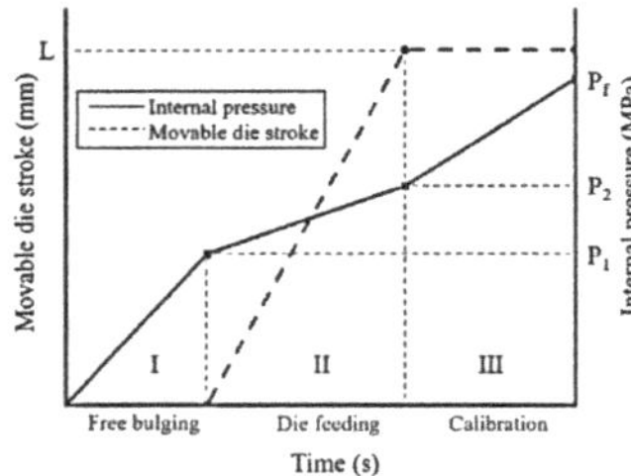

Figure 3. Designed loading path in hydroforming process with movable dies.

The geometric configurations of the dies and tube for the three stages in the hydroforming process are shown in Figure 4. The initial geometric configuration between the punch, dies and tube is shown in Figure 4a. At the first stage, only internal pressure is applied into the tube to bulge the tube and make the tube undergo slight plastic deformation, as shown in Figure 4b. At the second stage, movable dies start to move forward to make the tube flow into the die corner, as shown in Figure 4c. At this stage, the internal pressure can be adjusted. At the final stage, the movable dies are closed completely, and a higher internal pressure is input to calibrate the tube to fill into the die corner completely, as shown in Figure 4d.

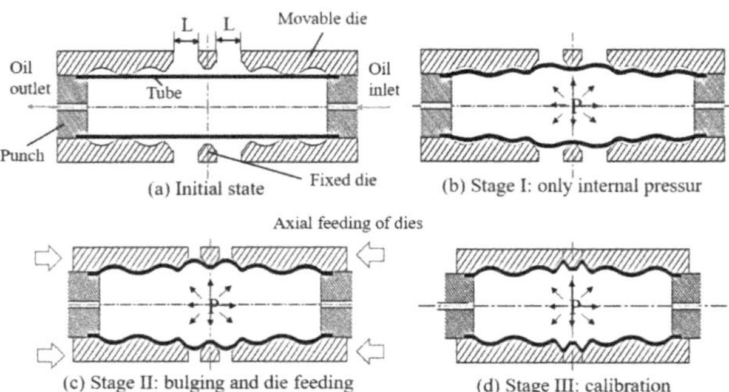

Figure 4. Geometric configurations of dies and tube at each stage. (**a**–**d**): Initial State to Stage III.

2.3. Movable Die Feeding Types

According to hydroforming machine categories, two kinds of movable die feeding types are proposed. For a hydroforming machine with two sets of hydraulic cylinders, one set of cylinders is used to push punches to seal the tube. The other set of cylinders is used to push the movable dies forward. This kind of feeding type is called feeding type 1. The geometric configurations of the movable die, tube, punch, and hydraulic cylinders for movable die feeding type 1 are shown in Figure 5. At the beginning, the tube is placed inside the die set and is sealed at both ends by the punch head pushed by cylinder set 1, as shown in Figure 5a. Then, hydraulic oil is input to bulge the tube, as shown in Figure 5b. During the die feeding stage (stage II), movable dies are pushed forward by cylinder set 2. The configurations of the tube and die after stage II are shown in Figure 5c. Finally, a calibration pressure is applied to make the tube material fill into the die cavities as much as possible, as shown in Figure 5d.

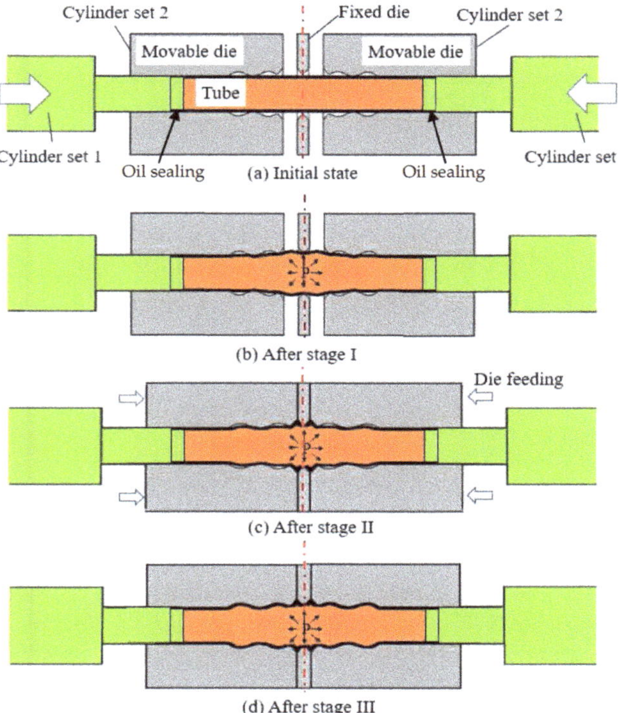

Figure 5. Forming procedures for feeding type 1 with two sets of cylinders (only movement of movable die). (**a**–**d**): Initial State to After Stage III.

For a hydroforming machine with only one set of hydraulic cylinders, the movable die movement and oil sealing function have to be accomplished by the only one set of cylinders simultaneously. This kind of feeding type is called feeding type 2. The forming procedures for feeding type 2 are also divided into three stages, as shown in Figure 6. At stage I, the cylinder set is used to seal the tube as the operation in feeding type 1. Initially, a small gap between the movable die and punch shoulder should be set. The gap is closed by the movement of the punch (cylinder) and the oil sealing function is accomplished accordingly. At stage II, only the one cylinder set is used to push punches to seal the tube and push the movable dies moving forward simultaneously. That is, the moving distance of movable die is the same as that of the punch, as shown in Figure 6c. At stage III, a higher internal pressure is input to calibrate the tube like the procedure in feeding type 1.

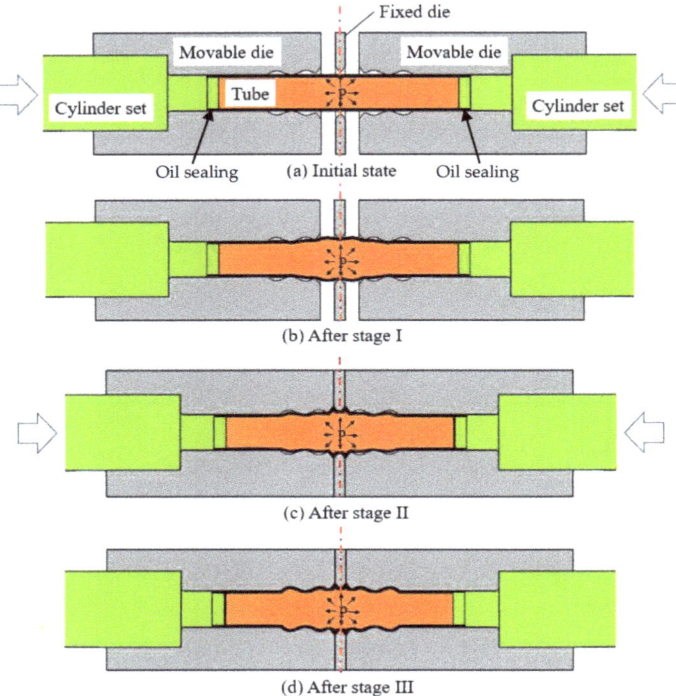

Figure 6. Forming procedures for feeding type 2 with one set of cylinders (Simultaneous movement of movable die and punch). (**a**–**d**): Initial State to After Stage III.

3. Finite Element Simulations

3.1. Finite Element Modeling

A commercial finite element software, DEFORM-3D, was used to simulate the hydroforming process. The tube material used in this study was stainless steel of SUS321. Tensile tests were conducted to obtain the mechanical properties and flow stress of SUS321. Table 1 shows the parameters used in the finite element simulations. Table 2 shows the dimensions of the irregular bellows simulated. The definition of the geometric variables is shown in Figure 1. To shorten the simulation time, only one quarter of the objects were used in the FE simulations. Convergence analyses were implemented. Finite element simulations with total element numbers of 20,000, 30,000, 40,000, 50,000, 65,000, and 75,000 were conducted. The simulation results of the maximal thinning ratio in the product were listed and the difference between two successive simulation results were compared. It was found that the relative differences in the maximal thinning ratios in the product decrease to within 1% as the total element number increases to 65,000. Accordingly, 75,000 tetrahedron elements were set for the tube in the subsequent finite element simulations. The minimal mesh length is about 0.09 mm. There are five layers of meshes in the thickness direction. The mesh configuration in the tube and the geometric configurations between the tube, punch, movable die and fixed die in FE simulations are shown in Figure 7.

Table 1. Forming parameters used in FE simulations.

Material	SUS321
Young's modulus (GPa)	178
Yield stress (MPa)	403
Flow stress (MPa)	$\sigma = 403 + 1270\ \varepsilon^{0.66}$
Poisson's ratio	0.33
Tube outer diameter, d_o (mm)	9.53
Tube thickness, t_0 (mm)	0.49
Punch velocity (mm/s)	0.1
Movable die velocity (mm/s)	0.1
Friction coefficient	0.05
Mesh number	75,000
Mesh layer number in thickness direction	4
Minimum mesh length (mm)	0.09

Table 2. Dimensions of simulated irregular bellows.

Variables	Values
d_1	10.08 mm
d_2	9.65 mm
d_3	13.25 mm
r_1	5.84 mm
r_2	7.62 mm
r_3	5.94 mm
r_4	1.27 mm
l	18.33 mm
θ	45°

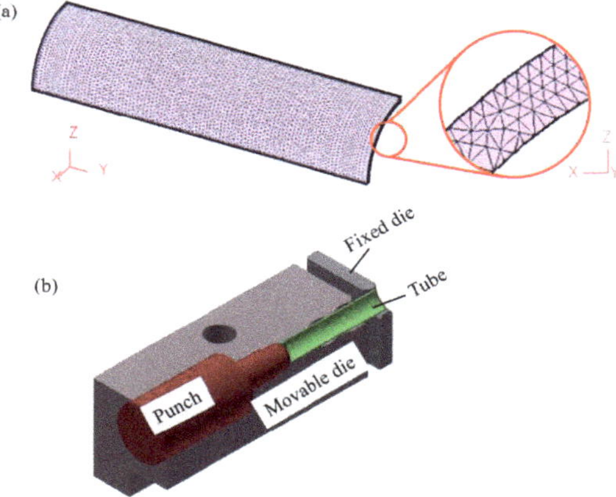

Figure 7. (**a**) Mesh configuration in tube, (**b**) geometric configurations between tube, punch, movable die, and fixed die in FE simulations.

3.2. Forming Windows for Different Feeding Types

In tube hydroforming of a product with small corner radii, a quite high hydraulic pressure is generally required and some defects probably occur if the loading path is not controlled appropriately [17]. The biggest advantage of applying movable dies is the reduction of the required maximal internal pressure. Figure 8 shows the comparisons of the product geometries and internal pressures needed with and without movables die design. Using a loading path of feeding type 1, the internal pressure is set as $P_1 = P_2 = 57.5$ MPa, $P_f = 120$ MPa, and the die gap width is set as L = 3.9 mm, an irregular bellows product with desired shape and a small corner radius of r = 0.63 mm was obtained, as shown in Figure 8a, Using a loading path of feeding type 2, a good bellows shape was also obtained with an internal pressures of $P_1 = P_2 = 50$ MPa, as shown in Figure 8b. However, using a conventional hydroforming process without movable die design, a bellows shape with a large corner radius of 4.7 mm was obtained by a quite large internal pressure of 200 MPa, as shown in Figure 8c. Clearly, using movable die design, a quite small corner radius in the hydroformed bellows could be obtained with a much smaller internal pressure.

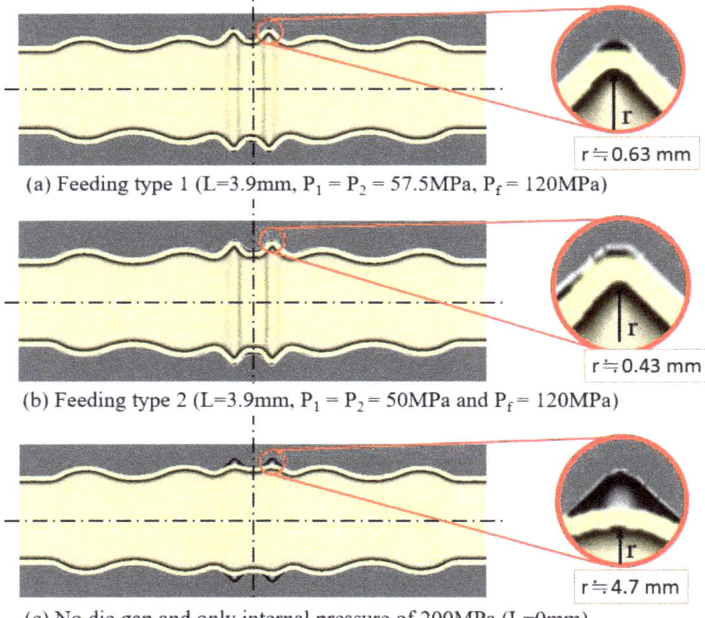

Figure 8. Comparisons of hydroformed bellows shapes with and without movable die design.

3.2.1. Forming Windows for Feeding Type 1

The process schedule for feeding type 1 is shown in Figure 5, in which only the movable die is pushed forward and the punch is used for oil sealing function only. A larger die gap L can make more tube material flow into the die cavity and can make the required bulging pressure smaller. However, if L is too large, the tube is probably bulged into the cavity too much at the free bulging stage (stage I), and the tube material would probably be clamped between the fixed and movable dies and this defect is called clamping. Thus, it is important to find the limitation of L and the relationship between L and the free bulging pressure, P_1, under which no clamping occurs.

Theoretically, there is no limitation for L in feeding type 1, because of no wrinkling or buckling occurring in feeding type 1. However, for a fixed die gap L, there is a constrained boundary for the bulging pressure P_1. The minimum bulging pressure depends on the tube yield stress, die contact geometries, and tube dimensions. In other words, the bulging pressure must be large enough to make the tube deform plastically. Generally, the time period at stage II of the loading path in Figure 3 is quite short. To simplify the loading path, let the internal pressure $P_i = P_1 = P_2$ [4,9]. From finite element simulations, the forming windows or formability ranges between L and P_i for feeding type 1 and the corresponding maximal thinning ratio curves are shown in Figure 9. Firstly, set the die gap L = 0.2 mm and divide the internal pressure P_i between 300 and 500 MPa into many small levels, with which the FE simulations are conducted. From the series of simulation results, the lower and upper critical internal pressures for sound products are determined. After that, L is increased with an increment of ΔL = 0.2 mm and a series of simulations are conducted to determine the corresponding lower and upper critical internal pressures. The simulations are repeated until L reaches 7 mm and all the lower and upper critical internal pressures for each L are obtained. Finally, curves C_1 and C_2 are drawn to represent the upper and lower boundaries of the safe region, respectively. Curve C_1 represents the upper critical values of P_i, over which clamping defects occur, and C_2 presents the lower critical values of P_i, under which the tube cannot fill up the die cavity after stage II. In other words, the region on the upper-right side of curve C_1 means the occurrence of tube clamping, and the region on the lower-left side of curve C_2 means that the die cavity cannot be filled up with the tube completely. The region between curves C_1 and C_2 is called the forming window, in which no clamping occurs and there is complete tube filling in the die cavity. From Figure 9, it is clear that the range of the allowable bulging pressure gets narrow as the die gap width L increases. When L is 3.3 mm, the allowable bulging pressure range is smaller than 5 MPa. At the maximal die gap, L = 7.1 mm, the bulging pressure can be reduced to 46 MPa, which is the minimum bulge pressure to make the tube deform plastically with a yielding stress of 403 MPa, an initial tube thickness of t_0 = 0.49 mm, and an initial tube diameter of d_o = 9.53 mm. The maximal thinning ratios of the formed bellows after calibration for curves C_1 and C_2 are also shown in Figure 9. Clearly, the mean value of the maximal thinning ratios decreases greatly with the die gap width L as 0.1 < L < 3 mm.

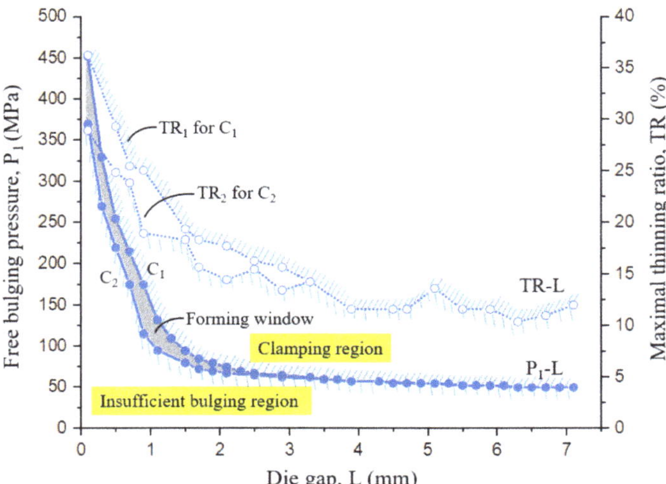

Figure 9. Forming windows and maximal thinning ratios for feeding type 1.

3.2.2. Forming Windows for Feeding Type 2

The process schedule for feeding type 2 is shown in Figure 6, in which the movable dies and tube are pushed forward simultaneously by one set of punches. The maximal die gap L is determined from the volume of the formed bellows and the initial tube dimensions. The geometrical configuration of a tube bulged to fill up the die cavities perfectly without thinning is shown in Figure 10. The radius functions of the outer and inner surfaces of the formed tube along the symmetric z axis are assumed as $f(z)$ and $g(z)$, respectively. The formed bellows is composed of seven arcs (zones 1–5, 7, and 9) and three straight lines (zones 6, 8, and 10), as shown in Figure 10. Arcs 1, 3, 5, and 9 curve upward, while arcs 2, 4, and 7 curve downward. A minus sign has to be input in front of the radius functions for arcs curving upward. Therefore, the radius functions at zones 1–5, 7, and 9 at the outer surface can be expressed as:

$$f_i(z) = (-1)^i\left[{R_i}^2 - (z - z_{ci})^2\right]^{1/2} + r_{ci},\ z_{i-1} \le z \le z_i,\ i = 1\text{–}5, \text{ and } 9 \tag{1}$$

$$f_7(z) = \left[{R_7}^2 - (z - z_{c7})^2\right]^{1/2} + r_{c7},\ z_6 \le z \le z_7 \tag{2}$$

where R_i is the radius of curvature at the outer surface of section i, i = 1–5, 7, and 9. z_{ci} and r_{ci} are the z and r coordinates of the arc centers at zones 1, 3, 5, and 9. z_{i-1} and z_i are the integration boundaries of zone i at the outer surface. The radius functions at zones 6, 8, and 10 are straight lines, and can be expressed as:

$$f_6(z) = z - 15.959,\ z_5 \le z \le z_6 \tag{3}$$

$$f_8(z) = 29.912 - z,\ z_7 \le z \le z_8 \tag{4}$$

$$f_{10}(z) = 5.684,\ z_9 \le z \le z_{10} \tag{5}$$

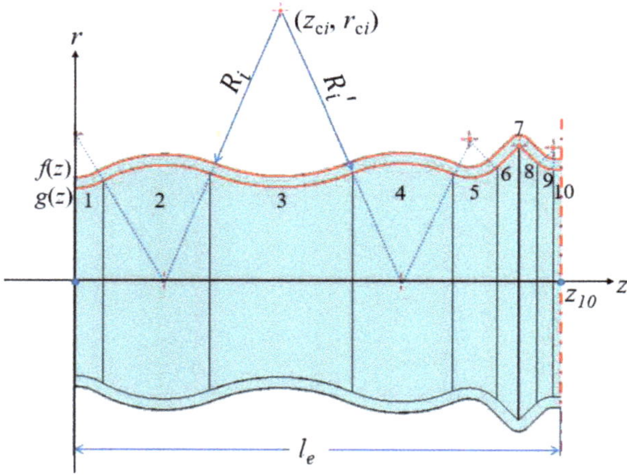

Figure 10. Geometric configurations of formed bellows. (Left-half part).

The inner surface is also composed of seven arcs and three straight lines. The thickness of the formed bellows is assumed to be uniform and the same as the initial thickness of the tube, t_0 = 0.49 mm. Thus, the arcs at zones 1–5, 7, and 9 have the same circular center as those of the outer surface, $f_i(z)$. The radius functions of each zone are listed below.

$$g_j(z) = (-1)^j\left[R_j'^2 - \left(z - z_{cj}\right)^2\right]^{\frac{1}{2}} + r_{cj},\ z_{j-1} \le z \le z_j,\ j = 1\text{–}5, \text{ and } 9 \tag{6}$$

$$g_7(z) = \left[R_7'^2 - (z - z_{c7})^2\right]^{1/2} + r_{c7},\ z_6 \le z \le z_7 \tag{7}$$

where R'_j is the radius of curvature at the inner surface of section j, j = 1–5, 7, and 9. z_j is the integration boundary of each zone of the inner surface. The straight lines at the inner surface have the same slopes as those at the outer surface, so the radius functions at zones 6, 8, and 10 can be expressed as

$$g_6(z) = z - 16.651,\ z_5 \le z \le z_6 \tag{8}$$

$$g_8(z) = 29.219 - z,\ z_7 \le z \le z_8 \tag{9}$$

$$g_{10}(z) = 5.194,\ z_9 \le z \le z_{10} \tag{10}$$

The relationship between the radii of curvature at the outer and inner surfaces can be expressed as:

$$\left|Ri - R'_i\right| = t_0 \tag{11}$$

The volume of the finally formed bellows, V_b, can be found using the disk integration method as below:

$$V_b = \pi \sum_{i=1}^{10} \int_{z_{i-1}}^{z_i} f^2(z)\, dz - \pi \sum_{j=1}^{10} \int_{z_{j-1}}^{z_j} g^2(z)\, dz \tag{12}$$

The tube volume can be written as below.

$$V_t = \frac{\pi}{4}\left(d_o{}^2 - d_i{}^2\right) l_0 \tag{13}$$

where d_o and d_i is the initial outer and inner tube diameters, and l_0 is the tube length. The volume of the formed bellows must be equal to that of the tube, i.e., $V_b = V_t$. From Equations (12) and (13), l_0 can be determined. The maximal die gap L_{max} is the difference between the die length l_e, as shown in Figures 2a and 10, and the initial tube length l_0 as below.

$$L_{max} = l_0 - l_e \tag{14}$$

According to the dimensions of the formed bellows shown in Table 2, the volume of the formed bellows by Equation (12) can be obtained as 378.65 mm^3. If d_o = 9.53 mm, d_i = 8.55 mm and l_e = 22.2 mm, the initial tube length is l_0 = 27.81 mm, and the maximal die gap is L_{max} = 5.61 mm.

Figure 11 shows the forming windows or formability ranges between L and P_i for feeding type 2 and the corresponding maximal thinning ratio curves. Generally, the range of allowable forming pressure becomes narrower as the die gap L increases. The maximal die gap L_{max} obtained by FE simulations is 5.5 mm which is quite close to the theoretically calculated value of 5.61 mm by Equation (14). In feeding type 2, wrinkling probably occurs at a large axial feeding and a low internal pressure. Therefore, the region on the upper-right side of curve C_1 means the occurrence of tube clamping just like the case for feeding type 1, whereas the region on the lower-left side of curve C_2 means that tube buckling occurs or the die cavity cannot be filled up with the tube completely. The region between curves C_1 and C_2 is the forming window, in which no clamping occurs and the die cavity is filled up with the tube completely. As L exceeds 4.7 mm, the allowable pressure becomes so low that tube wrinkling would probably occur due to axial feeding from the punch. However, because the movable dies also move forward along with the punches simultaneously, the tube wrinkling could be smoothed out by the movable die surfaces and the calibration pressure at the last stage.

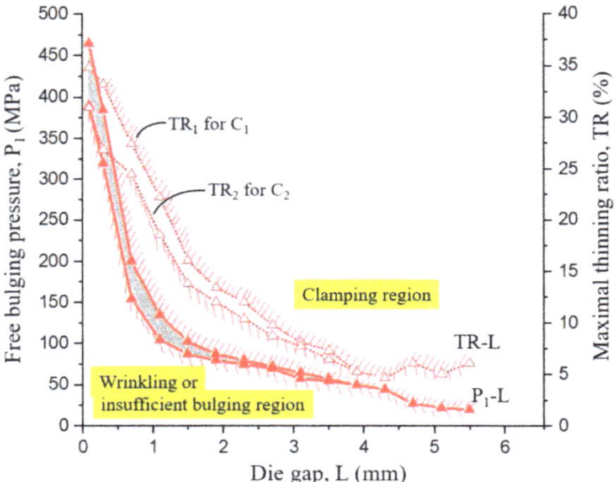

Figure 11. Forming windows for feeding type 2.

From Figures 9 and 11 for the forming windows using feeding types 1 and 2, it is known that the maximal thinning ratio can be reduced greatly by increasing the die gap L. Meanwhile, the bulging pressure can also be reduced greatly. For example, if the die gap is set as L = 2–4 mm, the maximum thinning ratios can be reduced to about TR = 11–15% and TR < 10% for feeding types 1 and 2, respectively. On the other hand, the bulging pressure required with L = 0.1 mm can be reduced to only approximately one sixth of the original pressure if the die gap is set as L > 3 mm.

Strain hardening exponents exhibit the formability of the tube material and the interface friction affects the flowability of the tube material during the forming process. The final total strain of the deforming tube is not so large and the relative movement between the tube and die is also not so severe in the hydroforming process with the proposed movable die concept. Thus, the effects of the strain hardening exponent and friction coefficient on the upper and lower boundaries of the safe region are considered to be small and limited.

4. Tube Hydroforming Experiments

A hydroforming machine with only one pair of cylinders was used to conduct tube hydroforming experiments of irregular bellows. Thus, movable die feeding type 2 was adopted in the experiments. This machine has a capacity of an internal pressure of 200 MPa, a die closing force of 200 ton, and an axial stroke of 150 mm. The geometric configurations of a self-designed die set is shown in Figure 12. As a result that two irregular bellows are formed at one pass, one pair of movable dies, one fixed die, two gaskets, and one pair of T-shape slideways in the upper track were designed and assembled in the upper die set. To avoid the upper movable dies dropping from the upper die holder, two T-shape slots were designed on the top surface of the movable dies. Four gaskets were inserted between the movable dies and the die holders to adjust the die gap width L between the movable die and fixed die. Four springs were used to keep the movable dies at the initial positions. As a result that the forming stage II in the loading path shown in Figure 3 is quite short, a constant internal pressure ($P_1 = P_2$) [5,14] was set in the hydroforming experiments. Stainless steel SUS321 seamless tubes with an outer diameter of 9.53 mm were used as the specimens in the hydroforming experiments. The thickness distributions of the tubes were measured and some thickness variations of 0.45–0.49 mm in the circumferential direction were found. The average thickness of the tubes was about 0.47 mm. After experiments, the hydroformed irregular bellows were cut into two halves by electric discharge wire cutting and the thickness distributions of the products were measured. Figure 13 shows the completed lower part of the movable die set for the hydroforming experiments.

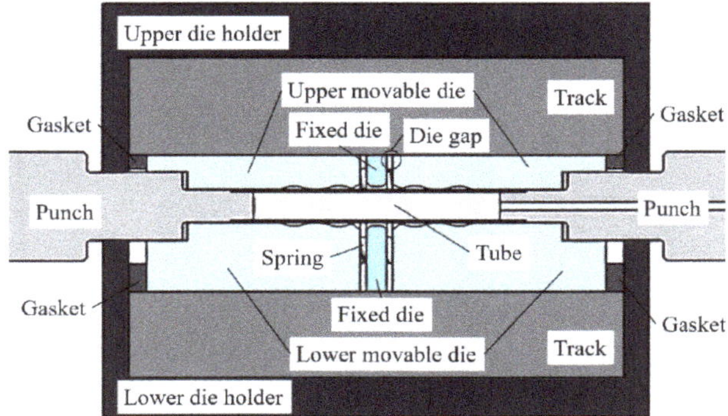

Figure 12. Geometric configurations of self-designed movable die sets.

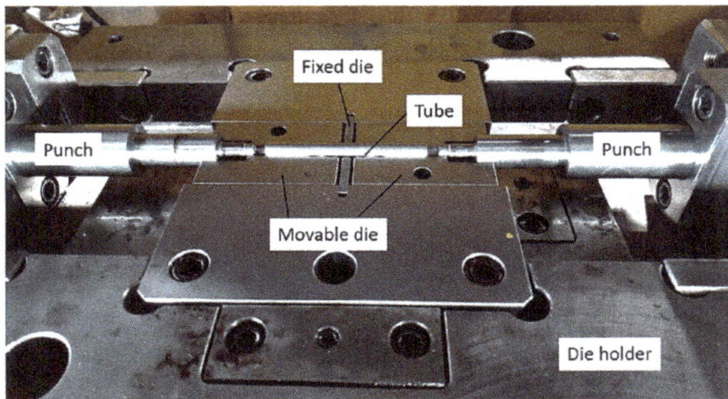

Figure 13. Completed lower part of movable die set for hydroforming experiments.

Hydroforming experiments of irregular bellows with a loading path of internal pressure $P_1 = P_2 = 70.2$ MPa and die gap width of L = 1.47 mm were carried out. The appearances at the outside and inner views of a hydroformed bellows after stage II of die feeding are shown in Figure 14a,b, respectively. Generally, a sound product with desired shape was obtained successfully. The central part and both ends of the product would be cut off and two irregular bellows with desired shape and dimensions could be obtained.

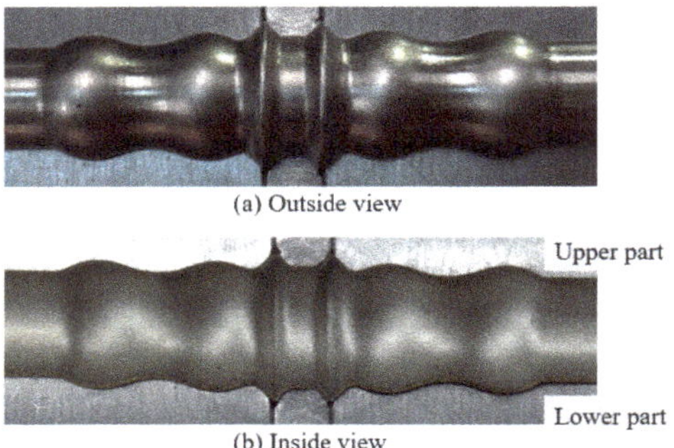

Figure 14. Appearance of hydroformed product after loading path stage II. (L = 1.47 mm, $P_1 = 70.2$ MPa).

The simulative and experimental thickness distributions at the upper and lower parts of the hydroformed products are shown in Figure 15. Generally, the thicknesses at the upper part are larger than those at the lower part of the hydroformed bellows, because there was some variation in the initial thickness distribution of the tube. Generally, the simulative thickness distribution was quite close to the experimental values. The thinnest positions occurred at points B_1, and B_2. Nevertheless, the simulative thinning ratio was only TR = 5.11%, and the experimental values were only TR = 4.47% at B_1 and TR = 5.53% at B_2.

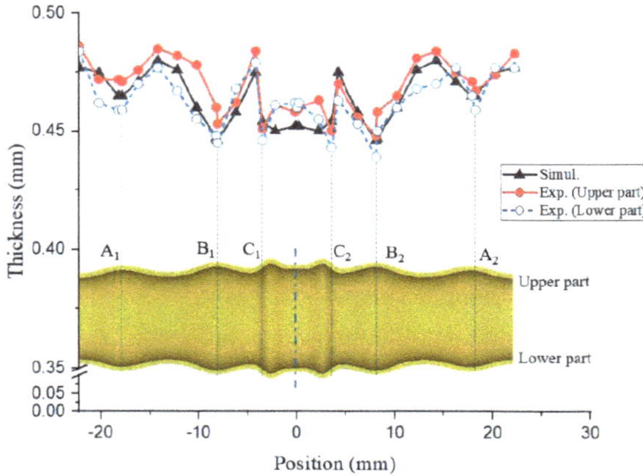

Figure 15. Thickness distributions of formed product after loading path stage II. (L = 1.47 mm, P_1 = 70.2 MPa).

The simulative and experimental thicknesses and diameters of the formed product at the left and right parts are summarized in Table 3. Even though there was some variation in the thickness at the left and right positions, the deviations are all below 2%. The diameters at positions A, B, and C were also measured to understand whether the die cavity is filled up with the tube material or not. The diameters at positions C_1 and C_2 are 13.40 and 13.45 mm, respectively, which are larger than 13.25 mm, the designed diameter at the tip of the taper part. The diameters at positions A and B are also quite close to the diameters at the corresponding die cavity. In other words, all the die cavities are filled up with the tube and two sound bellows with designed geometric dimensions were obtained. The maximal deviations in thicknesses and diameters between simulation and experimental values are 1.57% and 5.31%, respectively.

Table 3. Simulative and experimental thicknesses and diameters of formed product after stage II.

1	Experimental Results (mm)		Simulation (mm)
Average thickness	t_{A1} = 0.465 (Deviation = 0%)	t_{A2} = 0.468 (Deviation = 0.64%)	t_A = 0.465
	t_{B1} = 0.449 (Deviation = 0.67%)	t_{B2} = 0.444 (Deviation = 0.45%)	t_B = 0.446
	t_{C1} = 0.449 (Deviation = 1.11%)	t_{C2} = 0.447 (Deviation = 1.57%)	t_C = 0.454
Diameter	d_{A1} = 11.64 (Deviation = 4.47%)	d_{A2} = 11.64 (Deviation = 4.47%)	d_A = 11.12
	d_{B1} = 11.84 (Deviation = 0.42%)	d_{B2} = 11.84 (Deviation = 0.42%)	d_B = 11.79
	d_{C1} = 13.40 (Deviation = 5.31%)	d_{C2} = 13.45 (Deviation = 5.28%)	d_C = 12.74

5. Conclusions

In this paper, a tube hydroforming process using a novel movable die and loading path design was developed to manufacture irregular bellows with small thinning ratios in the formed product. A finite element simulation software "DEFORM 3D" was used to analyze the plastic deformation of the tube within the die cavity using the proposed movable die design concept. Two kinds of die feeding types were proposed to make the maximal thinning ratio in the formed bellows and the needed internal pressure as small as possible. Using the movable die design with an appropriate die gap width, the internal forming pressure needed can be reduced to only one sixth of the internal pressure needed without the movable die design. From the forming windows for the two proposed die feeding types, it is known that a larger die gap width is beneficial to reduce the free bulging pressure. However, a narrow range for the allowable forming pressure probably results in the occurrence of defects such as clamping and buckling if the forming pressure is not controlled precisely. Moreover, an excessively large die gap width may increase the maximal thinning ratio for die feeding type 1. Thus, a die gap width of 4 mm is an appropriate value that can improve the thickness distribution and reduce greatly the forming pressure for the both feeding types. Maximal thinning ratios of 11.63% and 5.31% could be obtained for feeding types 1 and 2, respectively. Finally, a movable die set was designed and manufactured and tube hydroforming experiments using feeding type 2 based on the forming conditions used in the finite element simulations were conducted. Comparisons of the diameters and the thickness distributions of the formed product between simulative and experimental results were carried out. The maximal deviations in thickness and diameter between simulative and experimental results were 1.57% and 5.31%, respectively, which validated the finite element modeling and the movable die design concept proposed in this paper.

Author Contributions: Conceptualization, methodology, research supervision, and writing—review and editing were conducted by Y.-M.H. Simulations, experiments, validation, and writing—original draft preparation were completed by Y.-J.T. Both authors have read and agreed to the published version of the manuscript.

Funding: The authors would like to extend their thanks to the Ministry of Science and Technology of the Republic of China under Grant no. MOST 106-2221-E-110 -029 -MY3. The advice and financial support of MOST are greatly acknowledged.

Conflicts of Interest: The authors declare no conflict of interest.

References

1. Ahmetoglu, M.; Altan, T. Tube hydroforming: State-of-the-art and future trends. *J. Mater. Process. Technol.* **2000**, *98*, 25–33. [CrossRef]
2. Kang, B.H.; Lee, H.Y.; Shon, S.M.; Moon, Y.H. Forming various shapes of tubular bellows using a single-step hydroforming process. *J. Mater. Process. Technol.* **2007**, *194*, 1–6. [CrossRef]
3. Faraji, G.H.; Mosavi Mashhadi, M.; Norouzifard, V. Evaluation of effective parameters in metal bellows forming process. *J. Mater. Process. Technol.* **2009**, *209*, 3431–3437. [CrossRef]
4. Lee, S.W. Study on the forming parameters of the metal bellows. *J. Mater. Process. Technol.* **2002**, *130–131*, 47–53. [CrossRef]
5. Bakhshi-Jooybari, M.; Elyasi, M.; Gorji, A. Numerical and experimental investigation of the effect of the pressure path on forming metallic bellows. *J. Eng. Manuf.* **2009**, *224*, 95–101. [CrossRef]
6. Ngaile, G.; Lowrie, J. Punch design for floating based micro-tube hydroforming die assembly. *J. Mater. Process. Technol.* **2018**, *253*, 168–177. [CrossRef]
7. Shinde, R.A.; Patil, B.T.; Joshi, K.N. Optimization of Tube Hydroforming Process (without Axial feed) by using FEA Simulations. In Proceedings of the 3rd International Conference on Innovations in Automation and Mechatronics Engineering, Vallabh Vidhyanagar, India, 5–6 February 2016; pp. 398–405.
8. Elyasi, M.; Bakhshi-Jooybari, M.; Gorji, A.; Hosseinipour, S.J.; Nourouzi, S. New die design for improvement of die corner filling in hydroforming of cylindrical stepped tubes. *J. Eng. Manuf.* **2009**, *223*, 821–827. [CrossRef]

9. Zhang, Q.; Wu, C.D.; Zhao, S.D. Numerical simulation on hydroforming tube-like eccentric shaft part by using movable die. In Proceedings of the 5th International Conference on Tube Hydroforming, Hokkaido, Japan, 24–27 July 2011; pp. 149–152.
10. Tomizawa, A.; Kubota, H.; Kurokawa, H.; Kojima, M. Development of tube hydroforming technology using axial movable dies. *JSTP* **2015**, *56*, 40–46. [CrossRef]
11. Hwang, Y.M.; Chen, Y.C. Study of compound hydroforming of profiled tubes. *Procedia Eng.* **2017**, *207*, 2328–2333. [CrossRef]
12. Hwang, Y.M.; Hsieh, S.Y.; Kuo, N.J. Study of large-expansion-ratio tube hydroforming with movable dies. *Key Eng. Mater.* **2016**, *725*, 616–622. [CrossRef]
13. Yuan, S.J.; Wang, X.S.; Liu, G.; Wang, Z.R. Control and use of wrinkles in tube hydroforming. *J. Mater. Process. Tchnol.* **2007**, *182*, 6–11. [CrossRef]
14. Xu, Y.; Zhang, S.H.; Cheng, M.; Song, H.W.; Zhang, X.S. Application of pulsating hydroforming in manufacture of engine. *Procedia Eng.* **2014**, *81*, 2205–2210. [CrossRef]
15. Imaninejad, M.; Subhash, G.; Loukus, A. Loading path optimization of tube hydroforming process. *Int. J. Mach. Tools Manuf.* **2005**, *45*, 1504–1514. [CrossRef]
16. Hama, T.; Ohkubo, T.; Kurisu, K.; Fujimoto, H.; Takuda, H. Formability of tube hydroforming under various loading paths. *J. Mater. Process. Technol.* **2006**, *177*, 676–679. [CrossRef]
17. Liu, G.; Yuan, S.; Teng, B. Analysis of thinning at the transition corner in tube hydroforming. *J. Mater. Process. Technol.* **2006**, *177*, 688–691. [CrossRef]

Article

Investigation of Effects of Strip Metals and Relative Sliding Speeds on Friction Coefficients by Reversible Strip Friction Tests

Yeong-Maw Hwang * and Chiao-Chou Chen

Department of Mechanical and Electro-Mechanical Engineering, National Sun Yat-sen University, No. 70, Lein-Hai Rd., Kaohsiung 804, Taiwan; imjj08800817@gmail.com

* Correspondence: ymhwang@mail.nsysu.edu.tw

Received: 12 September 2020; Accepted: 10 October 2020; Published: 14 October 2020

Abstract: Friction at the interface between strips and dies is an important factor influencing the formability of strip or sheet forming. In this study, the frictional behaviors of strips at variant speeds were investigated using a self-developed strip friction test machine with a dual tension mechanism. This friction test machine, stretching a strip around a cylindrical friction wheel, was used to investigate the effects of various parameters, including sliding speeds, contact angles, strip materials, and lubrication conditions on friction coefficients at the strip–die interface. The friction coefficients at the strip–die interface were calculated from the drawing forces at the strip on both ends and the contact angle between the strip and die. A series of friction tests using carbon steel, aluminum alloy, and brass strips as the test piece were conducted. From the friction test results, it is known that the friction coefficients can be reduced greatly with lubricants on the friction wheel surface and the friction coefficients are influenced by the strip roughness, contact area, relative speeds between the strip and die, etc. The friction coefficients obtained under various friction conditions can be applied to servo deep drawing or servo draw-bending processes with variant speeds and directions.

Keywords: strip friction test; friction coefficient; surface roughness; sliding speed; contact pressure

---

1. Introduction

Friction conditions between the workpiece and die influence the deformed sheet or strip material properties significantly. The surface grinding of the workpiece in contact with the die during the stretching process not only affects the finally stretched sheet surface quality, but also the formability of the forming process. The friction coefficient at the sheet–die interface is influenced by the surface roughness of the sheet and die, as well as the relative sliding speeds between the sheet and die and the contact pressure [1–3]. Tamai et al. [1] proposed a friction test apparatus, in which two identical compressive forces act on the test piece surfaces and a drawing force is imposed at one end of the test piece. The friction coefficient can be easily obtained from the force ratio. Tamai et al. [4] also developed a nonlinear friction coefficient model that considers contact pressure, sliding velocity, and sliding length to improve the accuracy of predictions of the formability of steel sheets. The effects of contact pressure and sliding velocity under mixed lubrication were estimated based on a friction test in which a long steel sheet was drawn between two dies. Servo press forming processes for a thin sheet utilize more complicated slide motion control, and the sliding direction of the material relative to the die is changed not only in sheet-forming but also in tube-forming processes. Such kinds of forming machines and forming processes are increasing [4]. Thus, it is important to investigate the friction characteristics when the slip direction is reversed in such a forming process.

Some researchers have proposed some friction test methods to measure friction coefficients in sheet-forming processes. For example, Weinmann et al. [5] proposed a friction coefficient measuring apparatus in which a metal strip is bent into a U shape through two fixed cylindrical friction pins. From the interaction of the pin and strip, coefficients of friction were calculated. Saha and Wilson [6] also conducted similar friction tests. They found that the friction coefficients of the steel sheet with the pins increased with the strip strain. That is because the plastic strain affected the strip surface roughness and the actual contact surface area. However, the friction coefficients from aluminum strip friction tests decreased with the strip strain because the severe plastic deformation made the contact area smaller. Hsu and Kuo [7] discussed the effects of dry friction and lubrication conditions on the friction coefficient and developed a boundary friction model, including plowing phenomena and bonding stresses. Kim et al. [8] measured the friction coefficient to investigate the friction characteristics of a coated metal and found that tool steel STD11 and copper alloys AMPCO have better surface roughness and formability than other metals. Lemu and Trzepiecinski [9] explored the friction behavior of steel, brass, and aluminum alloys with a self-developed friction test apparatus and discussed the effects of the deformation strains on the friction coefficient under dry and lubricated conditions. Ramezani et al. [10] used a steel pin as the counterpart under dry sliding conditions to investigate the friction coefficients of ZE10 and AZ80 magnesium alloys numerically and experimentally. The experimental results showed that increasing contact pressure lead to an increase in coefficient of friction for both alloys, while the effect of sliding speed was negligible. Fridmen and Levesque [11] investigated the effects of sonic vibrations on the coefficient of static friction for highly polished, ground, and sand-pitted steel surfaces. The coefficient of static friction could virtually be reduced to zero as a result of increased vibrations at frequencies between 6 and 42 kHz. Chowdhury et al. [12] carried out friction experiments under a normal load of 10–20 N and rotation speed of 500–2500 rpm. The experimental results showed that the friction coefficient decreased with the increase of sliding speed and normal load for aluminum sheets. They also found that the wear rates increased with the increase of sliding speed and normal load. Saha [13] developed a sheet tensile testing apparatus to measure the friction coefficient under various conditions. It was found that the friction force of the steel sheet increased with the plastic strain. That is because of a larger sheet roughness and a larger actual contact surface.

The level of generated vibrations is one of the most important exploitation parameters of rolling bearings. Adamczak and Zmarzły [14] examined five pieces of type 6304 ball bearings by measuring 2D and 3D roughness parameters of the bearings races with a contact method on a Form Talysurf PGI 1230 device made by Taylor Hobson. Statistical analysis based on the correlation calculation was used to evaluate the impact of 2D and 3D roughness parameters of active surfaces of rolling bearings on the level of generated vibrations. Ali et al. [15] adopted an experimental study to minimize the boundary friction coefficient via nanolubricant additives. The tribological characteristics of Al_2O_3 and TiO_2 nanolubricants were evaluated under reciprocating test conditions to simulate a piston ring/cylinder liner interface in automotive engines. The experimental results have shown that the boundary friction coefficient reduced by 35%–51% near the top and bottom dead center of the stroke for the Al_2O_3 and TiO_2 nanolubricants. Escosa et al. [16] evaluated the influence of both coating and austenitization treatment of $22MnB_5$ steel on friction and wear of tool steels. The results showed that Al–10%Si reduced the friction coefficient, while the hardening treatment resulted in an increase of friction coefficient due to Fe_2Al_5 brittle compounds. Wu et al. [17] investigated the friction and wear properties of the textured surfaces as well as the relationship between the tribological properties and the texture parameters by high-speed dry sliding tests. The results showed that the dimple textured titanium surfaces filled with molybdenum disulfide solid lubricants can effectively reduce the friction coefficient, as well as its fluctuation, compared with the untextured samples and textured samples without lubricants.

The above literature investigated the friction coefficients under only one pass of friction test with one kind of tested strip material. In this paper, a reversible friction test machine with a capacity of forward and backward moving directions was developed. A series of friction tests with multiple stages were conducted to make the friction situations similar to the loading path of a servo press. The effects of relative sliding speeds, different strip materials, strip contact angles, lubrication conditions, etc., on the friction coefficients are discussed.

2. Determination of Friction Coefficient

The geometric configurations between a strip and a friction wheel is shown in Figure 1. Bending and pulling of the strip are implemented by forces F_1 and F_2 acting on the two ends of the strip. A contact angle θ_c between the friction wheel and strip can be adjusted by changing the direction of F_2. As $F_2 > F_1$, the strip moves forward. On the contrary, as $F_2 < F_1$, the strip moves backward.

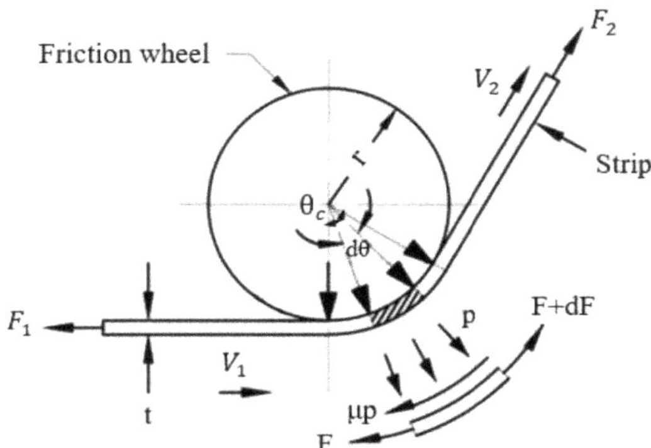

Figure 1. Free body diagram and geometric configurations between strip and friction wheel.

From the free body diagram shown in Figure 1, the force equilibrium in the radial direction yields:

$$pwrd\theta = F\sin\left(\frac{d\theta}{2}\right) + (F + dF)\sin\left(\frac{d\theta}{2}\right) \tag{1}$$

where w is the strip width and r is the friction wheel radius. The term of dFsin(dθ/2) can be ignored and sin(dθ/2) is approximately equal to dθ/2. Thus, the contact pressure p can be obtained as below:

$$p = \frac{F}{rw} \tag{2}$$

From the force equilibrium in the circumferential direction, we can get:

$$(F + dF)\cos\frac{d\theta}{2} = F\cos\frac{d\theta}{2} + \mu prwd\theta. \tag{3}$$

The above equation can be simplified as below:

$$dF = \mu prwd\theta. \tag{4}$$

Substituting Equation (2) into Equation (4) and after definite integration, we get:

$$\ln F = \mu\theta + c_1. \tag{5}$$

Taking exponential function on both sides yields

$$F = ce^{\mu\theta} \tag{6}$$

where c is the integral constant. From the boundary condition, $F = F_1$ as $\theta = 0°$, we get

$$c = F_1. \tag{7}$$

At $\theta = \theta_c$, F_2 can be obtained as follows:

$$F_2 = F_1 e^{\mu\theta_c}. \tag{8}$$

From the above equation, the friction coefficient at the interface of the trip and friction wheel can be obtained as a function of F_2, F_1 and θ_c as below:

$$\mu = \left| \frac{\ln \frac{F_2}{F_1}}{\theta_c} \right|. \tag{9}$$

For a thick sheet, the bending effect has to be considered. Please refer to Reference [18] for the friction coefficient formula derivation.

3. Experimental Apparatus and Friction Test Conditions

A self-developed friction test machine using a motor driving mechanism is shown in Figure 2. This apparatus consists of four main parts: (1) a driving system, including a motor, a work wheel, and a pneumatic cylinder, which is used to control the strip movement; (2) a measuring unit, including a load cell and a torque meter, which are used to measure force F_1 and torque T, respectively; (3) a friction mechanism, including a friction wheel and a strip, which generate a friction interface; and (4) a control panel, which is used to control the rotation speed of the work wheel and the strip moving speed. The advantages of this self-developed test machine are (1) the forward and backward movement of the strip can be easily controlled and (2) the contact angle can be easily arranged from 30° to 90°.

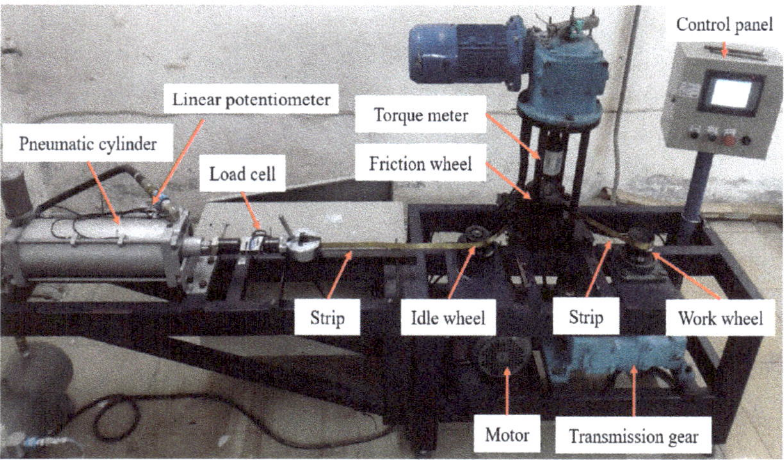

Figure 2. Appearance of self-developed friction test machine.

A series of friction tests for various friction conditions, such as the rotation speeds, interface conditions, contact angles, and different strip metals, were conducted. The friction wheel was fixed, whereas, the rotation speed of the work wheel was set as 5–9 rpm. The contact angles were 30°, 60°, and 90°. By adjusting the distance between the friction wheel and the center line connecting the idle wheel and work wheel, the contact angle could be easily arranged. Two kinds of surface conditions, dry friction and oil lubrication, at the interface between the friction wheel and strip metal were adopted. The friction wheel made of middle carbon steel had a surface roughness of Ra = 1.73 μm. It was difficult to install a load cell between the friction wheel and the work wheel to measure F_2 directly. For a thick sheet, F_2 is affected by the bending of the sheet. An alternative equation for thick sheet friction tests was given in the former publication Reference [18]. From drawing force F_1, measured from the load cell, drawing force F_2 could be obtained from the following equation:

$$F_2 = F_1 - \frac{T}{r} \tag{10}$$

where torque T is measured from the torque meter. Substituting F_2 into Equation (9), the friction coefficients for various friction conditions could be obtained.

Table 1 shows the friction test conditions for different strip materials, Vickers hardness, rotation speeds ω, contact angles θ_c, interface condition, and gauge pressure in the pneumatic cylinder. Carbon steel S25C, brass C2680, and aluminum alloy 6063T6, some kinds of metal materials commonly used in stamping or deep drawing processes, were adopted in this paper for friction tests. The strip thickness for different materials was 0.5–1.5 mm. The length of the test piece was 1800 mm. The radius of the friction wheel was 25 mm. The surface conditions at the strip and friction wheel interface were divided into dry friction and oil lubrication. The lubrication oil used was SAE 5W–30, the viscosity of which was 61.4 mm^2/s at 40 °C and density is 850 kg/m^3 at 20 °C. The pneumatic cylinder was used to act as a buffer or a brake while the strip was moved forward and acted as a driver as the strip was moved backward. A higher pneumatic cylinder pressure p_0 was set for a harder strip material used. The basis friction test conditions were rotation speed of ω = 5 rpm, contact angle of θ_c = 90°, and dry friction for the interface condition.

Table 1. Friction tests conditions.

	Strip Material	Vickers Hardness (HV)	Rotation Speed ω (rpm)	Contact Angle θ_c (°)	Interface Condition	Air Pressure p_0 (kPa)
Case 1	Carbon steel S25C (t = 1.1 mm)	130	5, 7, 9	90	Oil lubrication/Dry friction	103
Case 2	Brass C2680 (t = 0.5 mm)	128	5, 7, 9	90	Oil lubrication/Dry friction	69
Case 3	Aluminum 6063-T6 (t = 1.3 mm)	83	5, 6, 7	90	Oil lubrication/Dry friction	48
Case 4	Aluminum 6063-T6	83	5	30, 60, 90	Dry friction	48

Figure 3 shows strip movement variations during friction tests. The whole friction tests were composed of three stages. A linear potentiometer shown in Figure 2 was installed beside the pneumatic cylinder to monitor the displacement of the cylinder. Accordingly, the strip movement could be recorded. At stage 1, the work wheel was driven to pull the strip metal toward the work wheel side (forward) by 300 mm. The strip movement route or stage from 150 to 300 mm was called S_{1F}. As the work wheel was driven to rotate with inverse direction, the strip metal was pulled by the pneumatic cylinder and moved backward with a distance of 150 mm. This stage was called S_{1B}. The strip movement pattern was repeated two times, and the four stages were designated as S_{2F}, S_{2B}, S_{3F}, and S_{3B}. The strip movement pattern in Figure 3 was designed to be analogous to the punch movement in a servo press, so that the friction coefficients obtained by this reversible friction test machine can be applied to servo stamping or servo deep drawing processes.

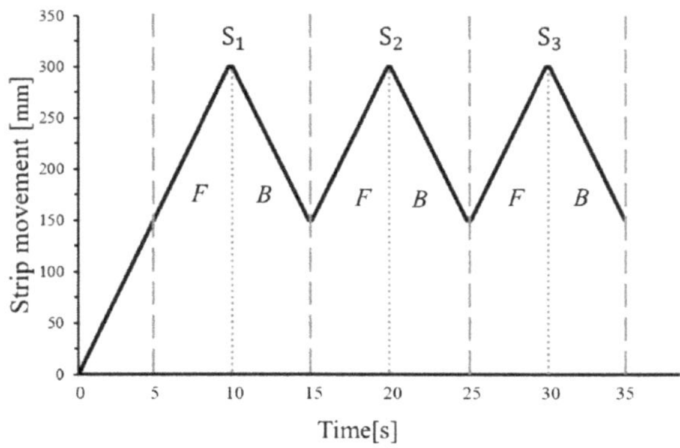

Figure 3. Strip movement variations during friction tests.

4. Friction Test Results and Discussion

Friction tests of carbon steel S25C strips were conducted repeatedly three times under identical rotation speeds ω, identical contact angles θ_c, identical pneumatic pressures p_0, and identical strip surface treatments. Figure 4 shows the drawing force variations F_1 and F_2 for the repeated three times with ± SD. The drawing force F_1 was measured from a load cell which was installed on the connecting rod in front of the pneumatic cylinder. From Figure 4, it is known that the measured drawing forces F_1 from the pneumatic cylinder were almost identical at all stages in the three repeated tests, because the gauge pressure in the pneumatic cylinder was quite steady. However, there was a slight difference in force F_2 at stages S_{1B} and S_{3F}. The maximal difference was about ±0.2 kN (±8%). That is because a variable frequency motor was used in this friction test machine, which could not control the rotation speed as accurately as a servo motor could.

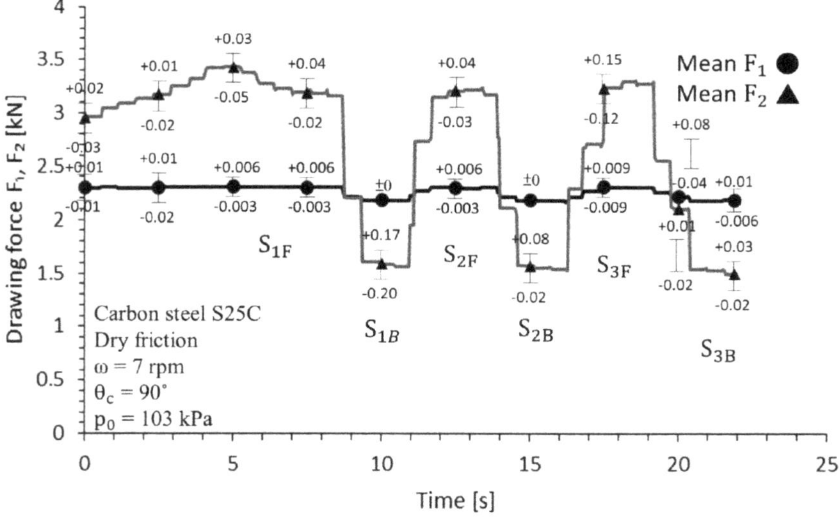

Figure 4. Drawing force variations in friction tests.

4.1. Friction Tests of Carbon Steel S25C Strip

Carbon steel S25C strips with a thickness of 1.1 mm were used as the test piece. The gauge pressure P_0 inside the pneumatic cylinder was set as 103 kPa. The contact angle of the strip at the friction wheel was 90°. The other testing conditions are given in Table 1 (case 1). The drawing force variations F_1 and F_2 with rotation speed of 5 rpm, equivalent to sliding speed of 30 mm/s at the interface of the strip and friction wheel, are shown in Figure 5. The drawing force F_1 was obtained from the load cell, which was installed between the pneumatic cylinder and the strip. The drawing force F_2 was obtained from Equation (10) and the torque meter, which was installed at the top of friction wheel. From Figure 5, it is known that a slightly larger F_2 was obtained at the very beginning of each stage. That is because static friction occurred as the work wheel changed its rotation direction. It is clear that F_1 values at the backward stages were smaller than those at the forward stages. That is because the pulling force from the pneumatic cylinder has to overcome the friction resistance at the piston ring as the pneumatic cylinder moves backward.

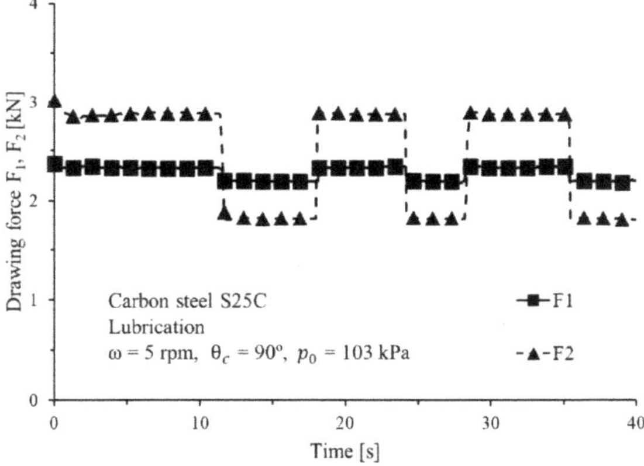

Figure 5. Drawing force variations with rotation speed of 5 rpm.

Figure 6 shows the contact pressure variations during the friction tests. The contact pressure was obtained from Equation (2), and the friction force F was regarded as the average value of F_1 and F_2. The drawing force variations of F_1 and F_2 are shown in Figure 5. It is clear that the contact pressures at the forward stages were larger than those at the backward stages. The maximal difference of the contact pressures at the forward and backward stages was about 1.2 MPa.

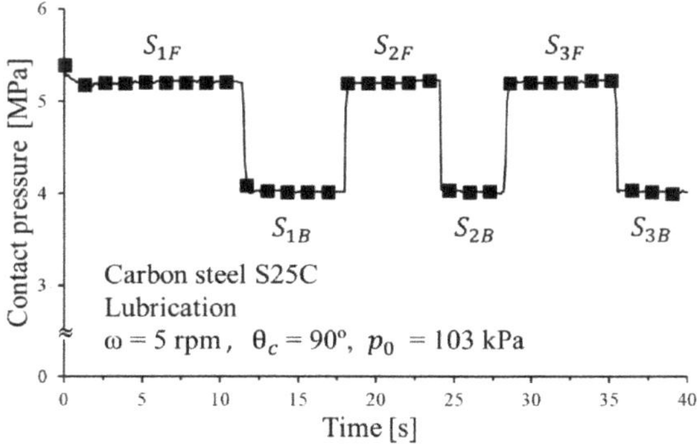

Figure 6. Contact pressure variations during friction tests.

Figure 7 shows the variations of the friction coefficient with rotation speeds of 5, 7, and 9 rpm under dry friction condition. The rotation speeds of 5, 7, and 9 rpm are equivalent to relative sliding speeds of 30, 40, and 50 mm/s, respectively, at the interface of the strip and friction wheel. The friction coefficient was determined from Equation (9) and the drawing forces F_1 and F_2 are shown in Figure 5. A smaller friction coefficient was obtained at the first stage (S_{1F} and S_{1B}), and a slightly larger friction coefficient was obtained at the third stage (S_{3F} and S_{3B}). Generally, the friction coefficient decreased as the rotation speed increases. The reason is probably that as the relative sliding speed at the interface increased, the strip momentum in the normal direction increased, which resulted in an increased separation force at the interface and reduced the real contact area. Accordingly, the friction coefficient decreased [9].

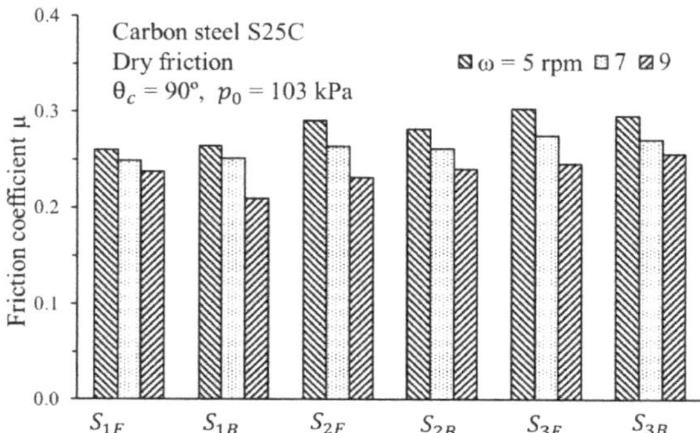

Figure 7. Effects of rotation speed on friction coefficient under dry friction condition.

Figure 8 shows the variations of friction coefficients with rotation speeds of 5, 7, and 9 rpm under lubricated conditions (SAE 5W–30). The friction test conditions are given in Table 1 (case 1). The friction coefficients obtained were 0.12–0.15. Generally, the friction coefficient increased at a later stage, like the tendency in the case under dry friction condition shown in Figure 7. This is because the strip surface became rougher after reciprocating friction test at the late stage. It was also found that the friction coefficients at the forward stages were larger than those at the backward stages, because of a larger contact pressure at the forward stages. The friction coefficient increased as the rotation speed increased, which is opposite to the tendency in the case under dry friction conditions. The reason is probably that for a dynamic viscous flow at the interface between the strip and friction wheel, a larger drawing force is needed at a higher relative sliding speed.

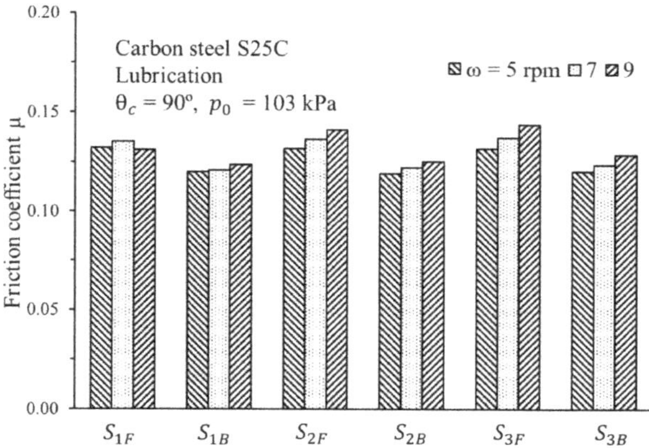

Figure 8. Effects of rotation speed on friction coefficient under lubricated condition.

4.2. Friction Tests of Brass C2680

In this section, brass C2680 strips with a thickness of 0.5 mm were used as the test piece. The Vickers hardness of brass C2680 is 128 MPa, quite close to 130 MPa, the hardness of carbon steel S25C strips. The gauge pressure P_0 in the pneumatic cylinder was set as 69 kPa. The contact angle of the strip at the friction wheel was 90°. The other friction test conditions are given in Table 1 (case 2). Figure 9 shows the variations of friction coefficients with rotation speeds of 5, 7, and 9 rpm (equivalent to relative sliding speeds of 30, 40, and 50 mm/s, respectively) under the dry friction condition. Clearly, the friction coefficient decreased with the rotation speed and a larger friction coefficient could be obtained at a later stage. The tendency of the variations of friction coefficients for brass C2680 strips was the same as that in the friction tests of carbon steel strips under dry friction conditions.

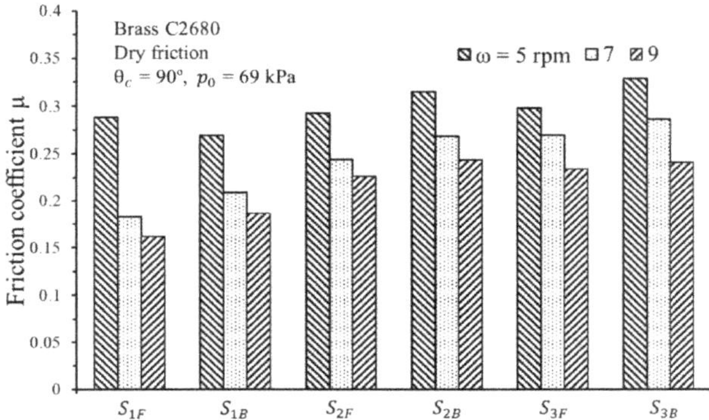

Figure 9. Effects of rotation speed on friction coefficient under dry friction conditions.

Figure 10 shows the variations of the friction coefficients with rotation speeds of 5, 7, and 9 rpm under lubricated conditions. It is known that the friction coefficient increases with the rotation speed, and a larger friction coefficient can be obtained at a later stage. The tendency of the variations of friction coefficients for brass C2680 strips was the same as that in the friction tests of carbon steel strips under lubricated conditions.

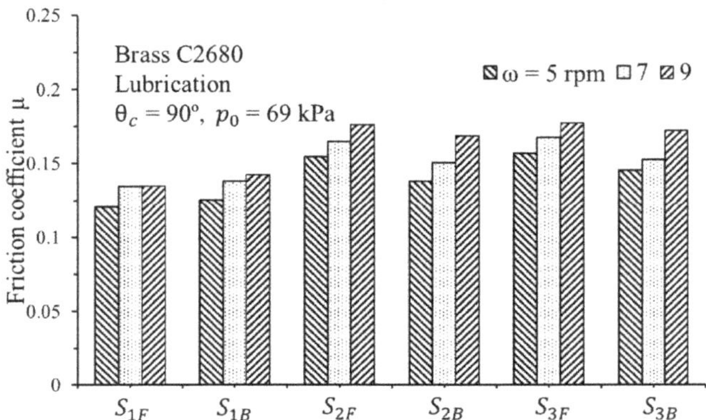

Figure 10. Effects of rotation speed on friction coefficient under lubricated conditions.

The friction coefficients obtained from friction tests of brass C2680 strips with 5 rpm at forward stages under dry friction and lubricated conditions are summarized in Table 2. μ_1, μ_2, and μ_3

correspond to the friction coefficients at S_{1F}, S_{2F}, and S_{3F} stages, respectively. The friction coefficient differences between two successive stages under dry friction conditions increased slightly, and the differences were less than 2%. Whereas, under lubricated conditions, the friction coefficient difference between μ_1 and μ_2 reached 22.1%, and the friction coefficient difference between μ_2 and μ_3 was only 1.9%. That is because the lubrication oil was greatly squeezed out after the first stage S_{1F}.

Table 2. Friction coefficients of brass C2680 at different forward stages with 5 rpm.

	Dry Friction		Lubrication	
	Friction Coefficient μ	$\frac{\mu_{i+1}-\mu_i}{\mu_i}\times 100\%$	**Friction Coefficient μ**	$\frac{\mu_{i+1}-\mu_i}{\mu_i}\times 100\%$
μ_1	0.288		0.120	
μ_2	0.292	1.4%	0.154	22.1%
μ_3	0.298	2.0%	0.157	1.9%

4.3. *Friction Tests of Aluminum 6063-T6*

In this section, aluminum alloy 6063-T6 strips with a thickness of 1.3 mm were used as the test piece. The aluminum alloy 6063-T6 strips with a Vickers hardness of 83 MPa is a relatively softer material compared with carbon steel and brass strips used in the previous sections. Friction tests were conducted under dry friction conditions. The contact angle of the strip with the friction wheel was 90°, and the other testing conditions are given in Table 1 (case 3). Figure 11 shows the variations of friction coefficients with rotation speeds of 5, 6 and 7 rpm, which correspond to relative sliding speeds of 30, 40, and 50 mm/s, respectively. From Figure 11, it is clear that a larger friction coefficient was obtained at forward stages and at a lower rotation speed. The tendency is the same as that in friction tests of carbon steel strips under dry friction conditions. Whereas, the friction coefficients decreased at a later stage, which is different from the tendency in the case of carbon steel strip. That is because the initial roughness of aluminum alloy 6063-T6 was Ra = 0.248 μm and the roughness after the first stage became Ra = 0.448 μm, which was larger than 0.410 and 0.365 μm, the roughness after the second and third stages, respectively. Due to the fact that only arithmetic mean surface roughness Ra values were used to evaluate the friction coefficient at the sheet-die interface in literature [1,6,12], only Ra values, no Rt values, were recorded during the surface roughness measurements.

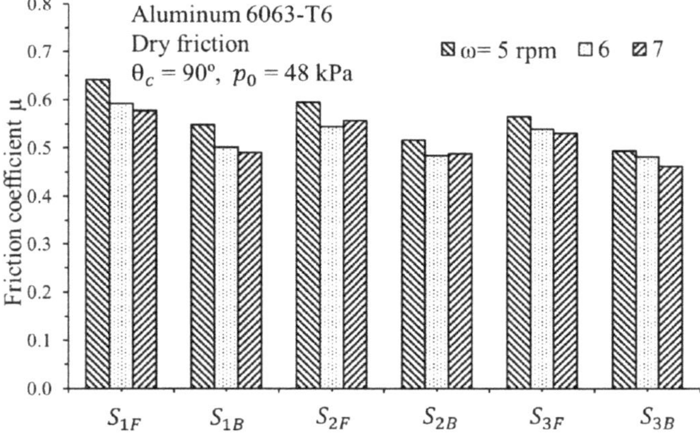

Figure 11. Effects of rotation speed on friction coefficient under dry friction conditions for aluminum alloy 6063-T6 strips.

Figure 12 shows the variations of friction coefficients with rotation speeds of 5, 6, and 7 rpm under lubricated conditions. The other testing conditions were the same as those shown in Figure 11 and are given in Table 1 (case 3). The friction coefficients became smaller at a later stage, and the friction coefficients at the forward stages were larger than those at the backward stages. The friction coefficients under lubricated conditions were significantly lower than those under dry friction conditions, shown in Figure 11. Generally, the friction coefficients became smaller as the rotation speed increases. That resulted from the surface roughness reduction after each stage. The surface roughness after stage 1 at rotation speeds of ω = 5, 6, and 7 rpm were Ra = 0.376, 0.338 and 0.289 μm, respectively.

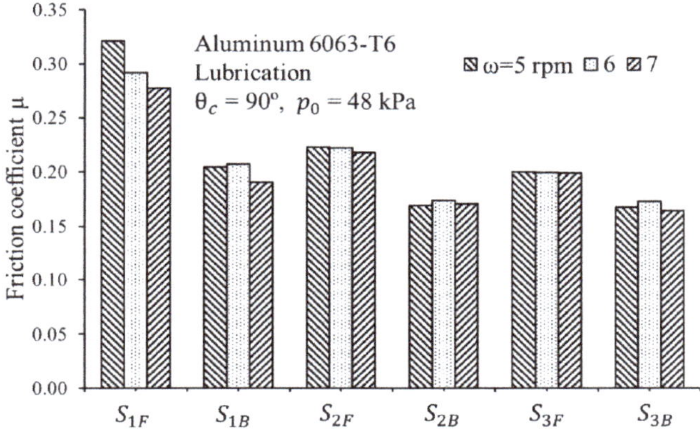

Figure 12. Effects of rotation speed on friction coefficient under lubricated conditions for aluminum alloy 6063-T6 strips.

Figure 13 shows the contact pressure variations with contact angles of 30°, 60°, and 90° under dry friction conditions. The gauge pressure p_0 in the pneumatic cylinder was set as 48 kPa, and rotation speed was 5 rpm. Friction test conditions are given in Table 1 (case 4). Clearly, the contact pressure increased with the contact angle at the forward stages, whereas the contact pressure decreased slightly with the contact angle at the backward stages.

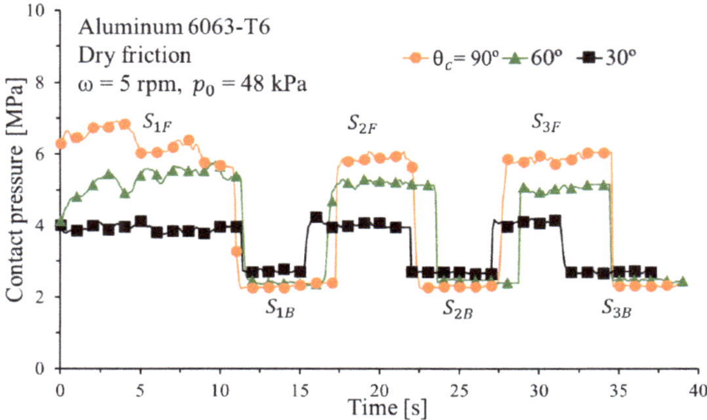

Figure 13. Contact pressure variations with different contact angles for Al6063-T6.

Figure 14 shows of friction coefficient variations with different contact angles under dry friction conditions. Clearly, the friction coefficients increased with the contact angle. That is because the contact pressure and contact area between the strip and friction wheel became larger at a larger contact angle.

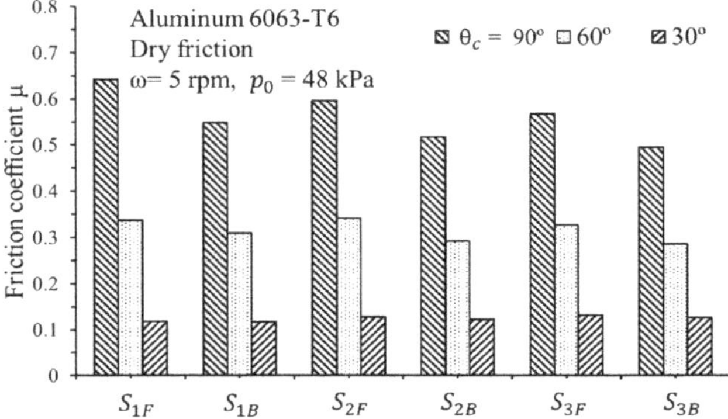

Figure 14. Friction coefficient variations with different contact angles for Al6063-T6.

Table 3 summarizes the friction coefficients obtained at different stages under dry friction with various friction test conditions given in Table 1. The values in square brackets denote the friction coefficients under lubricated conditions. It is clear that the friction coefficients under lubricated conditions were much smaller than those under dry friction conditions. Large friction coefficients of about 0.6 for aluminum alloy A6063T6 were obtained at a large contact angle of 90° and a low rotation speed of 5 rpm under dry friction conditions. The magnitude of the friction coefficients under dry friction is quite close to those obtained in Reference [12]. The friction coefficients obtained from the friction tests of carbon steel were about 0.1–0.14, which is quite close to the range of 0.1–0.18 obtained by Tamai et al. [1]. The friction coefficients decreased with increasing sliding speed under dry friction, the tendency of which was the same as that in Reference [1].

Table 3. Average friction coefficients μ under dry friction, values in square brackets denote μ under lubricated conditions.

Material	**Carbon Steel S25C**			**Al6063-T6**					**Brass C2680**		
Rotation Speed (rpm)	5	7	9	5	5	5	6	7	5	7	9
Contact angle θ_c	90°	90°	90°	30°	60°	90°	90°	90°	90°	90°	90°
S_{1F}	0.248 [0.132]	0.248 [0.135]	0.238 [0.131]	0.191	0.478	0.642 [0.321]	0.593 [0.292]	0.578 [0.277]	0.288 [0.120]	0.183 [0.135]	0.161 [0.135]
S_{1B}	0.264 [0.120]	0.251 [0.121]	0.209 [0.124]	0.191	0.407	0.548 [0.204]	0.501 [0.207]	0.491 [0.190]	0.269 [0.125]	0.209 [0.138]	0.186 [0.142]
S_{2F}	0.29 [0.131]	0.263 [0.136]	0.231 [0.141]	0.209	0.434	0.595 [0.223]	0.544 [0.222]	0.557 [0.218]	0.292 [0.154]	0.244 [0.165]	0.226 [0.176]
S_{2B}	0.282 [0.119]	0.261 [0.122]	0.24 [0.125]	0.198	0.373	0.517 [0.169]	0.484 [0.174]	0.489 [0.171]	0.29 [0.138]	0.243 [0.150]	0.219 [0.168]
S_{3F}	0.303 [0.131]	0.275 [0.137]	0.246 [0.144]	0.214	0.412	0.566 [0.200]	0.54 [0.199]	0.531 [0.199]	0.298 [0.157]	0.27 [0.168]	0.234 [0.177]
S_{3B}	0.295 [0.120]	0.271 [0.124]	0.255 [0.128]	0.205	0.361	0.494 [0.168]	0.483 [0.173]	0.463 [0.164]	0.304 [0.145]	0.261 [0.152]	0.221 [0.172]

5. Conclusions

A reversible friction test machine was developed and a series of friction tests with three stages were conducted to investigate the effects of contact angles, strip materials, surface lubrication, sliding speeds etc., on the friction coefficients at the interface of the strip and friction wheel. Generally, the friction coefficients increased with the contact angle between the strip and friction wheel and the friction coefficients decreased with increasing rotation speed under dry friction conditions. For harder strips of carbon steel and brass, the friction coefficients increased with increasing rotation speed under lubricated conditions, whereas, the friction coefficients decreased with increasing rotation speed for softer strips of aluminum alloy. Generally, the friction coefficients obtained in the friction tests of the three kinds of strip metals under dry friction conditions were approximately twice the values obtained in the friction tests with oil lubricant at the strip–die interface. From a series of friction tests, it is known that the friction coefficients at the forward stages during the friction tests were different from those at the backward stages and the friction coefficients at the earlier stages were different from those at the later stages. It is noteworthy that a variant, not a constant friction coefficient model, at the sheet–die interface should be considered in a servo press forming process with repeated forward and backward punch motions. In the future, the friction parameters or conditions can be adjusted in the reversible friction tests to extend the application scopes, such as shortening stroke distances to resemble the movement of a servo press, increasing force F_1 to increase the contact pressure, rotating the friction wheel to increase the relative sliding speed between the strip and die, adopting different lubricants, coating the surface of the friction wheel with a chromium film to decrease the roughness of the friction wheel, and so on. The effects of the surface texture on the friction behavior will be explored. Furthermore, a nonlinear friction coefficient model for servo press forming processes will be developed.

Author Contributions: Conceptualization, methodology, research supervision and writing—review and editing were conducted by Y.-M.H.; experiments, validation and writing—original draft preparation were completed by C.-C.C. All authors have read and agreed to the published version of the manuscript

Funding: The authors would like to extend their thanks to the Ministry of Science and Technology of the Republic of China under Grant no. MOST 107-2622-E-110-006-CC2. The advice and financial support of MOST are gratefully acknowledged.

Acknowledgments: The authors would like to extend their thanks to the Ministry of Science and Technology of the Republic of China under grant no. MOST 107-2622-E-110-006 -CC2. The advice and financial support of MOST are greatly acknowledged.

Conflicts of Interest: The authors declare no conflict of interest.

References

1. Tamai, Y.; Fuzita, T.; Inazumi, T.; Manabe, K. Effects of contact pressure, sliding velocity and sliding length on friction behavior of high-tensile-strength steel sheets. *J. JSTP* **2012**, *54*, 537–541.
2. Hwang, Y.M.; Huang, L.S. Friction tests in tube hydroforming. *Proc. IMechE Part B J. Eng.Manuf.* **2005**, *219*, 587–594. [CrossRef]
3. Fratini, L.; Lo Casto, S.; Lo Valvo, E. A technical note on an experimental device to measurefriction coefficient in sheet metal forming. *J. Mater. Process. Technol.* **2006**, *172*, 16–21. [CrossRef]
4. Tamai, Y.; Inazumi, T.; Manabe, K. FE forming analysis with nonlinear friction coefficient model considering contact pressure, sliding velocity and sliding length. *J. Mater. Process. Technol.* **2016**, *227*, 161–168. [CrossRef]
5. Weinmann, K.; Bhonsle, S.; Gerstenberger, J. On the determination of the coefficient of friction and the friction factor by the strip-tension friction test. *CIRP Ann.* **1990**, *39*, 263–266. [CrossRef]
6. Saha, P.K.; Wilson, W.R. Influence of plastic strain on friction in sheet metal forming. *Wear* **1994**, *172*, 167–173. [CrossRef]
7. Hsu, T.C.; Kuo, S.G. Boundary friction model in punch friction test. *J. Mater. Process. Technol.* **1994**, *45*, 601–606. [CrossRef]

8. Kim, H.; Hwang, B.; Bae, W. An experimental study on forming characteristics of pre-coated sheet metals. *J. Mater. Process. Technol.* **2002**, *120*, 290–295. [CrossRef]
9. Lemu, H.G.; Trzepiecinski, T. Numerical and experimental study of frictional behavior in bending under tension test. *Stroj. Vestn. J. Mech. Eng.* **2013**, *59*, 41–49. [CrossRef]
10. Ramezani, M.; Neitzert, T.; Pasang, T.; Selles, M.A. Dry sliding frictional characteristics of ZE10 and AZ80 magnesium strips under plastic deformation. *Tribol. Int.* **2015**, *82*, 255–262. [CrossRef]
11. Fridman, H.D.; Levesque, P. Reduction of static friction by sonic vibrations. *J. Appl. Phys.* **1959**, *30*, 1572–1575. [CrossRef]
12. Chowdhury, M.A.; Khalil, M.K.; Nuruzzaman, D.M.; Rahaman, M.L. The effect of sliding speed and normal load on friction. *Mechatron. Eng.* **2011**, *11*, 45–49.
13. Saha, P.K. Factors affecting the accuracy and control of the sheet metal forming simulator. *Trans. NAMRI/SME* **1994**, *22*, 47–54.
14. Adamczak, S.; Zmarzły, P. Research of the influence of the 2D and 3D surface roughness parameters of bearing raceways on the vibration level. *J. Phys. Conf. Ser.* **2019**, *1183*. [CrossRef]
15. Ali, M.K.A.; Xianjun, H.; Elagouz, A.; Essa, F.A.; Abdelkareem, M.A.A. Minimizing of the boundary friction coefficient in automotive engines using Al_2O_3 and TiO_2 nanoparticles. *J. Nanopart. Res.* **2016**, *18*, 377. [CrossRef]
16. Escosa, E.G.; García, I.; Damborenea, J.J.D.; Conde, A. Friction and wear behaviour of tool steels sliding against 22MnB5 steel. *J. Mater. Res. Technol.* **2017**, *6*, 241–250. [CrossRef]
17. Wu, Z.; Xing, Y.; Huang, P.; Liu, L. Tribological properties of dimple-textured titanium alloys under dry sliding contact. *Surf. Coat. Technol.* **2017**, *309*, 21–28. [CrossRef]
18. Hwang, Y.M.; Yu, H.P.; Chen, C.C. Study of friction tests of strips with variant relative speeds, Tube Hydroforming Technology. In Proceedings of the 9th International Conference On Tube Hydroforming, Kaohsiung, Taiwan, 18–21 November 2019; pp. 178–183.

Article

Crash Characteristics of Partially Quenched Curved Products by Three-Dimensional Hot Bending and Direct Quench

Atsushi Tomizawa [1,*], Sanny Soedjatmiko Hartanto [2], Kazuo Uematsu [3] and Naoaki Shimada [4]

1 Faculty of Production Systems Engineering and Sciences, Komatsu Univercity, Nu 1-3 Shichomachi, Komatsu, Ishikawa 923-8511, Japan
2 Department of Mechanical Systems Engineering, Okinawa National College of Technology, 905 Henoko, Nago-shi, Okinawa 905-2192, Japan; SannySoeHartanto@gmail.com
3 Steel Research Laboratories, Nippon Steel Corporation, 1-8 Fuso-cho, Amagasaki, Hyogo 660-0891, Japan; uematsu.h5h.kazuo@jp.nipponsteel.com
4 Steel Research Laboratories, Nippon Steel Corporation, 20-1 Shintomi, Futtsu, Chiba 293-0011, Japan; shimada.57y.naoaki@jp.nipponsteel.com
* Correspondence: atsushi.tomizawa@komatsu-u.ac.jp; Tel.: +81-(0)-761-48-3145

Received: 18 August 2020; Accepted: 30 September 2020; Published: 2 October 2020

Abstract: Recently, improvement of hybrid and electric vehicle technologies, equipped with batteries, continues to solve energy and environmental problems. Lighter weight and crash safety are required in these vehicles body. In order to meet these requirements, three-dimensional hot bending and direct quench (3DQ) technology, which enables to form hollow tubular automotive parts with a tensile strength of 1470 MPa or over, has been developed. In addition, this technology enables to produce partially quenched automotive parts. In this study, the crash characteristics of 3DQ partially quenched products were investigated as the fundamental research of the design for improving the energy absorption. Main results are as follows: (1) for partially quenched straight products in axial crash test, buckling that occurs at nonquenched portion can be controlled; (2) for the nonquenched conventional and overall-quenched curved products, buckling occurs at the bent portion at the initial stage in axial crash tests, and its energy absorption is low; (3) by optimizing partially quench conditions, buckling occurrence can be controlled; and (4) In this study, the largest energy absorption was obtained from the partially quenched curved product, which was 84.6% larger than the energy absorption of the conventional nonquenched bent product in crash test.

Keywords: crash safety; hot bending; partial-quench; FEM

1. Introduction

Recently, improvement of hybrid and electric vehicle technologies, equipped with batteries, continues to solve energy and environmental problems. Lighter weight and crash safety are required in the vehicles' bodies [1]. In order to satisfy these requirements, it has been promoted to apply ultrahigh tensile steel [2]. In the case of sheet metal, application of hot stamping has been promoted to achieve over 1500 MPa tensile strength [3,4]. However, assembling closed-sectional-shaped automotive parts from sheet metal requires spot welding of two or more parts. Therefore, flange for spot welding is indispensable. So, it is difficult to reduce the weight of the components. In addition, ideal rigidity cannot be obtained because the spot welds are intermitted. On the other hand, hydroforming technology has been developed to manufacture high-rigidity hollow components from steel tube. However, generally hydroforming cannot form over 980 MPa strength tubes, and manufacturing facility required for it is large and expensive [5,6]. In addition, as electrification and autonomous automobiles progresses, it is expected that newcomers of automotive manufacturer will increase, and they will demand more compact and low-cost manufacturing facility to depress initial investment cost. Resin body car prototype using 3D printer as the low-cost manufacturing facility even exists [7], but it is uncertain whether the crash performance is sufficient. Thus, the technology that can manufacture more high-strength tubular components with compact and low-cost facility has been desired. Therefore, three-dimensional hot bending and direct quench (3DQ) technology has been developed. 3DQ is a consecutive forming process, which allows bending and quenching at the same time. This technology enables to form hollow tubular structure components of a vehicle bodies with ultrahigh tensile strength and three-dimensional shapes using steel tube. The 3DQ technology is an innovative method, and the method makes it possible to fully enjoy the advantage of steel to manufacture high-strength automotive structural parts at lower costs with excellent recyclability, compared with other materials such as carbon fiber reinforced plastics (CFRP). In addition, 3DQ facility is more compact and low cost than hot press and hydroforming machine. In this study, the crash characteristics of partially quenched curved products by 3DQ and the suitable design are discussed for improving the energy absorption.

2. Outline of 3DQ Technology

2.1. Features of 3DQ Technology

Figure 1 shows the developed 3DQ machine. The main components of the 3DQ machine are axial feeding device, support guides or rolls, induction heater (IH), water cooling device, and bending robot. First, by inverse analysis from the three-dimensional product shape data, 3DQ operation data are determined (Figure 2). The straight tube is fed supported by the support guides or rolls. As shown in Figure 3, the straight tubes with various cross-sections, such as round, square, and odd-shaped, are available. Then, this tube is heated by the IH rapidly. The heat temperature is more than Ac3 transformation point. After heating, the tube is cooled to room temperature by water cooling device for quenching. At the same time, the tube is bent by robot. The deformation is concentrated in this narrow heating potion. By this process consecutively, the products which have complex three-dimensional bent shape and ultrahigh tensile strength are obtained (Figure 4).

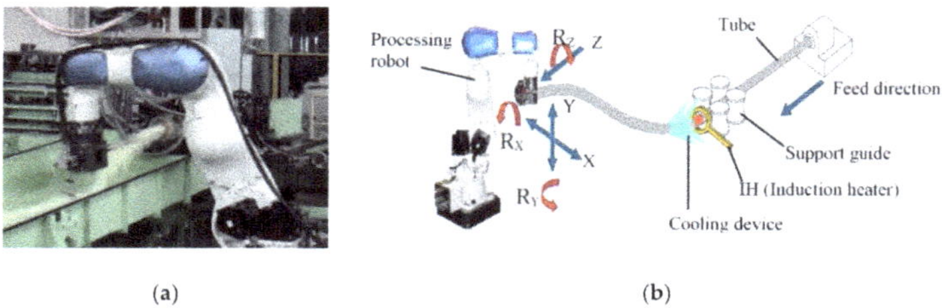

Figure 1. Three-dimensional hot bending and direct quench (3DQ) machine: (**a**) appearance photograph and (**b**) schematic illustration.

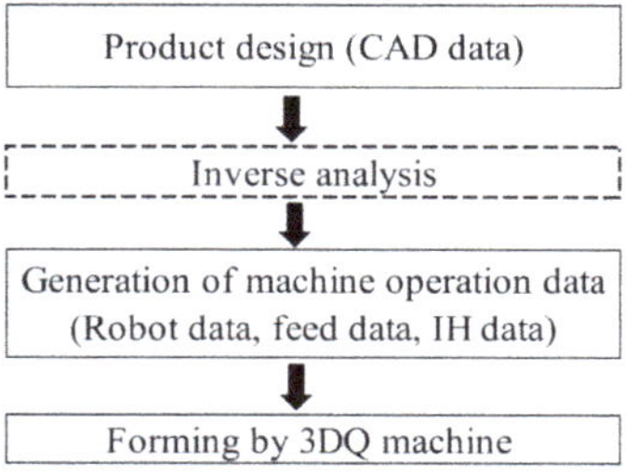

Figure 2. 3DQ production system.

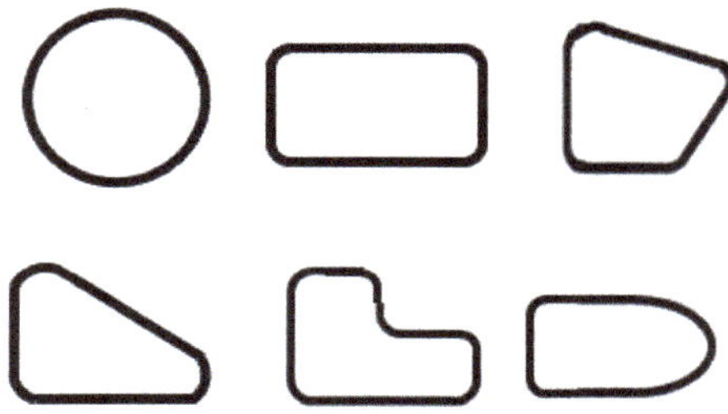

Figure 3. Examples of tube cross-section for 3DQ.

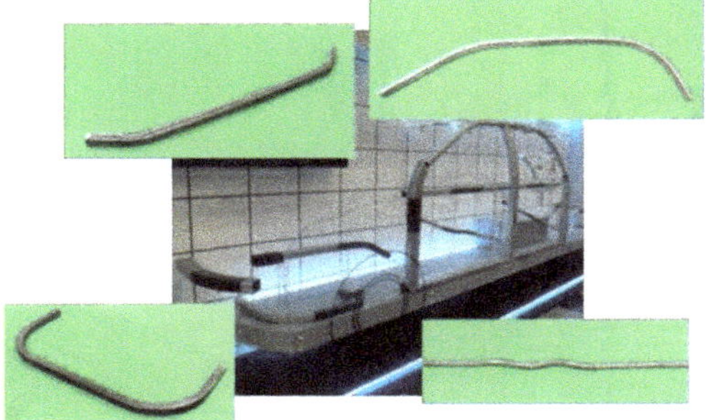

Figure 4. Examples of automotive products by 3DQ.

The 3DQ technology has achieved the following characteristics: [8–12]

1. Product tensile strength of 1470 MPa.
2. High shape fixability results in high forming precision.
3. Low residual stress.
4. Low die cost (die less forming).
5. No concern of delayed fracture.

In addition, it is possible to quenched products partially using the IH control.

2.2. Properties of 3DQ Products

In this study, the experiments were carried out using the 3DQ machine as shown in Table 1. In the experiment, electric welded steel tubes with a rectangular cross-section were used. Table 2 shows the chemical composition of the test tubes. Figure 5 shows an example of the temperature change during the 3DQ forming process of a square hollow section. It was measured with thermocouples on the internal surface of the material, which was fed at a rate of 80 mm/s. Here, the material was quickly heated to above the Ac3 temperature by the high-frequency heating coil, and then rapidly water-cooled to room temperature.

Table 1. Main specifications of experimental three-dimensional hot bending and direct quench (3DQ) machine.

Maximum Induction Heating Output	300 kW
Frequency of induction heater	9.8 kHz
Maximum feed stroke	1700 mm
Maximum feed speed	120 mm/s (at load 5 kN)
Payload of robot	165 kg

Table 2. Chemical compositions of material (mass %).

C	Si	Mn	B
0.21	0.22	1.20	0.0015

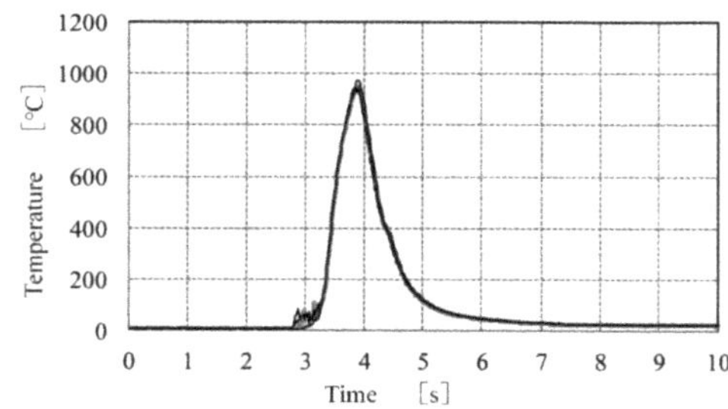

Figure 5. Example of tube temperature of in 3DQ process (feed speed 80 mm/s).

Figure 6 shows the hardness distribution of the product by the 3DQ machine. Vickers hardness of 450 HV, which is equivalent to tensile strength of 1470 MPa, is obtained in all portions of the product, and the microstructure of the product became uniform martensitic structure, as shown in Figure 7.

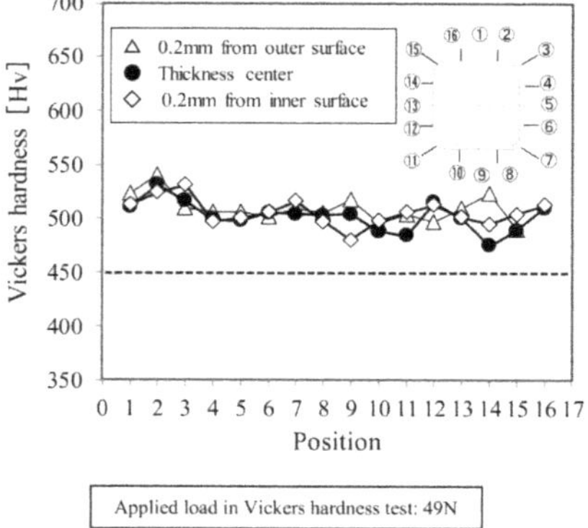

Figure 6. Circumferential hardness distribution of 3DQ product (40 mm × 40 mm and thickness 1.8 mm).

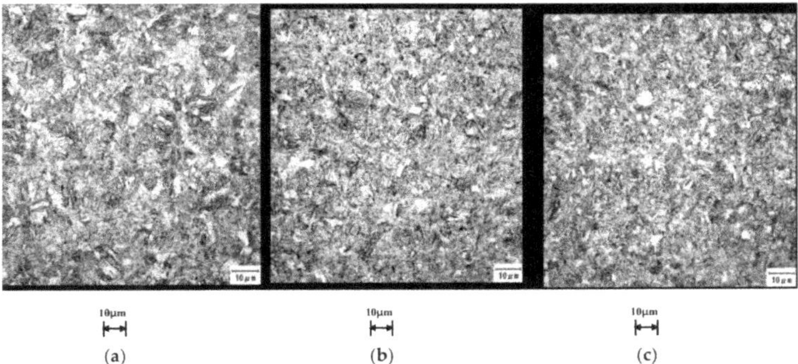

Figure 7. Microstructure of product by 3DQ (40 mm × 40 mm and thickness 1.8 mm): (**a**) 0.2 mm from outer surface, (**b**) thickness center, and (**c**) 0.2 mm from inner surface.

3. Crash Characteristics of Partially Quenched Curved Product

3.1. Deformation Behavior of the Partially Quenched Straight Product

Partial quenching by 3DQ allows strengthening of only the areas of an automobile component that need high-tensile strength. To clarify the deformation behavior of the partially quenched straight product by 3DQ, the axial crash tests were carried out as shown in Table 3 [13]. In the crash test, the load is measured by the load cell. The stroke signal was measured by a laser displacement meter. Hence, the load-stroke diagram is obtained by synchronizing these signals. Figure 8 shows longitudinal hardness distribution of the partially quenched straight product in part of this test tube. Figure 9 shows the results of the axial crash test of the nonquenched and the partially quenched specimens. The nonquenched product deforms sequentially from top of the specimen, whereas the portion which is not quenched deformed preferentially in the partially quenched specimen. In other words, it is suggested that the buckling mode can be controlled by partial quenching. Figure 10 shows the energy absorption of the specimen in this test. For the straight product, the energy absorption of the partially quenched product is a little larger than that of the nonquenched product or in the same level, because the deformation occurs in the nonquenched area.

Table 3. Conditions of axial crash test.

Cross-section of Specimen	50 mm × 70 mm
Specimen thickness	1.8 mm
Partially quenched specimen (see Figure 9)	Partially quench area: 30 mm × 4 Nonquenched area: 10 mm ×5
Impactor weight	430 kg
Impactor speed	7–10 m/s

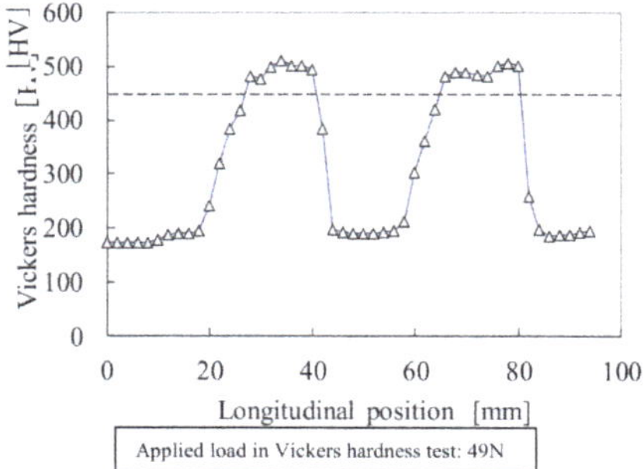

Figure 8. Longitudinal hardness distribution of the partially quenched straight product.

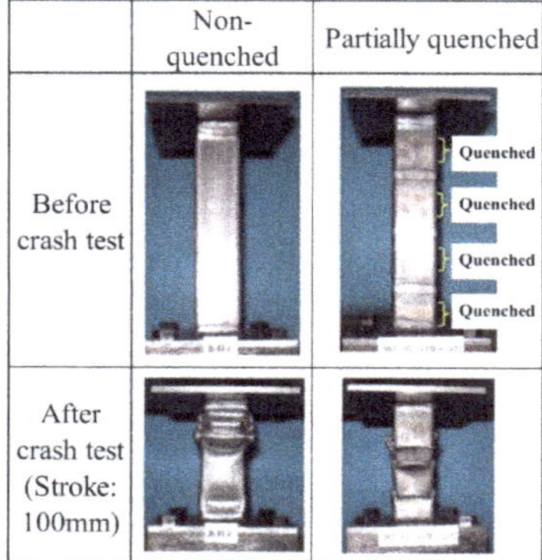

Figure 9. Deformation of the partially quenched straight tube in the axial crash test.

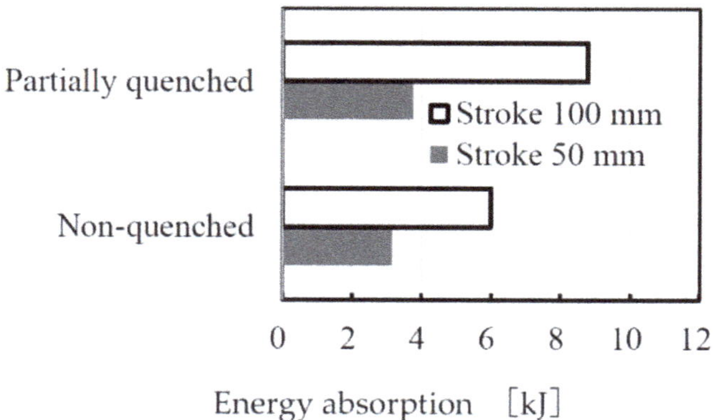

Figure 10. Energy absorption of the partially quenched straight tube in the axial crash test.

3.2. Deformation Behavior of Partially Quenched Curved Product

The shapes of automotive parts are generally curved. To investigate the effects of some major factors on axial crash behavior of partially quenched curved products by 3DQ, the finite element analysis has been conducted with RADIOSS 13.0 (Altair Engineering, Inc., Troy, MI, USA) using the dynamic explicit method. Table 4 shows the analytical conditions. During the plastic deformation, the material expressed the strengthening phenomenon, so called work hardening, where the stress of material increases exponentially to the strain as shown in Equation (1). In this calculation model, the transition region of hardness, as shown in Figure 8, was not considered.

$$\sigma = a + b \cdot \varepsilon^{n}. \quad (1)$$

Table 4. Analytical conditions (see Equation (1))

Symbol	Item	Nonquenched Area	Quenched Area
ρ	Initial density (kg/mm^3)	7.85×10^{-6}	
E	Young's modulus (GPa)	210	
ν	Poisson's ratio (−)	0.3	
a	Yield stress (GPa)	0.6	1.6
b	Strength coefficient (GPa)	1.754	3.978
n	Strain hardening exponent (−)	0.3663	0.3377

Figure 11 shows the example of the curved product shapes in the crash test: (a) the overall-quenched product and (b) the partially quenched product. In this simulation, the bending radius and its quenched area were varied to investigate the crash characteristics of the curved design products.

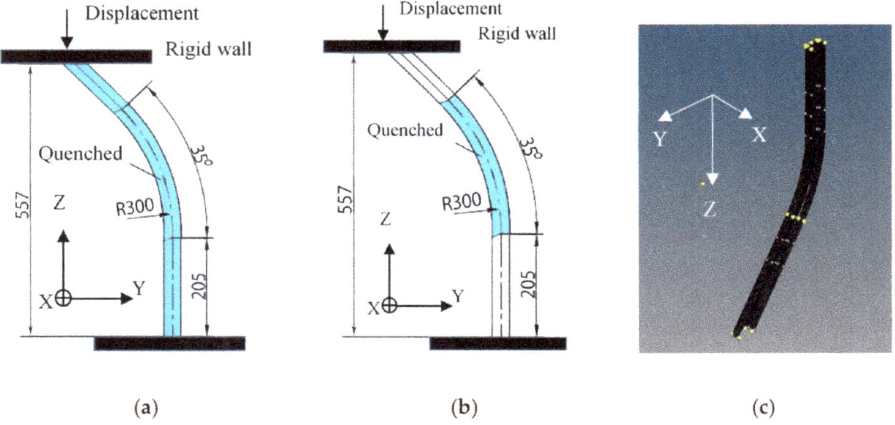

(a) (b) (c)

Figure 11. Initial product shape in axial crash test: (**a**) overall-quenched, (**b**) partially quenched, and (**c**) finite element method (FEM) model.

In the finite element method (FEM) model, the top of the products is restricted from translation of both X and Y axes and rotation of Z axis, and is deformed by a rigid wall as shown in Table 5. Under the consideration of the symmetry of the test product, half analytical model is applied to the simulation. The bottom of the product is restricted from both displacement and rotation of all axes. Meshing size is 2.0 mm × 2.0 mm.

Table 5. Boundary conditions.

Location of Surface	Restriction of Translation			Restriction of Rotation		
	X axis	Y axis	Z axis	X axis	Y axis	Z axis
Top	Fixed	Fixed	Free	Fixed	Fixed	Fixed
Bottom	Fixed	Fixed	Fixed	Fixed	Fixed	Fixed

The initial product shapes are shown in Figure 11a,b. In the crush test, two 3DQ curved products are used as shown in Figure 12a and top and bottom plates are welded to these curved products. The crash behavior of overall-quenched product is shown in Figure 12. The buckling occurs at the bending section during the initial stage. Figure 13 shows the crash behavior of partially quenched product. The first buckling occurs at the nonquenched upper part of the product and the second buckling occurs at the nonquenched lower part of the product. In this study, the simulation results are in good agreement with the experimental results. Figures 14 and 15 show the relationship between load and stroke of overall-quenched product and partially quenched product in crash test for both experimental and simulation results. Both graphs are in good agreement practically. These absolute values difference seems to be due to difference between their boundary conditions. Since the energy absorption represents the area of stroke-load diagram in crash test, there is a possibility of improving crash performance by applying partially quenched products by 3DQ, as seen in Figures 14 and 15.

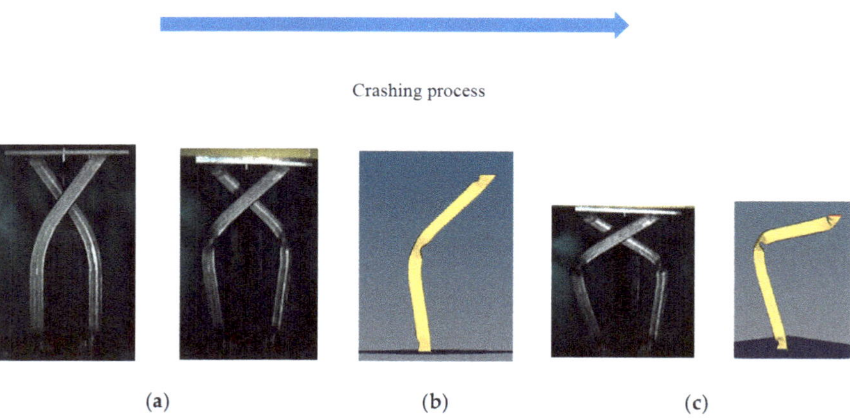

(a) (b) (c)

Figure 12. Comparison of crash deformation of overall-quenched product between experiment and simulation in axial crash test: (**a**) initial product shape, (**b**) axial stroke = 15 mm, (**c**) axial stroke = 150 mm.

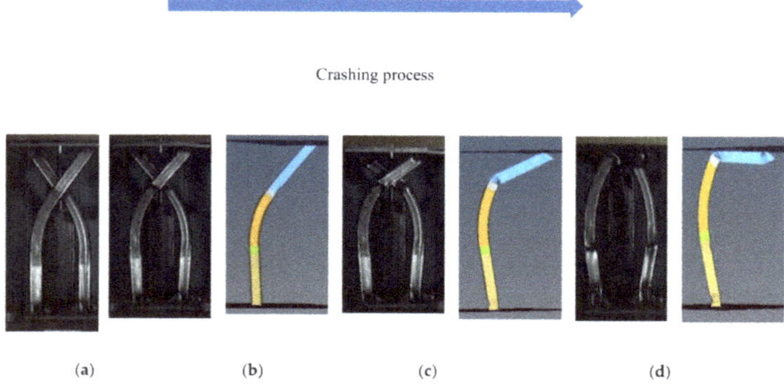

(a) (b) (c) (d)

Figure 13. Comparison of crash deformation of partially quenched product between experiment and simulation in axial crash test: (**a**) initial product shape, (**b**) axial stroke = 15 mm, (**c**) axial stroke = 85 mm, and (**d**) axial stroke = 177 mm.

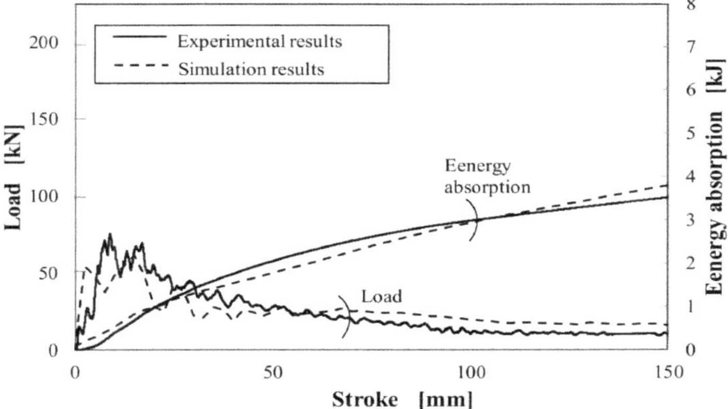

Figure 14. Relationship between load and stroke of overall-quenched product in axial crash test.

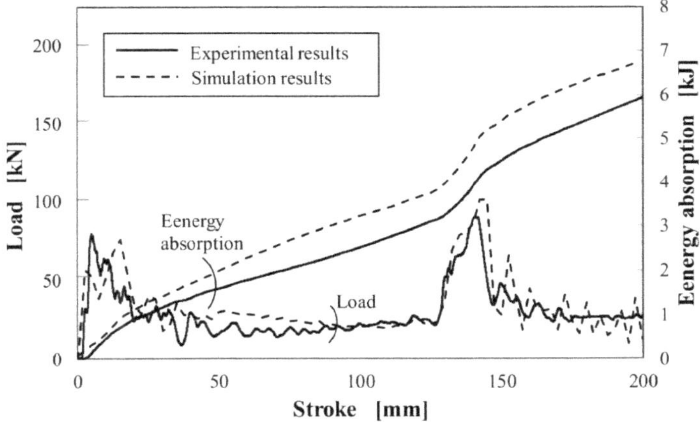

Figure 15. Relationship between load and stroke of partially quenched product in axial crash test.

3.3. *Influence of the Quenched Area on Crash Deformation*

To investigate effect of the quenched area on crash behavior, calculations were carried out using previous finite element (FE) analysis. Table 6 and Figure 16a show the analytical conditions. The calculations were carried out under the bending radius R = 400 mm with extended quenched area δ ranging from 0 to 75 mm. Calculation results are shown in Figure 16b. Figure 16 shows the stroke and buckling location. Buckling occurs at the center portion (C), upper portion (U), and lower (L) portion of curved product. The arrows indicate the order of buckling when the buckling occurred multiple times. As seen in Figure 16, buckling behavior depends on the quenched area. In the case of the products with 50 and 75 mm extended quenched area, the buckling occurrence is same as the crash test of overall-quenched products. In the case of smaller quenched area products, such as the products with 25 and 37.5 mm extended quenched area, the local buckling occurs two times in the crash test. The first buckling occurs at the nonquenched upper part of the product and the second buckling occurs at the bending section. This kind of buckling at bending section leads to rupture in some case. In the case of products with 0 and 12.5 mm extended quenched area, local buckling occurs three times in the crash test. The first buckling occurs at the nonquenched upper part of the product. Then, the second buckling occurs at the nonquenched lower part of the product. Finally, the last buckling occurs at the bending section.

Table 6. Analytical conditions.

Bending Angle θ	35 (Deg)
Bending radius R	300, 400, 500 (mm)
Extended quenched area δ	0, 12.5, 25, 37.5, 50, 75 (mm)

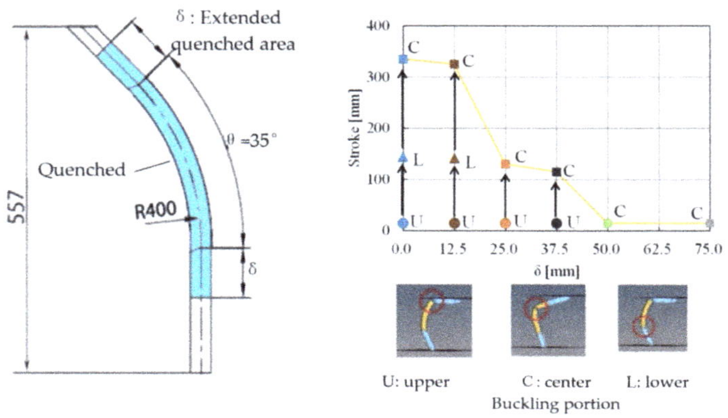

(a) (b)

Figure 16. Relationship between extended quenched area and the stroke at buckling occurred: (**a**) initial product shape and (**b**) calculation results.

3.4. Influence of Bending Radius on Crash Deformation

To investigate effect of bending radius on crash behavior, calculations were carried out under the same condition (extended quenched area δ = 25 mm) with bending radius R ranging from 300 to 500 mm. Figure 17a shows initial product shape in calculation. The calculation results are shown in Figure 17b. As can be seen from the figure, for the partially quenched products with bending radii greater than 400 mm, buckling easily occurs at the initial stage owing to its larger moment at the bent portion. Thus, it exhibits smaller energy absorption. Contrary, for those partially quenched products with bending radius below 400 mm, local buckling occurs three times.

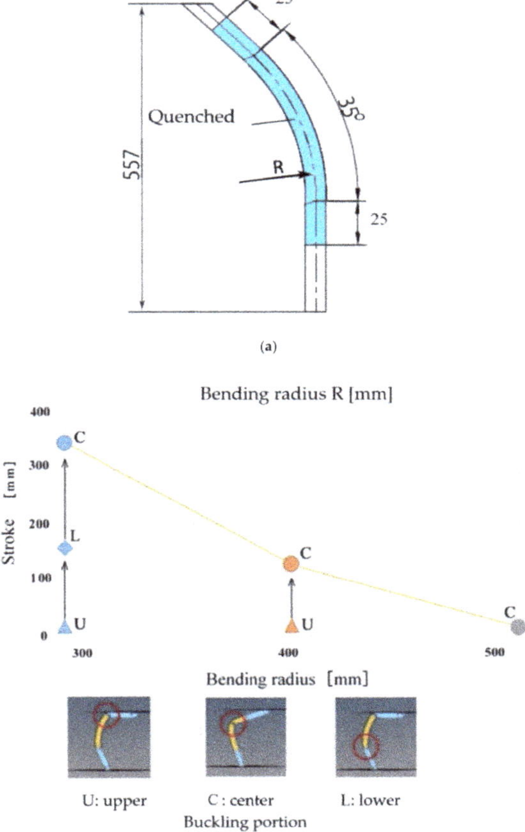

Figure 17. Relationship between bending radius and the stroke at buckling occurred for partially quenched product with extended quenched area δ = 25 mm: (**a**) initial product shape and (**b**) calculation results.

4. Discussion

From the FEM analysis in the previous section, it is made clear that the buckling behavior significantly changed by changing the partially quenched area. The deformation behavior in the crash test is classified into the following three types: (1) buckling occurs only at the bent portion (C); (2) buckling occurs at the upper nonquenched portion, and then buckling occurs at the center bent portion (U-C); (3) buckling occurs at the upper nonquenched portion, then buckling occurs at the lower nonquenched portion, and finally, third buckling occurs at the center bent portion (U-L-C). Figure 18 shows relationship between the extended quenched area and energy absorption, correspond to the buckling deformations.

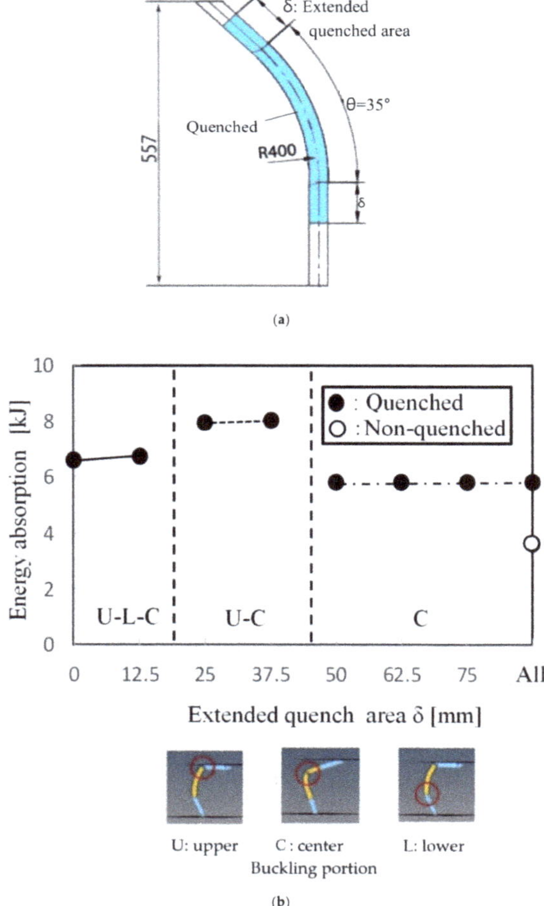

Figure 18. Relationship between the extended quenched area and energy absorption in crash test: (**a**) initial product shape and (**b**) calculation results.

The energy absorption of the conventional curved product is 3.5 kJ, whereas the overall-quenched product exhibits a larger value, whose buckling occurs at the bent portion during the initial stage. As shown in Figure 17, in the case of the products with 50 and 75 mm extended quenched area, the buckling occurrence are the same as that in the crash test of the overall-quenched products. For products with a smaller quenched area, such as those with 25 and 37.5 mm extended of the quenched area, the buckling occurs twice. The first buckling occurs at the nonquenched upper portion, and the second buckling occurs at the quenched bent portion. Although this type of buckling exhibits such a large value of energy absorption, as shown in Figure 17, rupture easily occurs, hence its buckling load is varied.

As shown in Figure 18, the most suitable crash characteristic is obtained when local buckling occurs three times during the crash process in the following condition: 0 and 12.5 mm extended quenched area. The first buckling occurs at the nonquenched upper part of the product and the second buckling at the nonquenched lower part of the product and then, the third buckling at the bent portion.

In Figure 19, the effect of partially quenched products on energy absorption is shown. The conventional nonquenched curved product has the smallest energy absorption in the crash test. Due to the effect of its quenching characteristics, the energy absorption of overall-quenched product became 59.3% larger than that of the conventional nonquenched curved product. Furthermore, in this study, the crash characteristics of partially quenched products by 3DQ are investigated. The energy absorption increased 84.6% compared with the energy absorption of conventional nonquenched curved product.

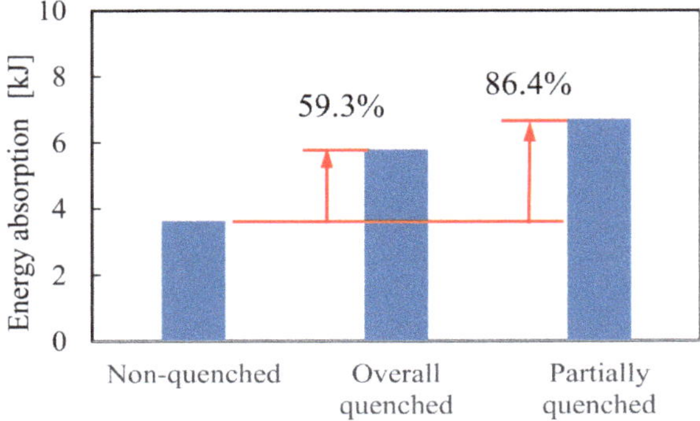

Figure 19. Energy absorption of product with optimized partially quenched area (see Figure 18).

5. Conclusions

In this study, the characteristics of 3DQ curved products were investigated as the fundamental research of the most suitable automotive design for improving the energy absorption capability. Main results are concluded as follows:

1. For partially quenched straight products in the axial crash test, buckling behavior occurred at nonquenched portion can be controlled.
2. For the conventional nonquenched curved product and overall-quenched curved products, buckling occurs at the bent portion at the initial stage in axial crash tests, and its energy absorption was low.
3. The crash deformation of curved partially quenched 3DQ products were performed by FE analysis.
4. For overall-quenched products, buckling occurred at the bent portion and, its energy absorption increased 59.3% compared to the conventional nonquenched curved product.
5. By optimizing the partially quenched area, buckling can be controlled. In this study, the largest energy absorption was obtained from the partially quenched curved product, which is 84.6% larger than the energy absorption of the conventional nonquenched curved product during the crash test.

Author Contributions: Conceptualization, A.T.; Methodology, A.T.; Resources, A.T., K.U. and N.S.; Material provision, K.U. and N.S.; Experimental work, A.T., K.U. and N.S.; Numerical analysis, A.T. and S.S.H.; Data curation, A.T.; Writing—original draft preparation, A.T.; Supervision, A.T.; Project administration, A.T.; Funding acquisition, A.T. All authors have read and agreed to the published version of the manuscript.

Funding: This work was supported by JSPS KAKENHI Grant Number JP20K05172.

Conflicts of Interest: The authors declare no conflict of interest.

References

1. Watabnabe, K.; Kuriyama, Y.; Inazumi, T.; Fukui, K. World Auto Steel Program Future Steel Vehicle (Third Report). *Trans. JSAE* **2013**, *44*, 523–528.
2. Takahashi, M. Current Status and Future Prospects of High-Strength Steel Sheet for Automobiles. *J. JSTP* **2017**, *58*, 105–109. [CrossRef]

3. Kojima, N.; Nishibata, T.; Uematsu, K.; Uchihara, M.; Imai KAkioka, K.; Kikuchi, K. The Effect of Process Factors on Performance of Hot Stamped Parts. *Trans. JSAE* **2007**, *38*, 321–326.
4. Asai, T.; Iwaya, J. Hot Stamping Drawability of Steel. *Proc. IDDRG* **2004**, 344–354.
5. Dohmann, F.; Hartl, C. Tube hydroforming—Research and practical application. *J. Mater. Process. Technol.* **1997**, *71*, 174–186. [CrossRef]
6. Hasegawa, Y.; Fujita, H.; Endo, T.; Fujimoto, M.; Tanabe, J.; Yoshida, M. Development of Front Pillar for Visibility Enhancement. *Honda R D Tech. Rev.* **2008**, *20*, 106–113.
7. LSEV World's First Mass Produced 3D Printed Car. Available online: https://polymaker.com/lsev-worlds-first-mass-produced-3d-printed-car/ (accessed on 9 March 2020).
8. Tomizawa, A.; Shimada, N.; Kubota, H.; Okada, N.; Sakamoto Yoshida, M.; Yamamoto, K.; Mori, H.; Hara, M.; Kuwayama, S. Development of three-dimensional hot bending and direct quench technology. *Nippon Steel Sumitomo Met. Tech. Rep.* **2013**, *397*, 83–89.
9. Tomizawa, A.; Shimada, N.; Hara, M.; Kuwayama, S.; Okahisa, M.; Suyama, T. Development of Three-Dimensional Hot Bending and Direct Quench (3DQ) Mass Processing Technology. In Proceedings of the 5th International Conference on Tube Hydro Forming, Hokkaido, Japan, 24–27 July 2011; pp. 137–140.
10. Tomizawa, A.; Shimada, N.N.; Kubota, H.; Okada, N.; Hara, M.; Kuwayama, S. Forming characteristics in three-dimensional hot bending and direct quench process- Development of three-dimensional hot bending and direct quench technology. *J. JSTP* **2015**, *56*, 961–966. [CrossRef]
11. Kubota, H.; Tomizawa, A.; Yamamoto, K.; Okada, N.; Hama, T.; Takuda, H. Effect of cross-sectional shape and temperature distribution on formable range in three- dimensional hot bending and direct quench processes- Development of three-dimensional hot bending and direct quench technology. *J. JSTP* **2016**, *57*, 879–885. [CrossRef]
12. Uematsu, K.; Tomizawa, A.; Shimada, N.; Mori, H. Development of Three-Dimensional Hot Bending and Direct Quench Using Robot, Journal of Advance Control. *Autom. Robot.* **2015**, *1*, 18–24.
13. Sumitomo Metals Develops High-Precision Drop Weight Impact Test Machine. Available online: https://www.nipponsteel.com/en/news/old_smi/2010/news2010-10-20-01.html (accessed on 25 September 2020).

Publisher's Note: MDPI stays neutral with regard to jurisdictional claims in published maps and institutional affiliations.

Article

Conductive Heating during Press Hardening by Hot Metal Gas Forming for Curved Complex Part Geometries

Mirko Bach [1,*], Lars Degenkolb [2], Franz Reuther [1], Verena Psyk [1], Rico Demuth [1] and Markus Werner [1]

1 Fraunhofer Institute for Machine Tools and Forming Technology IWU, 09126 Chemnitz, Germany; franz.reuther@iwu.fraunhofer.de (F.R.); verena.psyk@iwu.fraunhofer.de (V.P.); rico.demuth@iwu.fraunhofer.de (R.D.); markus.werner@iwu.fraunhofer.de (M.W.)

2 Salzgitter Hydroforming Verwaltungs GmbH, 08451 Crimmitschau, Germany; L.Degenkolb@szhf.de

* Correspondence: mirko.bach@iwu.fraunhofer.de; Tel.: +49-371-5397-1244

Received: 26 June 2020; Accepted: 14 August 2020; Published: 17 August 2020

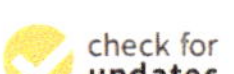

Abstract: Climate targets set by the EU, including the reduction of CO_2, are leading to the increased use of lightweight materials for mass production such as press hardening steels. Besides sheet metal forming for high-strength components, tubular or profile forming (Hot Metal Gas Forming—HMGF) allows for designs that are more complex in combination with a lower weight. This paper particularly examines the application of conductive heating of the component for the combined press hardening process. The previous Finite-Element-Method (FEM)-supported design of an industry-oriented, curved component geometry allows the development of forming tools and process peripherals with a high degree of reliability. This work comprises a description regarding the functionality of the tools and the heating strategy for the curved component as well as the measurement technology used to investigate the heat distribution in the component during the conduction process. Subsequently, forming tests are carried out, material characterization is performed by hardness measurements in relevant areas of the component, and the FEM simulation is validated by comparing the resulting sheet thickness distribution to the experimental one.

Keywords: tube hydroforming; lightweight structure; bending; formability; numerical methods; processing technology

1. Introduction

According to Yang [1], vehicles and road transportation produce more than 20% of greenhouse gas emissions. Here, the use stage causes approximately 85% of a passenger car's global warming potential [2]. In this phase, the vehicle weight is essential with regard to fuel consumption, therefore implementing lightweight design approaches is a key in order to reduce fuel consumption and CO_2 emissions, respectively. Yang [1] shows that the CO_2 emissions of new passenger cars have been steadily decreasing in the EU over the last 25 years. Nevertheless, additional effort is necessary in order to achieve the climate targets fixed by the EU, which include a 40% reduction of CO_2 by 2030 compared to the state of 1990 [3]. Aspects of material and design must be optimized in order to minimize component weight. Material optimization includes the application of typical lightweight materials such as aluminum and magnesium alloys or composites. However, economic aspects also have to be taken into account, especially in mass production. In addition, the application of high-strength steels, including press hardening steels, is superior to the other lightweight materials [4]. Weight optimized design strategies such as structuring [5] or functional integration usually result in very complex geometries. Thus, innovative forming technologies are needed that are capable of producing demanding geometries from hardly formable materials. In automotive production, especially in the manufacturing of chassis components, press hardening (which is also referred to as hot stamping) of sheet metal is a well-established process which allows for the manufacturing of high-strength components with minimal springback [6]. However, when realizing complex components, the use of tubular or profile shaped semi-finished parts frequently allows replacing several sheet metal parts in total, leading to lower material use and consequently resulting in lower weight and high component stiffness. Hydroforming is a suitable technology here [7]. At the same time, joining operations can be eliminated from the process chain so that often the result is a shorter and more cost-efficient process chain, see [8], where complex titanium parts are superplastically formed by HMGF in such a way that complex manufacturing with several joining operations can be saved. An innovative process investigated at the Fraunhofer Institute for Machine Tools and Forming Technology (IWU) combines the advantages of both press hardening and hydroforming. In this process, a tube or hollow profile made of the typical press hardening steels such as 22MnB5 is heated to austenitizing temperature, and then an inner gas pressure is applied so that the part expands in the die cavity. This process is the so-called Hot Metal Gas Forming (HMGF). As soon as it aligns to the cold dies, the part cools down rapidly resulting in a martensitic microstructure [9]. The dies are water-cooled in order to guarantee sufficiently high cooling rates in the process. The principal feasibility of HMGF combined with press hardening is shown in [9].

Similar to the press hardening processes with sheet metal, heating can be realized in a furnace or by flames [10]. In this case, it is necessary to transfer the hot component to the forming die. The most important disadvantage is a temperature decrease of up to 200 K. Bach et al. [11] have shown that components, which were heated up to 950 °C (Ac_3-point for usual press hardening steels) in the furnace, had a residual temperature of approx. 850 °C after transfer into the tool, and the forming temperature dropped further to 750 °C during the closing of the dies. This drop in temperature led to a reduction in the formability of the base material and thus to the impossibility of fully forming of complex component geometries due to previously occurring cracks. Furthermore, scaling occurs due to the exposure of the hot part to atmospheric conditions. Conductive heating of the component is an alternative to avoid this transfer and the related disadvantages [12]. The authors of [13] describe the combination of HMGF with integrated conductive heating and press hardening for straight parts. Bach et al. [11] apply it to a curved component made of Docol PHS 1800 by SSAB (official current name and abbreviation of the former name Svenskt Stål AB) [14] for the first time and reveal that the heating strategy must be adapted in order to achieve homogeneous temperature in the curved part. The possibility of forming extremely complicated geometries with a high degree of functional integration, as they are often required in the industrial sector, can, under certain circumstances, justify the effort of extended process time and more complex tools. Based on these described methods, the aim of the current paper is to provide deeper insight into the heating process. It describes the tooling, including the conductive heating equipment, presents the heating strategy and shows its influence on the temperature distribution in the part and on the resulting component properties, specifically hardness and distribution of wall thickness.

2. Forming Task

The FEM-Simulation and investigation of the HMGF process are based on a complex demonstrator part geometry shown in Figure 1. It was designed in order to operate in the maximum forming ranges of the material in trials. It features 66° bending, different representative cross-section geometries of vehicle components and secondary form elements frequently occurring in typical hydroforming parts. The input geometry for the hydroforming process is a pre-bent tube made of Docol PHS 1800 [14] from SSAB with an initial diameter of 57 mm and a wall thickness of 1.5 mm. Detailed information on the chemical compound of the used tube material is shown in Table 1.

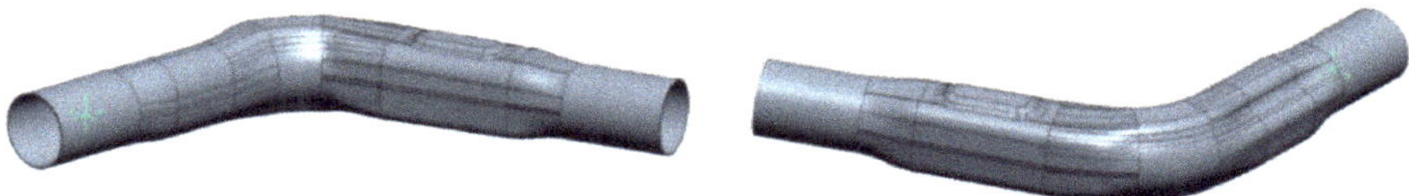

Figure 1. Demonstrator geometry.

Table 1. Chemical compound of Docol PHS 1800 by SSAB.

C (%)	Si (%)	Mn (%)	P (max%)	S (max%)	Cr (max%)	B (%)
0.27–0.33	0.15–0.35	1.00–1.45	0.025	0.010	0.35	0.0008–0.0050

According to [14], the HMGF process is expected to bring the material to a tensile strength of 1800 MPa at a failure strain of 6%. In the delivery condition, the material has a tensile strength of 500 MPa at an elongation of 27%. A thermomechanical forming simulation was carried out using LS-Dyna in order to estimate the principal feasibility of the component and to draw conclusions for necessary geometry adjustments. Here, models with all relevant boundary conditions and sub-steps were taken into consideration as well as earlier investigation results. For example, the test results with regard to pressure curves and forming, with which parts have already been successfully produced, see Figure 2 [11], were taken as a basis. This figure shows the deformation of the test part in relation to the pressure curve. As shown, the decisive shaping of the component is already completed at an internal pressure of <25 MPa.

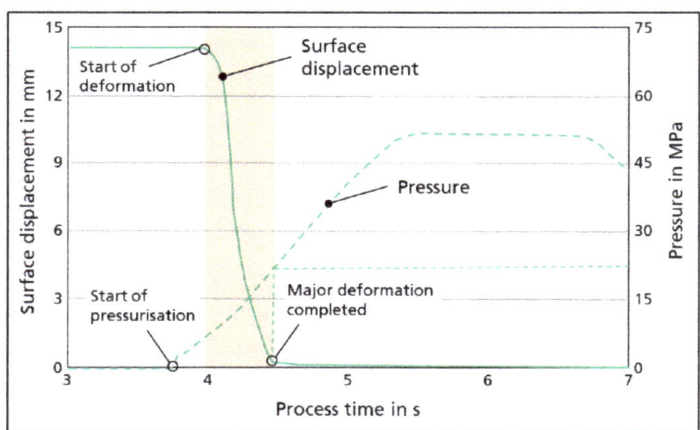

Figure 2. Forming via pressure curve in the Hot Metal Gas Forming (HMGF) process.

As usual for HMGF processes, also the tests within this study were carried out without axial feeding. The ends of the tube were fixed and simultaneously sealed by conical wedge elements. This sealing concept allows achieving high internal pressure. Approaches that work with axial feeding e.g., via a stepped sealing edge [15] the danger of leakage is high.

The achievement of the Ac_3 point of the tested material is mandatory for the calculation of the functionality of the process chain and the achievement of the desired material properties. In order to achieve a pragmatic and simplified representation of the HMGF process with sufficient accuracy, the simulation of the heating of the component has been omitted. Uniform temperature distribution in the component was assumed at the beginning of the thermomechanical-coupled simulation although during the experiments to heat the real component, locally different temperatures were expected due to its geometry. This circumstance is counteracted with the concept of pulsed power supply, see Section 4, in order to reach the Ac_3 point, in this case, 911 °C, at any point.

For the thermomechanical forming simulation, flow curve data was generated from tensile tests at seven different temperatures (950, 900, 850, 800, 750, 680, 600 °C) and for three different strain rates (0.5, 5, 50 s^{-1}). The individual derived and extrapolated flow curves were implemented into a temperature and strain rate-dependent isotropic material model in LS-DYNA.

In the simulation model, the tools were implemented in the form of rigid active surface meshes. In contrast, the tube was represented as an elastically plastically deformable shell with an initial element edge length of 0.75 mm and five integration points across the thickness. In the starting situation of the forming simulation, the tempered tube was positioned in the tool. The following sub-simulations up to the actual forming stage ensured representation of the process-specific boundary conditions in which the semi-finished product already cools down after conductive heating. This included the calculation of the time required to close the upper die, start up the sealing punches and apply the closing force onto the tool halves. Within these simulations, the heat transfer mechanisms heat conduction, heat transfer to air by convection and radiation as well as contact heat transfer to the cooled tool surfaces were taken into account based on [16]. In the subsequent simulation of the pressure generation phase, the forming of the semi-finished product in the tool cavity was realized by stepwise, linear application of a pressure load to the inner surface of the tube up to the target pressure level. A constant static coefficient of friction of $\mu = 0.35$ was assumed in the entire simulation steps, with which good experience has already been gained in previous work on the subject of tempered tube forming processes [11].

Results of the numerical simulation show that the circumferential expansion reaches values similar to a straight demonstrator regarded in earlier investigations [1], which could be successfully formed. Furthermore, the simulation predicts a minimum thickness of the tube material of approx. 0.85 mm after forming (Figure 3) which corresponds to a maximum thinning of 43.3%. This is also comparable with the successfully formed straight demonstrator mentioned above so that the currently regarded curved part can be expected to be feasible, too. However, it must be considered that the simulation disregards the strain results of the cold bending step, which is necessary before the HMGF process in order to allow positioning of the part in the hydroforming tool. This bending results in additional strain on the weld seam, but the corresponding impact on the resulting wall thickness distribution is expected to be small. This means that it was avoided to make the additional effort for implementing this preforming step in the numerical modeling for pragmatic reasons. Results of previous projects have shown that feasibility of hydroforming processes with pre-bent semi-finished parts will be the least affected if the weld seam is placed in the area of the neutral fiber and in zones featuring minor deformation during hydroforming. In the present case, this implies that the bottom of the component opposite of the dome is the most appropriate area for positioning the weld seam. Nevertheless, practical tests are indispensable in order to verify the simulation and to finally evaluate the feasibility.

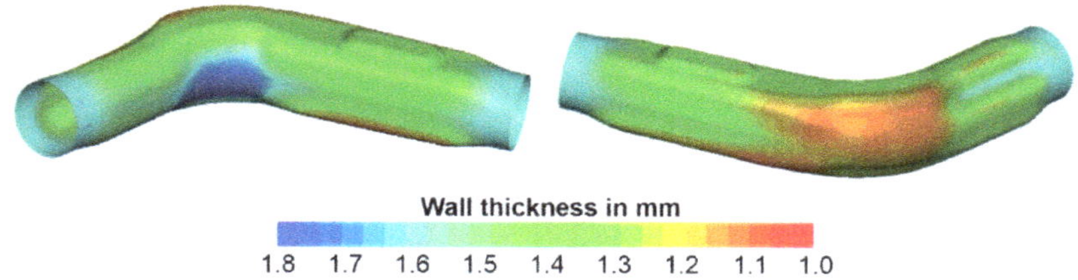

Figure 3. Numerically determined wall thickness distribution.

3. Tool Concept and Measurement Technique

For the experimental verification of the numerical simulation and the final proof of feasibility of the part, a tool was designed, which can be applied under conditions that are close to series production. As shown in Figure 4, this tool consists of

- The form elements providing the desired shape of the component with integrated cooling channels that are necessary to prevent accumulated heating of the dies over a number of tests; these elements are highlighted in Figure 4b,
- The electrodes for conduction heating, which are highlighted in Figure 4c,
- Guiding elements and force absorption elements, required for the functionality of the forming process, which are highlighted in Figure 4d,
- Axial punches, sealing the tube and applying the inner gas pressure, see overview in Figure 4a,
- An ejector that prevents the bending area from tilting into the tool engraving, see the green dot in the middle of the tool engraving in Figure 4c.

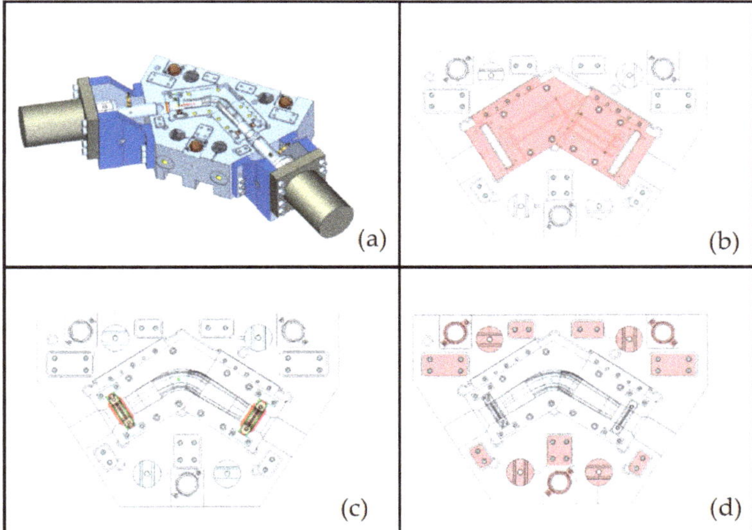

Figure 4. Bottom half of HMGF forming tool, CAD with details: (**a**) overview, (**b**) form elements, (**c**) ejector, (**d**) guiding and force absorption elements.

The electrodes are electrically insulated from the surrounding form inserts and spring-loaded so that they rise from the top and bottom when the tool is not completely closed. At the beginning of the process, the curved semi-finished part is clamped in these electrodes and lifted from the form inserts by the springs. Therefore, the part is electrically insulated from the metallic die elements so that the electrical current enters the tube at the electrode at one end and flows through the tube to the second electrode at the other end. The temperature of the tube rises due to resistive heating. As soon as the target temperature of the tube is reached and the tool halves start to get closed, the current is interrupted by an initiator. When the tool halves are closed, the axial punches seal the tube ends and the gas pressure is applied, leading to an expansion of the tube. All forming dies and electrodes are steadily water-cooled. In order to provide optimum cooling conditions, cooling channels must be located in direct proximity to the mold.

In general, the main application problem is gross cracking of the inserts due to insufficient toughness of the material if thermal shock occurs during the process. However, the mechanical strength of the tools and the leak tightness of the system must be guaranteed at the same time. A minimal distance between cooling channels and mold is required due to this fact. Extensive measurement technology was necessary for the thermal characterization of the component and process control. This measuring technology included a thermal camera, two pyrometers and type K thermocouples with direct measuring functions. The thermal camera (black in the lower part of Figure 5), whose orientation towards the component, provided detailed information about the local temperature distribution in the part at specific moments during the process. This information was used in the process analysis in order to characterize the overall heating behavior of the part and to identify relevant points of the workpiece, where more detailed information was needed and which were most suitable for process control. This measurement was complemented by the thermocouples and the pyrometers, which provided detailed information about the temperature as a function of the time for distinct points of the workpiece. The thermocouples served for analyzing the component heating behavior at six points in different regions of the part. However, due to the mechanical impact, this measurement technique turned out to be not suitable for reliable process control with the complete closing of the tool and the entire forming process. Therefore, contact-free optical measurement via pyrometers was used to control the conduction heating of the component at the two most relevant points.

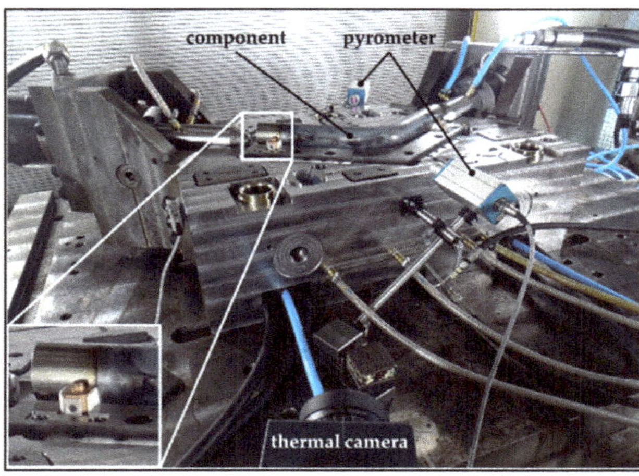

Figure 5. Bottom half of HMGF forming tool, experimental setup.

4. Heating Procedure

The thermal camera is directed towards the outer area of the pre-bent workpiece. Figure 6 shows the temperature of the part at different moments during the heating process (specifically after 30 s, left, after 50 s, middle) and the real component on the right.

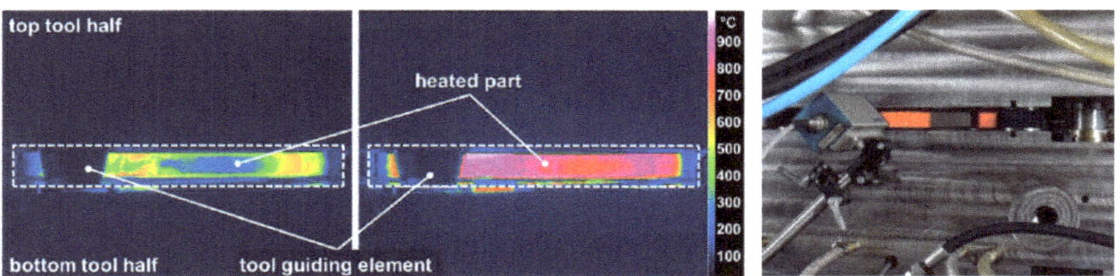

Figure 6. View of thermal camera after 30 s (l) and 50 s heating time (m), real heated component (r).

Obviously, only the middle section of the tube is visible since the component is partially covered by the tool halves that are not completely closed. In order to indicate the extent of the hidden sections, a dashed line represents the true contour of the part. After 50 s, large areas of the part reached the target temperature of 911 °C as assumed in the previous simulation. This especially concerns the straight end regions, while the temperature of the outer curved area is approx. 100 °C lower.

The explanation for this inhomogeneous temperature distribution is the current flow, which follows the shortest possible route through the component due to the smallest electrical resistance. Therefore, the current flow concentrates on the inner radius of the curved part. In order to counteract inhomogeneous temperature distribution and to achieve the required minimum temperature at any point, a suitable pulsed current was applied to the component. In the short time window without power supply in-between two subsequent pulses, areas that are relatively hot compared to their surroundings can dissipate warmth by heat flow into neighboring component areas, whereas regions that are relatively cool compared to their surroundings are heated by nearby hot areas. This results in the balancing of the temperature and in almost homogenous heating of the component to a targeted austenitization temperature of 911 °C. Measurement of the temperature distribution with attached thermocouples proved the suitability of this heating strategy as shown in Figure 7 without the HMGF-step. The thermal sensors 3, 4, 5 and 6 at the beginning and at the end of the pipe (top, bottom, inside and outside) are heated to similar temperature levels. The highest temperature is detected by sensor 1, which is positioned on the inner curve, i.e., directly in the area where the current flows and the resistive heat is generated. Due to the pulsing of the current, the lower temperature at the outer bend of the component (i.e., at thermal sensor 2) raises to austenitization temperature without overheating the other areas. When the thermal sensor at the inner bend (1) reaches the pre-set maximum temperature, the power supply from the conductor is automatically switched off, followed by an interruption of a few seconds. A straight orange line highlights the theoretical start of the HMGF-process after reaching Ac_3-point in all part areas. The conducting system has the following four adjustable parameters for controlling the pulsing current: maximum temperature, heating or pulse time, maximum current and pause time. The temperature is measured by the two mentioned pyrometers during the heating process, see Figure 5. Due to the inhomogeneous heating of the component, it was decided to focus one pyrometer onto the inner bend and the other onto the outer bend of the component in order to control the conductor. This was to ensure that the temperature in the faster-heating inner region of the part does not exceed a pre-set maximum temperature and that the outer region reaches the set minimum temperature. Figure 6, right, shows the glowing component in the almost closed tool during the heating process. As mentioned above, the top and the bottom sections of the part are hidden by the tool. First trials at Fraunhofer IWU were carried out to determine the optimized conduction heating process with the focus on reaching the A_{C3} temperature at any point of the component. Several tests have additionally been done with different currents. With lower current levels compared to the maximum of 2.500 A on the equipment side, only the heating time has been extended too much or the desired temperature has not been reached. Table 2 shows the final corresponding parameters characterizing the ideal heating process.

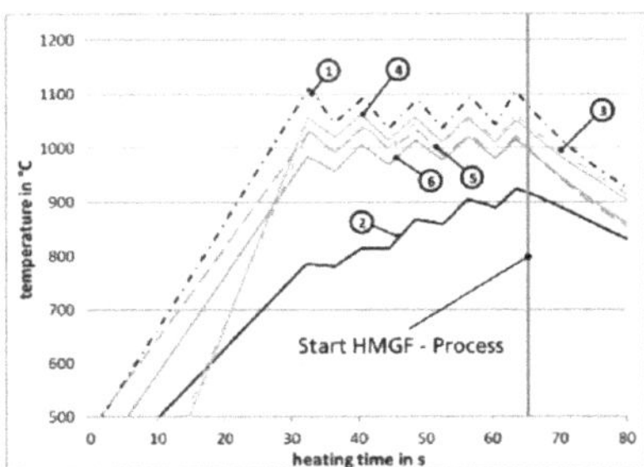

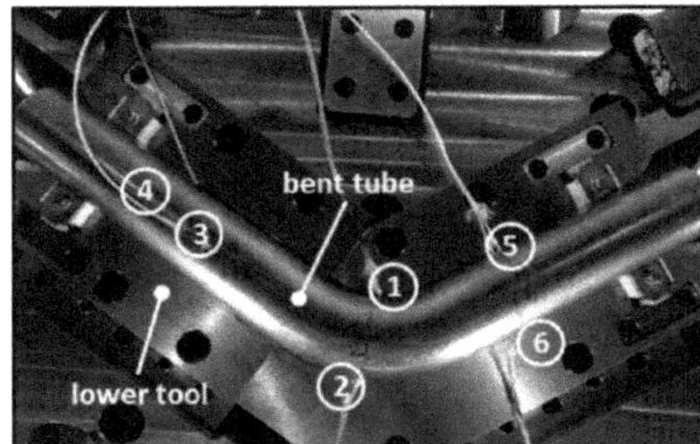

Figure 7. Heating curves during pulsing current; the numbers 1–6 indicate the positions of the thermal sensors.

Table 2. Times and temperatures for optimized component heating.

Target Temperature	Pulsed Current Amplitude	Heating Time	Maximum Temperature	Pause Time	No. of Cycles	Total Heating Time
980 °C	2.500 A	5 s	1070 °C	3 s	5	35 s

5. Forming Tests

The first tests aimed at demonstrating the feasibility of manufacturing curved, press-hardened components with conductively heated preforms. Figure 8 exemplarily shows the finished part.

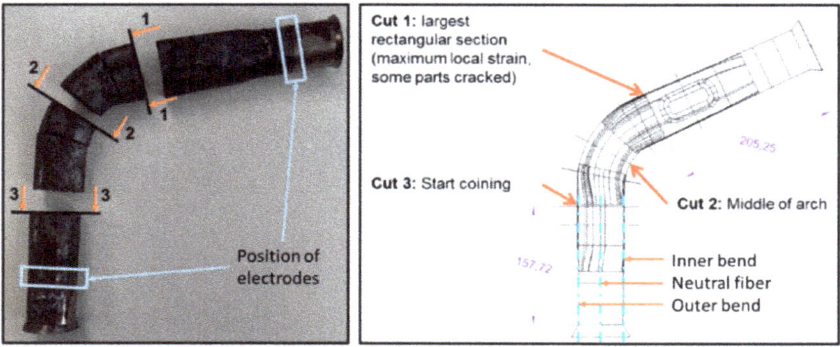

Figure 8. Component formed by HMGF and position of cuts.

This result serves as a first verification of the simulation and proves the importance of combined numerical and experimental feasibility evaluation described in Section 2 of this paper. The darkened area between the positions of contacted electrodes (blue squares), through which the current flew, was heated up to austenitization temperature Ac_3. A significantly reduced scaling on the surface is immediately noticeable due to the elimination of component transfer and the correspondingly reduced exposure of the part to atmospheric conditions. This is particularly remarkable as this part was manufactured without surface coating or protective gas. In hot forming processes, the sheet is usually protected against surface scaling by AlSi® or X-tec® coating. These are aluminium-based corrosion-protective and passivating coatings, which consist in the case of AlSi 85–95% of Aluminum and 5–11% Silicon [17], in the case of X-tec of Aluminum in a special binding matrix [18]. In the investigated case, the excellent surface quality achieved by the use of conductive component heating offers the potential of shortening the process chain and of reducing the environmental impact by avoiding aggressive chemicals. Full forming is reached with an internal pressure of 60 MPa. By using the technology of conductive heating, a constant starting temperature was guaranteed for all components manufactured.

6. Characterizing Hardness and Wall Thickness of the Component

In order to evaluate the quality of the manufactured component and to provide a more detailed verification of the numerical simulation, the parts were characterized considering the distribution of hardness and wall thickness. For this purpose, exemplary components were cut at the three positions marked in Figure 8. The hardness was measured according to DIN EN ISO 6507-1:2006-03 on the outer and inner bend and on the neutral fiber on the cross-sectional area, illustrated by blue dotted lines. Figure 9 shows the results.

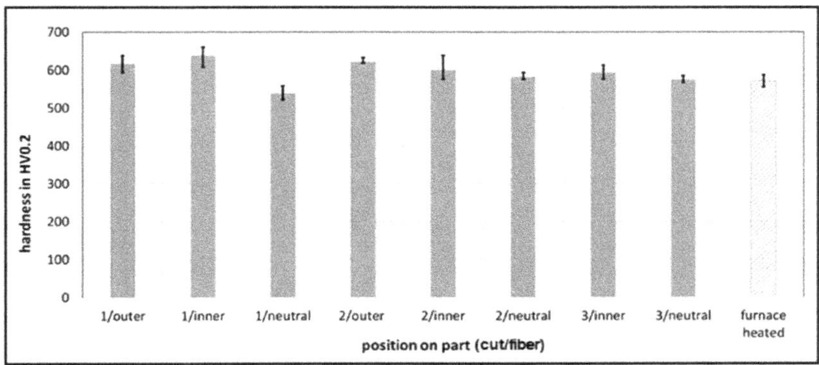

Figure 9. Comparison of hardness: part after conduction heating or furnace heating.

For comparison, the average hardness of a furnace-heated straight demonstrator of the same material is supplemented (striped bar in the diagram, Figure 8), compare DP3 in [11]. A minimum hardness of 540 HV was verified for all component areas. When converted according to DIN EN ISO 18,265, this value approximately corresponds to a tensile strength of 1775 MPa. This shows that the newly developed process allows for the design and heating procedure HMGF with integrated conductive component heating to be adapted to bent component geometries. Furthermore, the wall thickness of the manufactured component was measured along a circumferential path at the sections 1-1, 2-2 and 3-3 as shown in Figure 8. Figure 10 compares the result of the measurement with the simulation results. Each section is shown on the right with an initial arrow for the start of the measuring path. The curves at the different measurement positions feature different lengths since the cross-sections, and consequently, the circumferences, differ from each other. A good qualitative agreement exists between the respective simulation and real curves. In large parts, the quantitative agreement is also acceptable. Locally there are significant deviations in a range of about 30%, see for example curve 1-1, which may be caused by the inaccurate representation of the friction conditions at the high forming temperatures, which may be assumed as too low, and by the assumption regarding heat transfer. Furthermore, the results of the real tests are strongly influenced by the quality of the semi-finished product with strongly varying wall thickness in some places due to the manufacturing process.

The local thinning is suspected to be caused by the manufacturing process of the test tubes. These were produced, not as usual by roll forming, but because of their shorter dimensions by U-O-bending with prototype tools. Manufacturing deviations led to the formation of heels at radius transitions at the tool parts and thus to thinning of the raw sheet material while bent into tubes. If the material used for the pipes is established, it can be assumed that the quality will be improved in terms of wall thickness distribution.

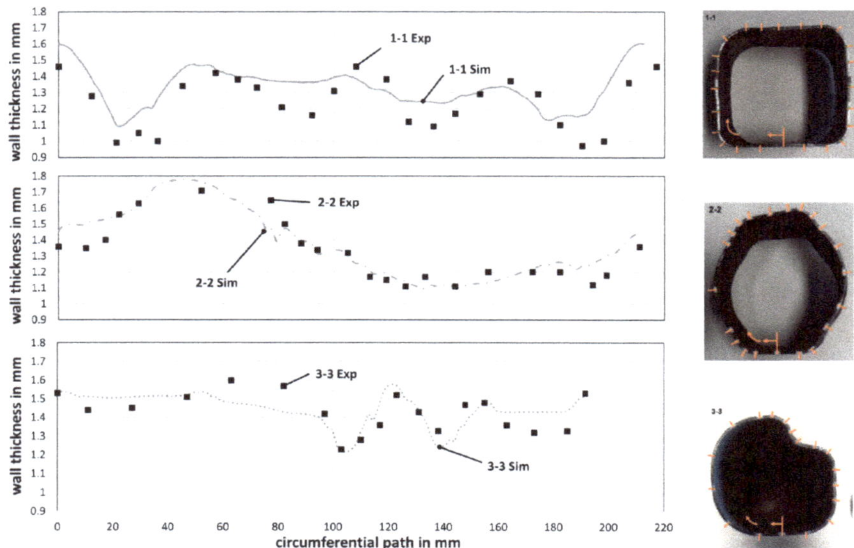

Figure 10. Wall thickness validation for the three sections, experiment (Exp) and simulation (Sim).

7. Summary

Based on earlier results of tests with tool-integrated component heating, a more complex production tool was designed with integrated conduction components. A pulsed current flow was successfully tested for heating and was used for the HMGF of complex bent parts in order to guarantee the complete heating of the part to the austenitization temperature without local overheating due to the curvature of the component and due to the correspondingly inhomogeneous current distribution in the part during conductive heating. The eliminated step of transferring heated preforms into the forming tool allows for a significant reduction in component scaling and ensures the same robust condition for cooling, thus influencing the material properties such as the desired minimum hardness and overall strength of the component areas. Furthermore, the additional technological expense of the integrated conduction device for medium component quantities can be justified since additional process steps for coating are avoided and no environmentally harmful chemicals are used. Finally, the experimental verification of the simulation results was realized by examining the sheet thickness curves in three-component sections. As a result, the FEM simulation for HMGF processes on curved components can be confirmed as a reliable design tool.

Author Contributions: Conceptualization, M.B., M.W. and V.P.; methodology, M.B. and V.P.; validation, M.B., F.R. and R.D.; investigation, F.R., L.D. and M.B.; resources, L.D. and M.B.; writing—original draft preparation, M.B., L.D. and F.R.; writing—review and editing, M.B. and V.P.; visualization, M.B., L.D., F.R. and R.D.; supervision, V.P.; project administration, V.P.; funding acquisition, M.W. All authors have read and agreed to the published version of the manuscript.

Funding: The content of this paper was developed within the framework of the following publically funded project: The EU-project "Development of energy-efficient press hardening processes based on innovative sheet and tool steel alloys and thermo-mechanical process routes" with the Grant Agreement No. RFSR-CT-2015-00019 was publically funded by the European Commission, Research Fund for Coal and Steel, Technical Group TGS 7. Fraunhofer IWU would like to extend its thanks to its project partners for their cooperation and support in this project.

Conflicts of Interest: The authors declare no conflict of interest.

References

1. Yang, Z. Overview of Global Fuel Economy Policies. In Proceedings of the 2018 APCAP Joint Forumand Clean Air Week Theme: Solutions Landscape for Clean Air, Bangkok, Thailand, 20 March 2018.
2. Koffler, C. Automobile Produkt-Ökobilanzierung. Ph.D. Thesis, Technische Universtität Darmstadt, Darmstadt, Germany, 2007.
3. European Commission. Communication from the Commission to the European Parliament, the Council, the European Economic and Social Committee and the Committee of the Regions. 2014. Available online: https://eur-lex.europa.eu/legal-content/EN/TXT/PDF/?uri=CELEX:52014DC0015&from=EN (accessed on 15 July 2020).
4. Taub, A.I.; Luo, A.A. Advanced lightweight materials and manufacturing processes for automotive applications. *MRS Bull.* **2015**, *40*, 1045–1054. [CrossRef]
5. Psyk, V.; Kurka, P.; Kimme, S.; Werner, M.; Landgrebe, D.; Ebert, A.; Schwarzendahl, M. Structuring by electromagnetic forming and by forming with an elastomer punch as a tool for component optimisation regarding mechanical stiffness and acoustic performance. *Manuf. Rev.* **2015**, *2*, 23. [CrossRef]
6. Karbasian, H.; Tekkaya, A.E. A review on hot stamping. *J. Mater. Process. Technol.* **2010**, *210*, 2103–2118. [CrossRef]
7. Bell, C.; Corney, J.; Zuelli, N.; Savings, D. A state of the art review of hydroforming technology. *Int. J. Mater. Form.* **2019**, *151*, 1–40. [CrossRef]
8. Paul, A.; Werner, M.; Trân, R.; Landgrebe, D. Hot metal gas forming of titanium grade 2 bent tubes. In *AIP Conference Proceedings*; AIP Publishing LLC: Melville, NY, USA, 2017; Volume 1896, p. 050009.
9. Paul, A.; Reuther, F.; Neumann, S.; Albert, A.; Landgrebe, D. Process simulation and experimental validation of Hot Metal Gas Forming with new press hardening steels. *J. Phys. Conf. Ser.* **2017**, *896*, 12051. [CrossRef]
10. Talebi-Anaraki, A.; Chougan, M.; Loh-Mousavi, M.; Maeno, T. Hot Gas Forming of Aluminum Alloy Tubes Using Flame Heating. *JMMP* **2020**, *4*, 56. [CrossRef]
11. Bach, M.; Degenkolb, L.; Reuther, F.; Mauermann, R.; Werner, M. Parameter measurement and conductive heating during press hardening by hot metal gas forming. In Proceedings of the 38th IDDRG Conference Enschede, Enschede, The Netherlands, 3–7 June 2019.
12. Reuther, F.; Mosel, A.; Freytag, P.; Lambarri, J.; Degenkolb, L.; Werner, M.; Winter, S. Numerical and experimental investigations for hot metal gas forming of stainless steel X2CrTiNb18. *Procedia Manuf.* **2019**, *27*, 112–117. [CrossRef]
13. Maeno, T.; Mori, K.-I.; Unou, C. Improvement of Die Filling by Prevention of Temperature Drop in Gas Forming of Aluminium Alloy Tube Using Air Filled into Sealed Tube and Resistance Heating. *Procedia Eng.* **2014**, *81*, 2237–2242. [CrossRef]
14. N.N. Docol PHS 1800 Presshärtender Stahl: Härtbarer Borstahl Für Die Automobilindustrie. Available online: https://www.ssab.de/produkte/warenzeichen/docol/products/docol-phs-1800?accordion=downloads (accessed on 15 July 2020).
15. Wu, Y.; Liu, G.; Wang, K.; Liu, Z.; Yuan, S. The deformation and microstructure of Ti-3Al-2.5V tubular component for non-uniform temperature hot gas forming. *Int. J. Adv. Manuf. Technol.* **2016**, *88*, 2143–2152. [CrossRef]
16. Shapiro, B. *Using LS-Dyna for Hot Stamping*; Dynamore, G.H., Ed.; LSTC: Livermore, CA, USA, 2009.
17. NANO-X GmbH. AlSi Coat 4001: Product Information. Available online: www.nano-x.de (accessed on 15 July 2020).
18. NANO-X GmbH. Verarbeitungsvorschrift X-Tec. Available online: www.nano-x.de (accessed on 15 July 2020).

Article

Deformation Property and Suppression of Ultra-Thin-Walled Rectangular Tube in Rotary Draw Bending

Kunito Nakajima [1], Noah Utsumi [2,*], Yoshihisa Saito [3] and Masashi Yoshida [4]

1 HVAC Systems Production Engineering Department, Marelli Corporation, 8 Sakae-cho, Sano, Tochigi 327-0816, Japan; kunito.nakajima@marelli.com
2 Faculty of Education, Saitama University, 255 shimo-Okubo, Sakura-ku, Saitama 338-8570, Japan
3 Graduates of Faculty of Education, Saitama University, 255 shimo-Okubo, Sakura-ku, Saitama 338-8570, Japan; y.saito.edu@gmail.com
4 Department of Advanced Science and Technology, Daido University, 10-3 Takiharu-cho, Minami-ku, Nagoya 457-0819, Japan; myoshida@daido-it.ac.jp
* Correspondence: utsumi@mail.saitama-u.ac.jp; Tel.: +81-48-858-9157

Received: 17 July 2020; Accepted: 7 August 2020; Published: 10 August 2020

Abstract: Recently, miniaturization and weight reduction have become important issues in various industries such as automobile and aerospace. To achieve weight reduction, it is effective to reduce the material thickness. Generally, a secondary forming process such as bending is performed on the tube, and it is applied as a structural member for various products and a member for transmitting electromagnetic waves and fluids. If the wall thickness of this tube can be thinned and the bending technology can be established, it will contribute to further weight reduction. Therefore, in this study, we fabricated an aluminum alloy rectangular tube with a height H_0 = 20 mm, width W_0 = 10 mm, wall thickness t_0 = 0.5 mm (H_0/t_0 = 40) and investigated the deformation properties in the rotary draw bending. As a result, the deformation in the height direction of the tube was suppressed applying the laminated mandrel. In contrast, it was found that the pear-shaped deformation peculiar to the ultra-thin wall tube occurs. In addition, axial tension and lateral constraint were applied. Furthermore, the widthwise clearance of the mandrel was adjusted to be bumpy. As a result, the pear-shaped deformation was suppressed, and a more accurate cross-section was obtained.

Keywords: ultra-thin walled tube; tube bending; laminated mandrel; rotary draw bending; Finite Element Analysis (FEA); deformation property

1. Introduction

Tubes are subjected to secondary forming processing such as bending and used as parts in the automobile and aerospace industries. To bend the tubes, press bending, rotary draw bending, and other bending methods have been developed and applied to actual product production. However, during bending of a thin-walled tube with a space in the cross-section, flattening, thickness deviation, wrinkling, folding, and other undesirable distortions occur [1–3]. Generally, it is effective to apply a mandrel to suppress undesirable distortions. A ball-type mandrel is often used [4–8], and the results of investigating the effects of the clearance between the pipe inner diameter and the mandrel on flattening [9,10] and wrinkling [11] have been reported. Moreover, there have been reports on the effect of mandrels on springback [12]. In addition, there have been reports on the effect of the friction coefficient of the ball mandrel on undesirable distortions and analysis of the stress state of the mandrel during bending [13–16]. In addition, other studies include the following: A study using a chain-link type mandrel, which is cheaper than a ball type one [17], a mandrel that combines hard and soft rubber [18], and a study that suppresses cross-sectional deformation by applying fluid pressure inside the pipe [19]. Moreover, the effects of mandrels on various materials such as steel, aluminum alloys, copper alloys, titanium, and high tensile strength materials have been reported [20–23].

However, these studies are mainly the results of investigations of circular pipes and materials with a wall thickness of 1.0 mm or more [24,25]. In contrast, there has been little research on clarifying and suppressing the cross-sectional deformation phenomenon during bending of a rectangular tube with a rectangular cross-section and ultra-thin materials with a wall thickness of 1.0 mm or less. If ultra-thin tubes can be applied to various components in the automobile and aerospace industries, it is expected that they will contribute significantly to the reduction of size and weight, which is a common issue in each industry in recent years. Among them, the rectangular tube is considered for application in the waveguide, which is a component for electromagnetic wave propagation in the aerospace industry. However, in order to improve the electromagnetic wave propagation efficiency, it is necessary to minimize the cross-sectional deformation after bending. We investigated the effect of a mandrel and a restraint plate on undesirable distortions in the cross section of an extruded square tube on press bending [26]. Furthermore, we also reported that wrinkling in rotary draw bending can be suppressed by axial tension, and that cracking is considerably affected by the bending radius and material. In addition, it has been clarified that pear-shaped cross-section deformation occurs when a mandrel is applied during the bending of ultra-thin tubes [27,28].

Therefore, in this study, in order to contribute to further weight reduction of the tubular material application parts in each industry by rendering the tubular material thinner, we investigated the deformation property during rotary draw bending using an ultra-thin aluminum alloy rectangular tube. As a result, the pear-shaped deformation peculiar to the ultra-thin rectangular tube was confirmed. The suppression method was then investigated. Specifically, axial tension was used to suppress wrinkling due to bending. A restraint jig was used to suppress convex distortion at the sides. To suppress the flattening, we applied a laminated mandrel to adjust the clearance between the tube shape and the mandrel. As a result, it was found that the deformation of the cross-section peculiar to the ultra-thin rectangular tube can be suppressed by controlling axial tension, applying side restraint, and adjusting the widthwise clearance of the mandrel on the compression and the tension sides of the tube.

2. Materials and Methods

2.1. Workpiece

The workpiece used in the experiments and simulations was an aluminum alloy with annealing (A6063-O). Table 1 shows the mechanical properties, and Figure 1 shows the shape and dimensions of the workpiece. The dimensions of the cross-sectional shape were set as follows: height H_0 = 20, 10 mm; width W_0 = 10, 20 mm; and thickness t_0 = 0.5 mm. The waveguide, which is an example of the application range of ultra-thin tubes, has a standard regarding the material, but the rectangular tube with a wall thickness of 0.5 mm applied in this study is not part of this standard. Moreover, it is not marketed because there is a limit to thinning in extrusion. Therefore, in this study, a commercial extruded tube was processed by drawing to achieve a wall thickness of 0.5 mm and applied to the experiment.

Table 1. Mechanical properties of the workpiece.

A6063-O	
Tensile Strength σ_B/MPa	91
Proof Stress $\sigma_{0.2}$/MPa	39
Elongation δ/%	24.3
Work-hardening Exponent n *	0.27
Plastic Modulus C */MPa	160

* Refer to JIS Z2201 $\sigma = C\varepsilon^n$.

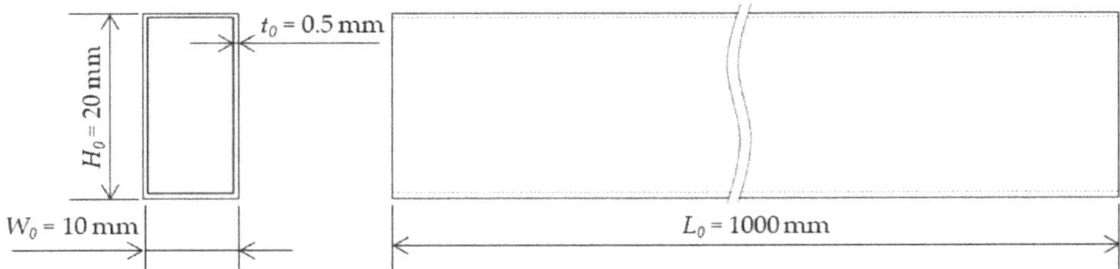

Figure 1. Shape and dimension of the workpiece.

2.2. Bend Radius

As shown in Figure 2, the bending radius R is the bending drum radius; more precisely, the inside of the rectangular tube is the bending radius. There are two types of rectangular tubes with outer height H_0 = 20, 10 mm, and thickness t_0 = 0.5 mm (H_0/t_0 = 40, 20).

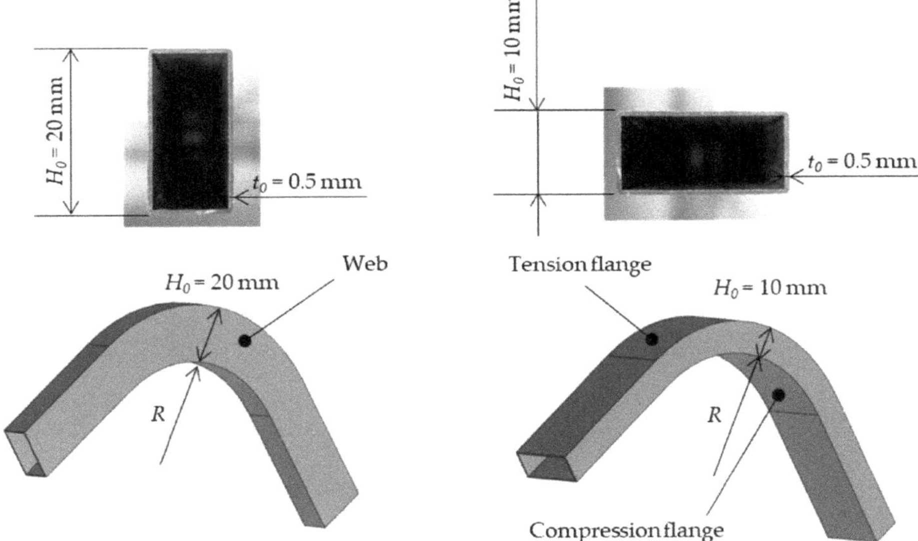

Figure 2. Bending radius and bending height.

2.3. Rotary Draw Bending System

Figure 3 shows a schematic of the rotary draw bending machine, and Figure 4 shows a schematic of the tool alignment of rotary draw bending. This device has a mechanism with a chuck that fixes the material, a bending drum that determines the bending radius, and a bending load rod that generates a bending moment. In addition, it is also equipped with a structure that can apply axial tension. The bending drum can be selected from those with a radius R of 20 to 200 mm, and in this experiment, the workability R/H_0 = 1, 2.5. Moreover, a plate made of MC nylon was installed as a side restraint jig.

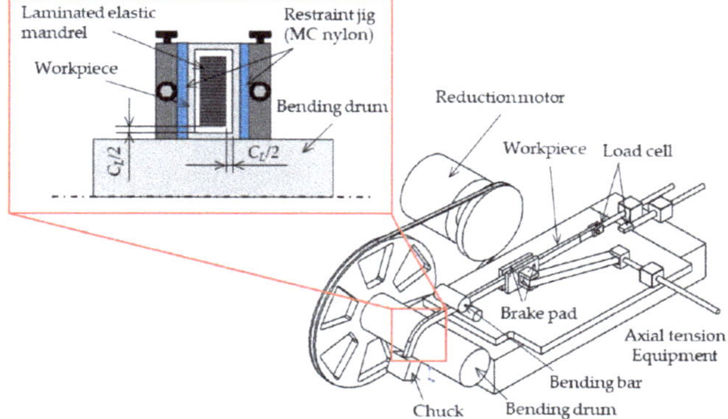

Figure 3. Schematic representation of the rotary draw bending device.

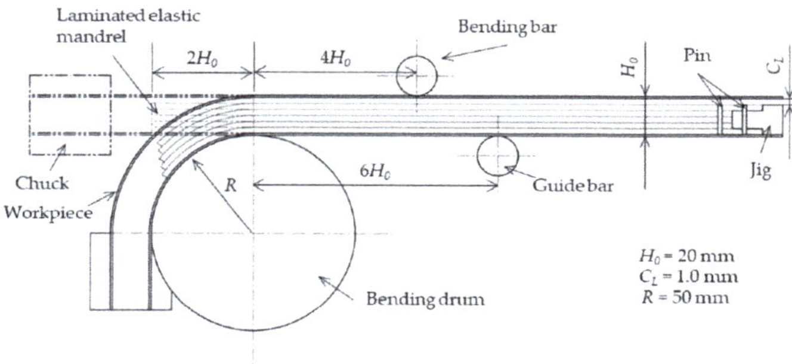

Figure 4. Tool alignment of rotary draw bending.

2.4. Mandrel

Figure 5 shows a schematic of the mandrel. The core material has a structure in which 0.01 mm to 2.0 mm PVC (polyvinyl chloride) and stainless-steel thin plates are bundled by a jig. The flexural rigidity of the laminated mandrel was adjusted to minimize the adverse effect on the workpiece during bending. In addition, the clearance with the workpiece was adjusted to C_L = 1.0 mm in both width and height.

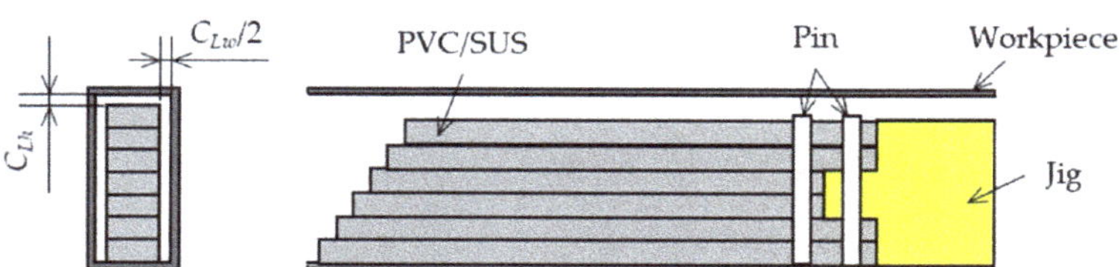

Figure 5. Schematic of the laminated elastic mandrel.

2.5. Finite Element Analysis (FEA) Model

Figure 6 and Table 2 show a schematic of the Finite Element Analysis (FEA) model and the boundary conditions of FEA. LS-DYNA 3D (Ver.R10.1.0, ANSYS Inc. Canonsburg, PA, USA), a commercially available finite explicit element analysis software package, was used for the simulation. The laminated mandrel is a solid element and the rest are shell elements. The mesh size was 2 mm and the number of elements of the workpiece was 17,250. The friction coefficient between the inner surface of the workpiece and the mandrel was set to 0.01 on the assumption that there was almost no friction because the lubricant was applied. In addition, the workpiece was analyzed as an elasto-plastic (Von-Mises) according to the exponentiation hardening rule, the mandrel as an elastic body, and the rest as a rigid body. The workpiece in the model was assumed to be an isotropic elastoplastic body with the power law hardening rule. The constitutive equation is given as Equation (1).

$$\sigma = C\left(\varepsilon_{yp} + \overline{\varepsilon}^{p}\right)^{n} \quad (1)$$

where, ε_{yp} is an elastic strain and $\overline{\varepsilon}^{p}$ is an equivalent plastic strain. Moreover, the elastic strain is calculated by the following Equation (2).

$$\varepsilon_{yp} = (E/C)^{(1/(n-1))} \quad (2)$$

where, E is Young's modulus. Accordingly, the constitutive equation of the workpiece can be expressed by Equation (3).

$$\sigma = 160(0.002424 + \overline{\varepsilon}^{p})^{0.27} \quad (3)$$

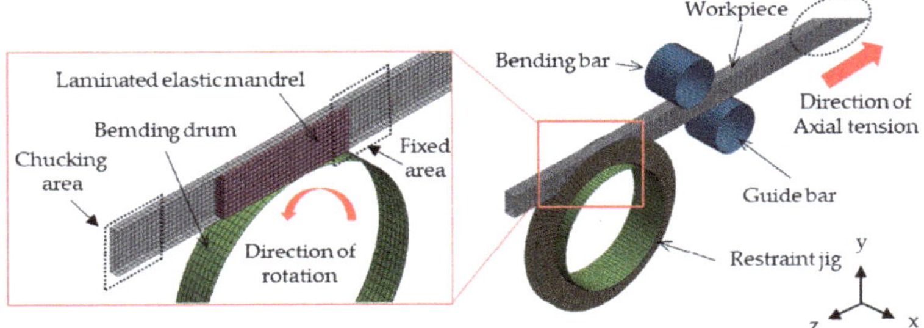

Figure 6. Schematic representation of the simulation model in rotary draw bending.

Table 2. Boundary conditions of Finite Element Analysis (FEA).

	Workpiece	Mandrel	Bending Drum	Restraint Jig	Bar
Element Type	**Elasto-Plastic/ShellElastic/Solid**		**Rigid/Shell**	**Rigid/Shell**	**Rigid/Shell**
Number of Elements	17,250	1200	1515	H_0 = 10:1010 H_0 = 20:2020	765
Restraint condition	free	free *	90° rotation	fixed	fixed

* Fixed area of mandrel.

3. Results and Discussion

3.1. *The Mechanism of Cross-Sectional Deformation*

Figures 7 and 8 show the mechanism of cross-sectional deformation. Here, M is the Bending moment, σ_t is the tensile stress, σ_c is the compression stress, P_T is the circumferential force of the tension side in the flattening, P_C is the circumferential force of the compression side, P_{TN} is the flattening component of the tension side, and P_{CN} is the flattening component of the compression side. The flange falls due to the component force generated in the compression flange and the tension flange by the bending moment. Subsequently, due to the moment generated by the component force, the component force acts on the web, resulting in flattening. The purpose of applying the mandrel is to restrain the component force P_{TN} generated in the compression flange and the tension flange. In addition,

P_{TN} is expressed as Equations (4) and (5), using bending stress σ_{TE} in the elastic state. However, it is qualitatively the same as that in plastic region.

$$P_{TN} = 2\sigma_{TE}t_0W_0\sin\left(\frac{d\theta}{2}\right) \fallingdotseq \sigma_{TE}t_0W_0d\theta = \frac{t_0W_0EH_0d\theta}{2\rho} \tag{4}$$

$$dx_T = \left(\rho + \frac{H_0}{2}\right)\cdot d\theta \fallingdotseq \rho d\theta \tag{5}$$

where t_0 is the thickness of the workpiece, W_0 is the width of the workpiece, dx_T is the length between points A and C on the tension flange, E is the Young's modulus, and ρ is the bending degree ($R + H_0/2$) [29].

Distributed flattening force w_{TN} on the tension flange is expressed as Equation (6), using Young's modulus, E, in the elastic state. In the plastic state, the tangent modulus E_t is used in Equation (7).

$$w_{TN} = \frac{P_{TN}}{dx_TW_0} = \frac{H_0t_0E}{2\rho^2} \tag{6}$$

$$E_t = \frac{d\sigma}{d\varepsilon} = C\varepsilon^n \tag{7}$$

Figure 9 shows the graph of w_{TN} calculated under various conditions from Equation (4). It demonstrates that the w_{TN} increased as the wall thickness increased; thus, flattening was more likely to occur. Additionally, w_{TN} increases as ρ/H_0 decreases. In this study, $R/H_0 = 1.5$ ($H_0 = 20$, $R = 50$), and $t_0 = 0.5$ mm. Under this condition, the w_{TN} is small because the wall thickness is thin, but ρ/H_0 is also smaller than that of a square tube with a general wall thickness; accordingly, so the w_{TN} is larger and the deformation after bending is expected to be larger. Furthermore, when the axial tension P_a is applied, P_{TN} (w_{TN}) is further increased because P_a is added to P_T.

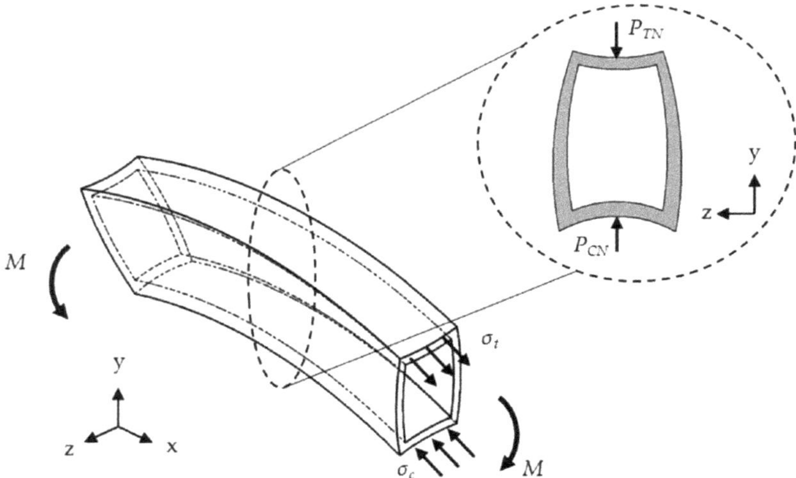

Figure 7. Schematic of flattening distortion of a square tube.

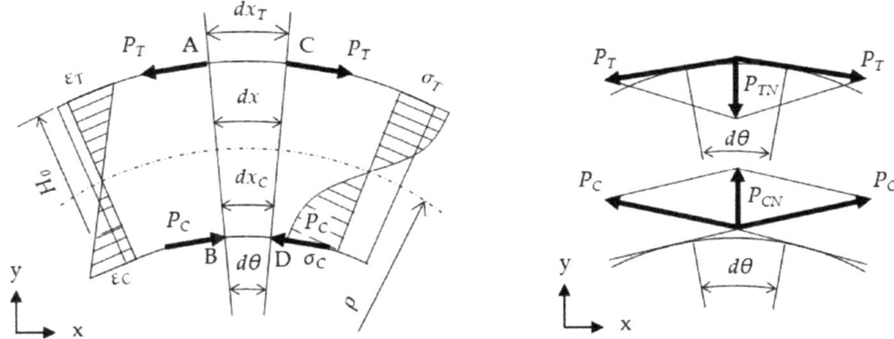

Figure 8. Mechanism of flattening components.

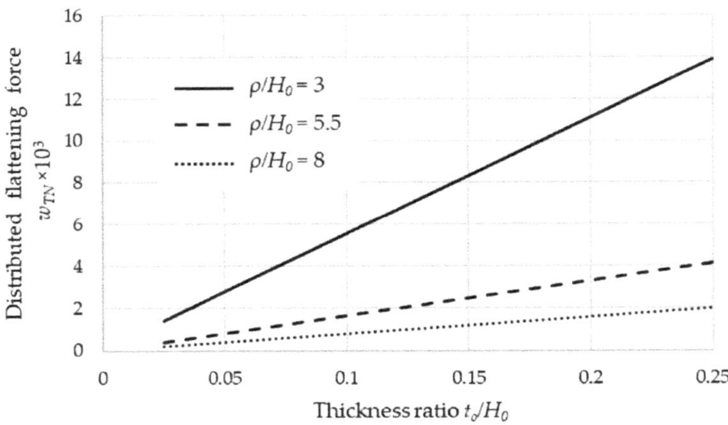

Figure 9. Relationship of flattening component of the tension side and thickness.

3.2. Method for Evaluating Deformation

The deformation ratio was defined as $(H-H_0)/H_0$, $(W-W_0)/W_0$ (measured values H, W; initial values H_0, W_0), and the cross-sections before and after processing were compared. In addition, the maximum height and width of the workpiece are H_{max} and W_{max}, and the minimum workpiece width is W_{min}. Furthermore, the load factor of axial tension was defined and evaluated as Pa/F_{max} (axial tension Pa, tensile strength F_{max} of the workpiece).

3.3. Effect of the Laminated Mandrel and Axial Tension

Figure 10 shows the appearance of the workpiece after rotary draw bending without a material, restraint jig, and axial tension. At $R/H_0 = 1.5$ ($H_0 = 20$, $R = 50$), folding occurred, and at $\mathbf{R}/H_0 = 5.0$ ($H_0 = 10$, $R = 50$), wrinkling occurred. In addition, Figure 11 shows the effect of axial tension. As shown in Figure 11, the wrinkling can be suppressed by axial tension. In contrast, the fall of the flange became large. The axial tension reduced the compression region of the tube, suppressed buckling deformation, and suppressed wrinkling. However, the component force P_{TN} increased as the tensile force increased, causing the tensile flange to fall greatly, resulting in a large expansion of the web. From this result and Equation (6), it was found that the following factors increased influence the cross-section deformation:

- Axial tension
- $1/R$
- Cross-section outline
- Thickness
- Work-hardening exponent
- Plastic modulus (Young's modulus)

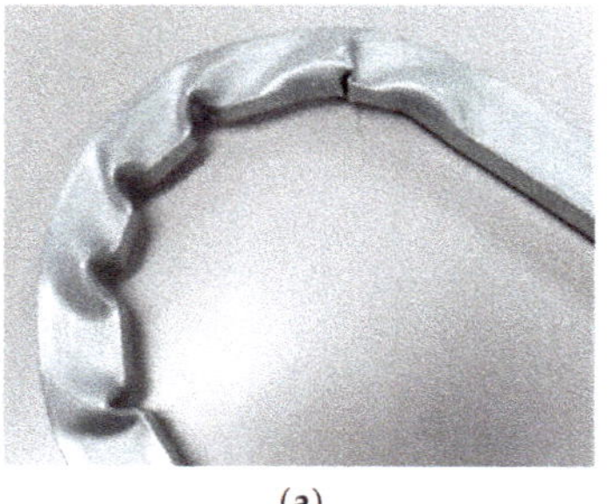

(a)

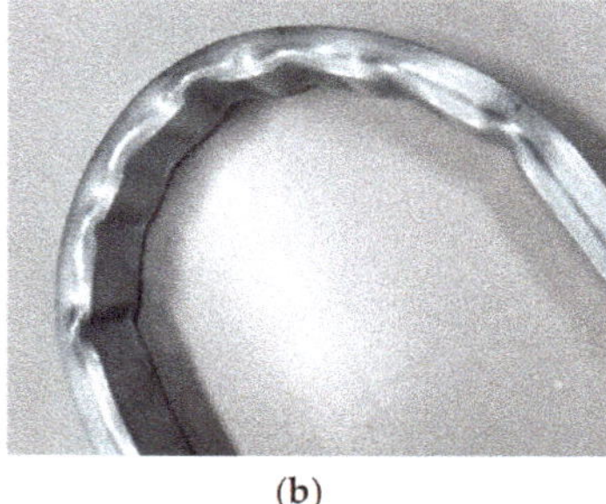

(b)

Figure 10. Result of rotary draw bending (Without a mandrel and axial tension); (**a**) Folding (H_0 = 20, R = 50); (**b**) Wrinkling (H_0 = 10, R = 50).

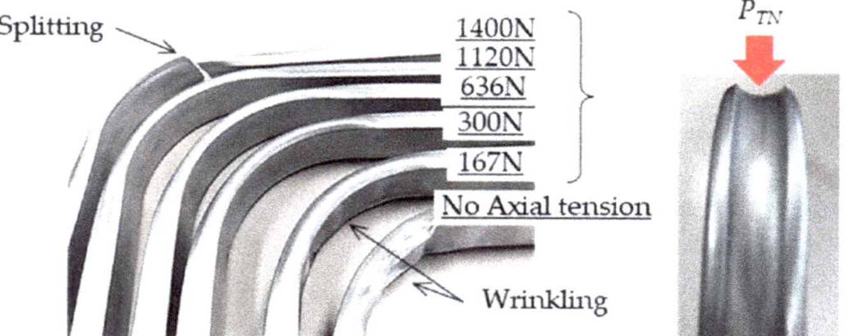

Figure 11. Effect of axial tension (H_0 = 10, R = 50).

Figure 12 shows the effects of the mandrel and axial tension. R/H_0 = 2.0 (H_0 = 10, R = 20) was possible bending without wrinkling and splitting, whereas splitting occurred in the condition of R/H_0 = 1.0 (H_0 = 20, R = 20). Wrinkling and splitting did not occur at R/H_0 = 1.5 (H_0 = 20, R = 50). In contrast, the cross-section deformation was large for both H_0 = 10 and 20. Especially at R/H_0 = 1.5 (H_0 = 20, R = 50), a pear-like deformation occurred in which the deformation of the web side enlarged. Considering this deformation as buckling, and considering one side of the web as buckling of a long column, the end on the tension side was constrained and the end on the compression side is constrained to rotate, which is similar to the deflection of a beam. In addition, the effect of the mandrel was confirmed by the small deformation of both the tension and compression side flanges.

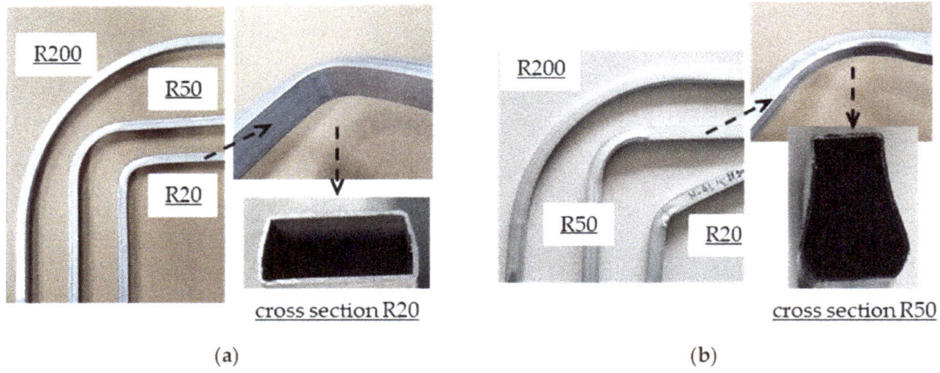

(a) (b)

Figure 12. Effect of axial tension and the mandrel; (**a**) H_0 = 10; (**b**) H_0 = 20.

Additionally, the convex distortion of the web is larger when H_0 = 20 than when H_0 = 10. It is assumed that this is because the mandrel was compressed by P_{TN} by axial tension and deflected to the web. In this experiment, the number of laminated mandrels with H_0 = 20 was larger than that with H_0 = 10. This is because by lowering the bending rigidity of the mandrel, it is possible to prevent adverse effects such as splitting and convex distortion in the tensile flange of the tube. Therefore, as the number of laminated layers increases, it is possible that minute voids are generated between the plates. It is considered that the voids were compressed by P_{TN} and became substantially larger than the set clearance, resulting in larger deformation than H_0 = 10.

Furthermore, Figure 13 shows the results of FEA. The same phenomenon as in the experiment was confirmed in FEA. Therefore, it was confirmed that the pear-shaped deformation is a deformation peculiar to the ultra-thin rectangular tube. The analysis conditions were R/H_0 = 1.5 (H_0 = 20, R = 50), the mandrel was applied, and axial force was applied. To reduce the analysis time, the laminated mandrel in FEA has a smaller number and a larger thickness than in the experiment. The laminated plate is an elastic body, and Young's modulus was set small to match the moment of inertia with the experiment. Thus, the deformation of the mandrel in the plate thickness direction was larger than in the experiment, resulting in depression or convex distortion in the tension and compression flanges.

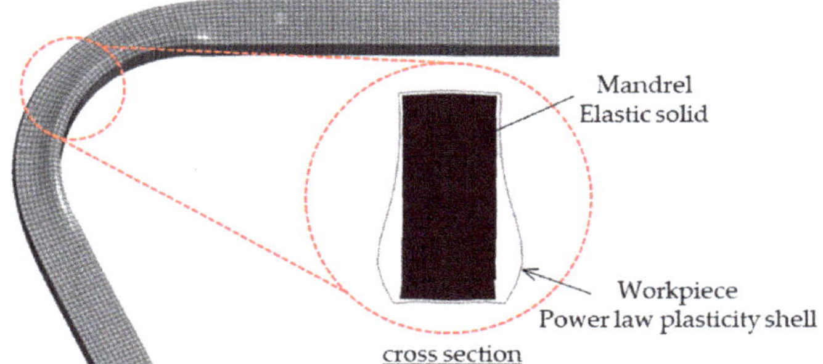

Figure 13. Result of FEA; (H_0 = 20, R = 50, using the mandrel and axial tension).

3.4. Effect of Restraint Jig

An experiment was conducted by applying a jig that restrains the web surface in order to suppress the outward convex distortion of the web. Figure 14 shows a schematic of the jig and FEA model. In bending with H_0 = 10, the height of the restraining ring was 10 mm. At H_0 = 20, the deformation of the compression side is particularly large. Therefore, two types of conditions were applied: one for constraining only the compressed part of the web and one for constraining the entire web. Namely, there are two types of ring height h = 10, 20 mm.

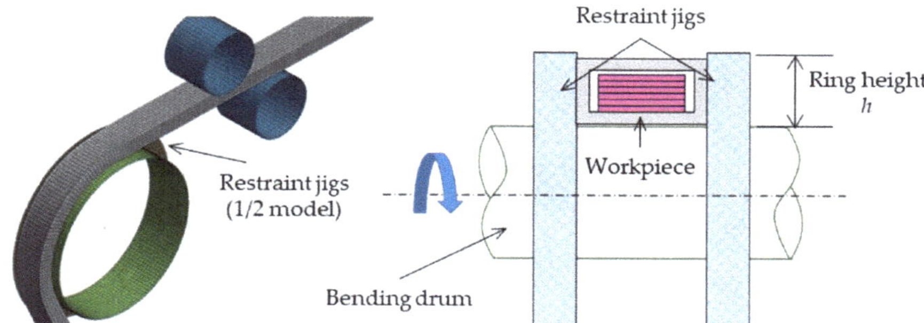

Figure 14. Schematic representation of restraint jigs and simulation model.

Figure 15 shows the experimental results for R/H_0 = 2.0 (H_0 = 10, R = 20). It was found that the convex distortion of the web was suppressed to less than 5% by applying the restraint jig. In addition, at the minimum part (tensile flange), the width W decreased due to axial tension applied to suppress wrinkling. Therefore, it was found that the accuracy of the width tends to deteriorate slightly compared to the maximum part (compression flange).

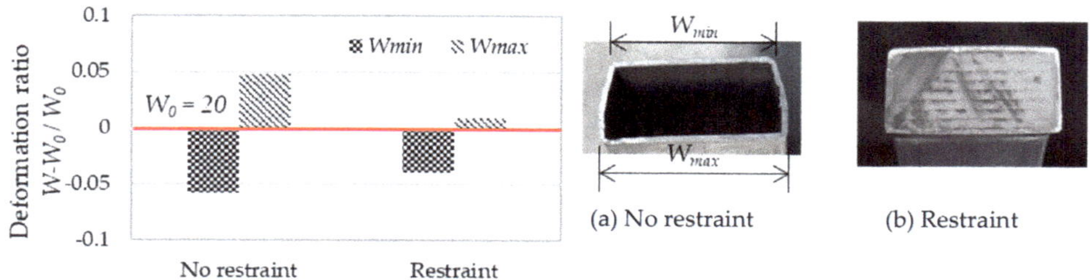

Figure 15. Effect of restraint jigs (H_0 = 10, R = 20).

In addition, Figure 16 shows the experimental and FEA results for R/H_0 = 1.5 (H_0 = 20, R = 50), and restraint jig ring height h = 10. As shown, the web compression side is inward, and the web tension side is outward. As mentioned above, it is known that convex distortion occurs outward during bending on the compression side. However, in this case, a restraint jig was provided on the side surface, and accordingly, outward convex distortion was suppressed. Nevertheless, since there is clearance (0.5 mm × 2 on each side) between the tube and the mandrel, it is presumed that the web is deformed inward by that amount. In addition, since the restraint jig was not provided on the tension side, it is considered that the web collapsed outward and convex distortion due to the above-mentioned deflection on the compression side.

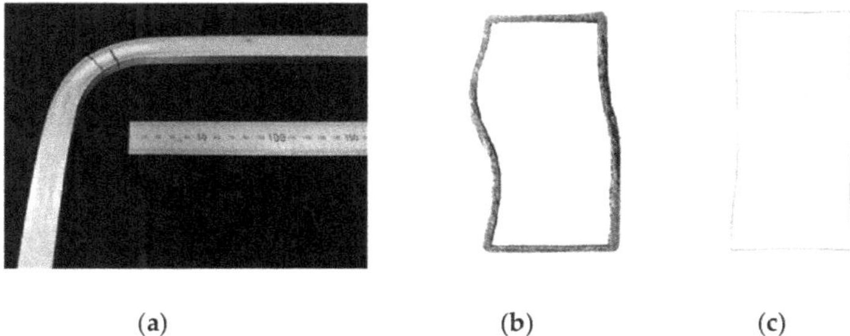

(a) (b) (c)

Figure 16. Effect of restraint jigs (H_0 = 20, R = 50, h = 10); (**a**) Appearance after bending; (**b**) Experiment; (**c**) FEA.

Furthermore, Figure 17 shows the results when the restraint jig ring height h = 20. Deformation was suppressed compared to h = 10 in both experiments and FEA. However, it was not possible to completely suppress the waveform deformation of the web. This is attributed to the fact that the clearance between the tube and the mandrel was set to 1.0 mm in both height and width. It is considered that the P_{TN} during bending compressed the web by the amount of clearance, buckled, and caused wavy deflection.

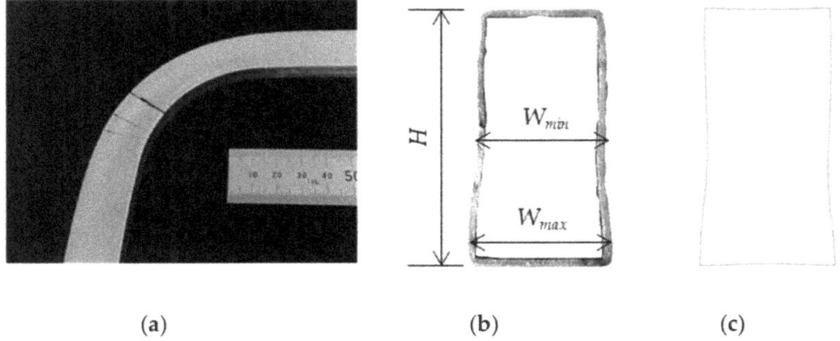

(a) (b) (c)

Figure 17. Effect of restraint jigs (H_0 = 20, R = 50, h = 20); (**a**) Appearance after bending; (**b**) Experiment; (**c**) FEA.

3.5. *Effect of the Bumpy Laminated Elastic Mandrel*

Based on the results obtained so far, Figure 18 shows a comparison of the cross-sectional shapes acquired by unconstrained and constrained bending. As shown in Figure 18, the maximum width $W_{\max}$ was significantly improved by applying the constraint. In addition, the height H was reduced by 4.4%, and the minimum width $W_{\min}$ was reduced by 9.8%. A decrease in height H is considered to be due to the clearance between the tube and the mandrel, and the decrease in minimum width W_{min} is considered to be due to the axial tension.

In addition, Figure 19 shows the result of FEA using a model in which the wall thickness of the rectangular tube was changed. The wall thickness of the rectangular tube increased; the above-mentioned deformation in which the web collapses inward and wavy deformation was not observed, but the deformation was mainly trapezoidal. Thus, it is considered that the pear-shaped or corrugated cross-section deformation is peculiar to the ultra-thin rectangular tube. In order to suppress this waveform deformation, it is necessary to adjust the clearance between the rectangular tube and the core material.

Therefore, bending was performed by applying a stepped laminated mandrel, as shown in Figure 20. Table 3 shows the experimental conditions. The height of the rectangular tube was H_0 = 20 and the bending radius was R = 50. Axial tension was applied, and the height of the restraint jig was set to h = 20. The clearance between the rectangular tube and the core material was C_{hL} = 0 to 0.5 in the height direction and 0.5 mm in the width direction. Moreover, the effect was confirmed by adding a step of 1.0 mm on the compression side and 0.5 mm on the tension side with the center line as the boundary. Figure 21 shows the appearance after bending under each condition. Wrinkling occurred in Figure 21a. This is considered to be because the clearance in the height direction was as large as 0.5 mm and the depression of the compression side flange surface, P_{TN}, could not be suppressed. In addition, in Figure 21b, splitting occurred. In contrast, in Figure 21c, a cross-sectional shape without wrinkling or splitting was obtained. In order to examine these results, the post-bending cross-sections were compared under the condition that the clearance was 1.0 mm in both height and width, and the condition where the bumpy mandrel was applied. Figure 22 shows a comparison of deformation ratio and Figure 23 shows a comparison of changes in wall thickness. Figure 20 demonstrates $H_{\max}$ was significantly suppressed in the bumpy mandrel. It is considered that this is because the bending in the case of using the bumpy mandrel was C_{Lh} = 0 mm; thus, the deformation in the height direction could be suppressed more than C_{Lh} = 1.0 mm. Accordingly, it was found that the clearance in the height direction needs to be as small as possible to suppress the deformation in the height direction. In contrast, the deformation of the width W was suppressed to some extent, but its effect was smaller than that of H. As shown in Figure 21b, if the clearance on the tension side in the width direction is set small (0.5 mm), a decrease in the width direction is suppressed, but conversely, cracking occurs. It is considered that this is because the tension of the mandrel was increased and the friction between the mandrel and the tube was increased by decreasing the clearance. Therefore, it was found that it is necessary to adjust the clearance in the width direction of the mandrel on the tension side to allow the tension and friction to relief in order to obtain a highly accurate cross-section after bending in thin-walled tubes.

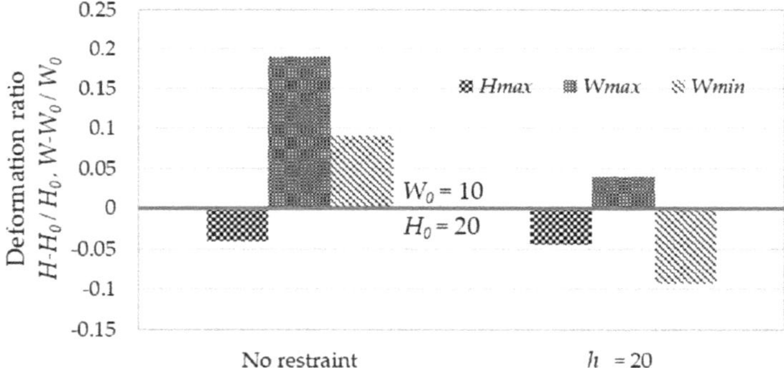

Figure 18. Comparison of the existence of restraint jigs.

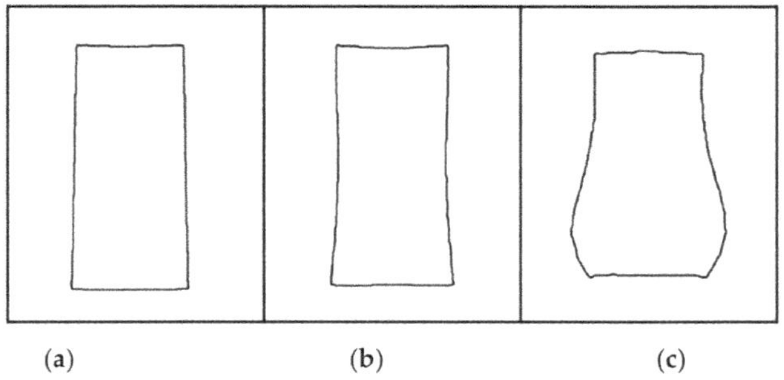

Figure 19. Effect of thickness; (**a**) t_0 = 2.0 mm; (**b**) t_0 = 1.0 mm; (**c**) t_0 = 0.5 mm.

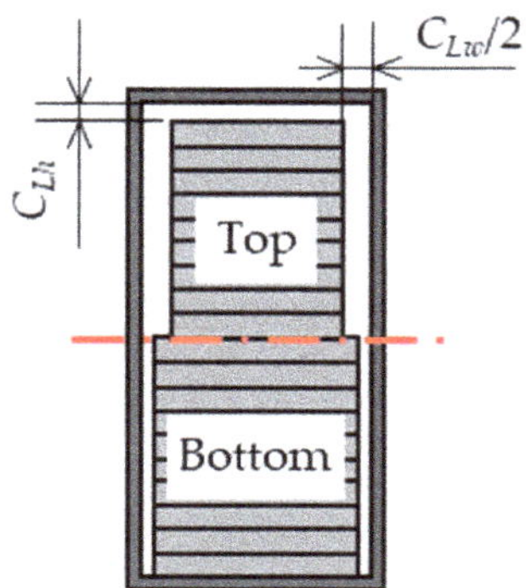

Figure 20. Schematic of the bumpy laminated elastic mandrel.

Table 3. Conditions of experiment and simulation.

Height of Workpiece H_0		20 mm		
Bending drum radius R			50 mm	
Axial tension α			Apply	
Height of restraint jigs h			20 mm	
Laminated Elastic Mandrel	Height C_{Lh}	0.5 mm	0 mm	0 mm
	Width C_{Lw} (Top/Bottom)	0.5/0.5 mm	0.5/0.5 mm	1.0/0.5 mm

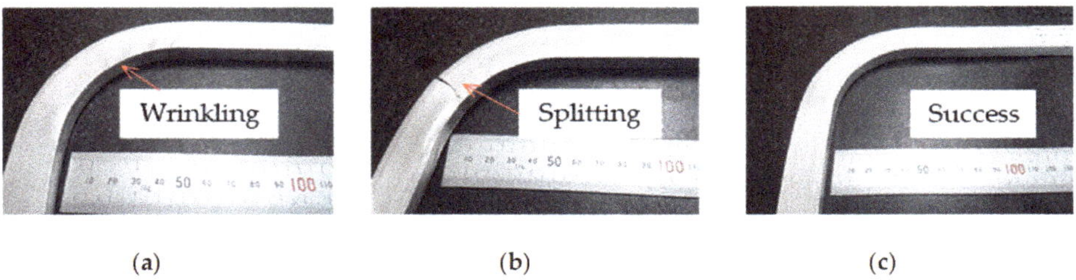

Figure 21. Effect of the bumpy laminated elastic mandrel; (**a**) C_{Lh} = 0.5 mm, C_{Lw} = 0.5/0.5 mm, Axial tension; (**b**) C_{Lh} = 0 mm, C_{Lw} = 0.5/0.5 mm, Axial tension; (**c**) C_{Lh} = 0 mm, C_{Lw} = 1.0/0.5 mm, Axial tension.

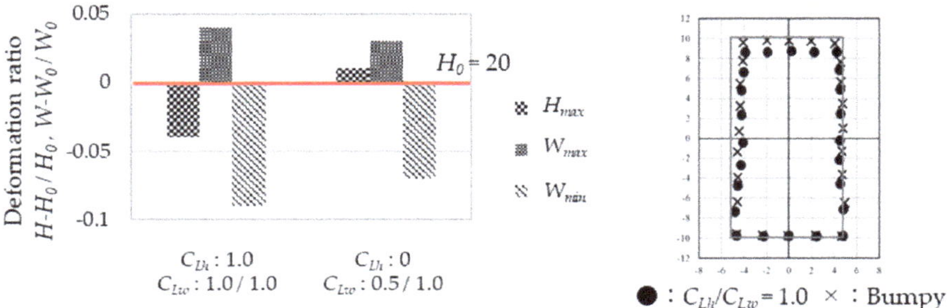

Figure 22. Effect of the bumpy laminated elastic mandrel on deformation ratio.

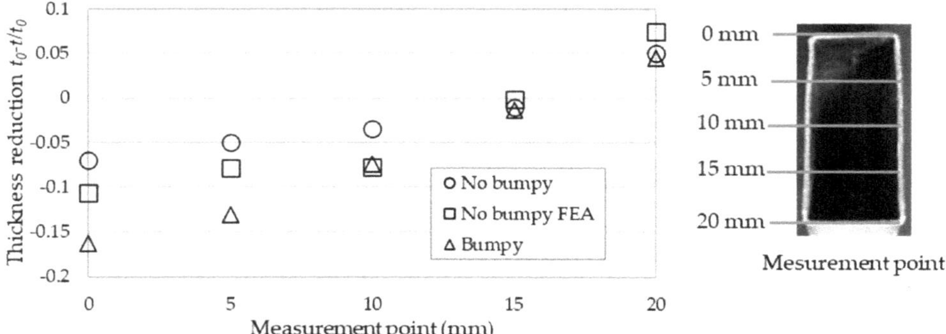

Figure 23. Effect of the bumpy laminated elastic mandrel in thickness.

4. Conclusions

The results of investigating the deformation characteristics and suppression of the ultra-thin rectangular tube in rotary draw bending are as follows:

1. The component flattening forces P_{TN} and the distributed flattening force w_{TN} increase as bending degree ρ (bending radius R), thickness t_0, height H_0, and axial tension P_a increase.
2. Wrinkling tended to occur when bending an ultra-thin wall tube. However, wrinkling can be suppressed by applying axial tension.
3. By applying the mandrel to a tube with R/H_0 = 1.5 (H_0 = 20, R = 50), it was possible to suppress cross-sectional deformation in the height direction. In contrast, pear-shaped deformation peculiar to ultra-thin wall tube occurred.
4. The pear-shaped deformation could be suppressed to $W_{\text{min}} = -4\%$, $W_{\text{max}} = 1\%$ by restraining the side surface of the ultra-thin wall tube with H_0 = 10, R = 20. In contrast, wrinkling and waveform deformation such as a long column buckling phenomenon occurred on the web of the tube with H_0 = 20, R = 50, h = 20.
5. By adjusting and stepping the clearance in the width direction of the mandrel on the tension side and the compression side of the ultra-thin wall tube with R/H_0 = 1.5 (H_0 = 20, R = 50, h = 20), along with the restraint on the side surface and the axial force, it was possible to suppress to $H = 1\%$, $W_{\text{max}} = 3\%$, and $W_{\text{min}} = -7\%$. Namely, it was possible to suppress the pear-shaped deformation peculiar to ultra-thin wall tube and waveform deformation such as a long column buckling phenomenon.
6. It was found that the deformation of the cross-section peculiar to the ultra-thin rectangular tube can be suppressed by applying axial tension, applying side restraint (h = 20), and adjusting the widthwise clearance of the mandrel on the compression and the tension sides of the tube (C_{Lh} = 0 mm, C_{Lw} = 1.0 mm/0.5 mm).

Author Contributions: Conceptualization, N.U.; Methodology, N.U.; Resources, N.U.; Material provision, N.U.; Experimental work, K.N. and Y.S.; Numerical analysis, K.N., N.U. and Y.S.; Data curation, K.N., N.U. and Y.S.; Writing—original draft preparation, K.N. and Y.S.; Supervision, N.U. and M.Y.; Project administration, N.U.; Funding acquisition, N.U. All authors have read and agreed to the published version of the manuscript.

Funding: This research was funded by AMADA Foundation grant number AF-2016026.

Acknowledgments: The authors would like to gratefully acknowledge the funding source for this study, the AMADA Foundation.

Conflicts of Interest: The authors declare no conflict of interest.

References

1. Takahashi, K.; Murata, M. Trend of tubular bending. *J. JPN Light Met.* **2006**, *56*, 283–289. [CrossRef]
2. Utsumi, N.; Sakaki, S. Countermeasures against undesirable phenomena in the draw-bending process for extruded square tubes. *J. Mater. Process. Technol.* **2002**, *123*, 264–269. [CrossRef]
3. Fuchs, F.J., Jr. Waveguide Bending Design Analysis: Theory of Bending and Formulae for Determination of Wall Thicknesses. *Bell Syst. Tech. J.* **1959**, *38*, 1457–1484. [CrossRef]
4. Gu, R.J.; Yang, H.; Zhan, M.; Li, H.; Li, H.W. Research on the springback of thin-walled tube NC bending based on the numerical simulation of the whole process. *Comput. Mater. Sci.* **2008**, *42*, 537–549. [CrossRef]
5. Zhang, H.; Liu, Y.; Liu, C. Multi-Objective Parameter Optimization for Cross-Sectional Deformation of Double-Ridged Rectangular Tube in Rotary Draw Bending by Using Response Surface Methodology and NSGA-II. *Metals* **2017**, *7*, 206. [CrossRef]
6. Strano, M. Automatic tooling design for rotary draw bending of tubes. *Int. J. Adv. Manuf. Technol.* **2005**, *26*, 733–740. [CrossRef]
7. Sherif, S.S.; Mostafa, S.; Mohammad, M.M. Limit load analysis of thin-walled as-fabricated pipe bends with low ovality under in-plane moment loading and internal pressure. *Thin Walled Struct.* **2019**, *114*, 106336.
8. Gantnerl, P.; Harrison, D.K.; De Silva, A.K.; Bauer, H. The Development of a Simulation Model and the Determination of the Die Control Data for the Free-Bending Technique. *Proc. Inst. Mech. Eng. Part B J. Eng. Manuf.* **2007**, *221*, 163–171. [CrossRef]
9. Masoumi, H.; Mirbagheri, Y. Effect of Mandrel, Its Clearance and Pressure Die on Tube Bending Process via Rotary Draw Bending Method. *Int. J. Adv. Des. Manuf. Technol.* **2012**, *5*, 47–52.
10. Hassan, H.R.; Abdulgaffar, S.M.; Mohsen, K.K.; Egab, K. Investigation of Material Properties Effect on the Ovalization Phenomenon in the Tube Bending Produced by RDB. *IOP Conf. Ser. Mater. Sci. Eng.* **2020**, *671*, 012010. [CrossRef]
11. Kourosh, H.; Mahmoud, B.; Behnaz, A.; Mehrdad, P. The effect of anisotropy on wrinkling of tube under rotary draw bending. *J. Mech. Sci. Technol.* **2013**, *27*, 783–792.
12. Xin, X.; Juan, L.; Gabriela, V.; Jose, G. Twist springback of asymmetric thin-walled tube in mandrel rotary draw bending process. *Procedia Eng.* **2014**, *81*, 2177–2183.
13. Christopher, H.; Rainer, S.; Bernd, E. Rotary-draw-bending using tools with reduced geometries. *Procedia Manuf.* **2018**, *15*, 804–811.
14. Linda, B.; Christopher, K.; Peter, F.; Bernd, E. Sensitivity analysis of the rotary draw bending process as a database of digital equipping support. *Procedia Manuf.* **2019**, *29*, 592–599.
15. Yang, H.; Li, H.; Zhan, M. Friction role in bending behaviors of thin-walled tube in rotary-draw-bending under small bending radii. *J. Mater. Process. Technol.* **2010**, *210*, 2273–2284. [CrossRef]
16. Muresan, C.; Achimas, G. Numeric Simulation of Material Behavior in Mandrel Bending. *Appl. Math. Mech. Eng.* **2014**, *57*, 455–460.
17. Mohammad, S.; Mahmoud, F.; Mahmoud, K.; Mohsen, N. A chain link mandrel for rotary draw bending: Experimental and finite element study of operation. *Int. J. Adv. Manuf. Technol.* **2015**, *79*, 1071–1080.
18. Kami, A.; Dariani, B.M. Prediction of wrinkling in thin-walled tube push-bending process using artificial neural network and finite element method. *Proc. Inst. Mech. Eng. Part B J. Eng. Manuf.* **2011**, *225*, 1801–1812. [CrossRef]
19. Lucian, L. Effect of internal fluid pressure on quality of aluminum alloy tube in rotary draw bending. *Int. J. Adv. Manuf. Technol.* **2013**, *64*, 85–91.
20. Kajikawa, S.; Wang, G.; Kuboki, T.; Watanabe, M.; Tsuichiya, A. Prevention of defects by optimizing mandrel position and shape in rotary draw bending of copper tube with thin wall. *Procedia Manuf.* **2018**, *15*, 828–835. [CrossRef]
21. Fang, J.; Lu, S.; Liang, C.; Ceng, L.; Zheng, D. Influences of mandrel types on forming quality of TA18 high strength tube in numerical control bending. *IOP Conf. Ser. Mater. Sci. Eng.* **2019**, *631*, 032032. [CrossRef]
22. Zhang, Z.; Yang, H.; Li, H.; Ren, N.; Tian, Y. Bending behaviors of large diameter thin-walled CP-Ti tube in rotary draw bending. *Prog. Nat. Sci. Mater. Int.* **2011**, *21*, 401–412. [CrossRef]
23. Fang, J.; Lu, S.; Wang, K.; Xu, J.; Xu, X.; Yao, Z. Effect of Mandrel on Cross-Section Quality in Numerical Control Bending Process of Stainless Steel 2169 Small Diameter Tube. *Adv. Mater. Sci. Eng.* **2013**, *2013*, 1–9. [CrossRef]

24. Frode, P.; Torgeir, W. Cross-sectional deformations of rectangular hollow sections in bending: Part I–experiments. *Int. J. Mech. Sci.* **2001**, *43*, 109–129.
25. Hong, Z.; Yuli, L.; Honglie, Z. Study on cross-sectional deformation of rectangular waveguide tube with different materials in rotary draw bending. *J. Mater. Res.* **2017**, *32*, 3912–3920.
26. Nakajima, K.; Utsumi, N.; Yoshida, M. Suppressing method of the cross-section deformation for extruded square tubes in press bending. *Int. J. Precis. Eng. Manuf.* **2013**, *14*, 965–970. [CrossRef]
27. Utsumi, N.; Saito, Y.; Yoshida, M. Effects of axial tension in rotary draw bending of thin-walled rectangular aluminum alloy tubes. In Proceedings of the 69th Japanese Joint Conference for the Technology of Plasticity, Kumamoto, Japan, 27–28 October 2018.
28. Saito, Y.; Kono, Y.; Kamimura, T.; Utsumi, N.; Yoshida, M. Deformation of property and working limit of thin-walled rectangular tube in rotary draw bending. In Proceedings of the 5th Asian Symposium on Materials and Processing, Bangkok, Thailand, 7–8 December 2018.
29. Sakaki, S.; Utsumi, N. Deformation property and working limit of extruded square sections in rotary-bending process. *Adv. Technol. Plast.* **2002**, *2*, 1763–1768.

Article

Circumferential Material Flow in the Hydroforming of Overlapping Blanks

Cong Han [1,2,*] and Hao Feng [1]

1 School of Materials Science and Engineering, Harbin Institute of Technology, Harbin 150001, China; kingmouseboca@126.com
2 National Key Laboratory of Precision Hot Processing of Metals, Harbin Institute of Technology, Harbin 150001, China
* Correspondence: conghan@hit.edu.cn; Tel.: +86-451-8641-7917

Received: 26 May 2020; Accepted: 24 June 2020; Published: 29 June 2020

Abstract: The hydroforming of the overlapping blanks is a forming process where overlapping tubular blanks are used instead of tubes to enhance the forming limit and improve the thickness distribution. A distinguishing characteristic of the hydroforming of overlapping blanks is that the material can flow along the circumferential direction easily. In this research, the circumferential material flow was investigated using overlapping blanks with axial constraints to study the circumferential material flow in the hydroforming of a variable-diameter part. AISI 304 stainless steel blanks were selected for numerical simulation and experimental research. The circumferential material flow distribution was obtained from the profile at the edge of the overlap. The peak value located at the middle cross-section. In addition, the circumferential material flow could be also reflected in the variation of the overlap angle. The variation of the overlap angle kept increasing as the initial overlap angle increased but the improvement of the thickness distribution did not. There was an optimal initial overlap angle to minimize the thinning ratio. An optimal thickness distribution was obtained when the initial angle was 120° for the hydroforming of the variable-diameter part with an expansion of 31.6%.

Keywords: hydroforming; overlapping blank; variable-diameter part; thickness

---

1. Introduction

In the past decades, the demand for lightweight structures and components has been increasing with the rapid development of the automotive industry [1]. A variety of tubular components, i.e., exhaust pipes, chassis systems and structural components are designed and assembled in automobiles. The manufacturing processes of these components should be straightforward, efficient and reliable [2]. Tube hydroforming has proved itself as an advanced metal forming process to manufacture these components because of its part consolidation, weight reduction and lower costs [3,4]. Variable-diameter components, as a kind of typical hydroformed parts, are widely used in the exhaust pipes of automotives. The diameter of the middle region of the variable-diameter tubular part is much larger than the end [5].

With regard to the tube hydroforming process of variable-diameter tubular components, there are certain limitations on the degree of deformation owing to the hydroformability of the material used [6]. Up to now, a number of studies on the hydroformability of tubular components have been reported [7]. A high strain hardening exponent and a high plastic anisotropy of the material were effective to improve the wall thickness [8,9]. Lubrication conditions had a significant effect on defects of hydroformed parts, such as wrinkling, buckling and cracking [10,11]. The thickness distribution was relatively uniform due to the low friction coefficient when an appropriate lubricant was used [12]. The loading path (the relationship of the internal pressure and the axial feeding) had a significant effect on the formability for a certain material [13]. On the basis of it, the useful wrinkles have been proposed and applied to enlarge the process window [14]. It has been demonstrated that useful wrinkles can improve the forming limit and thickness distribution by increasing axial feeding [15,16]. Wrinkling defects, however, may occur easily due to the axial feeding if the diameter-to-thickness ratio (*D*/*t*) of the part is extremely large [17]. It is not feasible to enhance the forming limit by increasing the axial feeding in this situation.

Therefore, a novel approach was proposed, in which overlapping tubular blanks were used instead of closed cross-section tubes [18]. A sound spherical part with an expansion of 60.0% was obtained by using this approach, whereas the maximum expansion was only 46.1% using tubes with closed cross-sections. The key to the enhancement of forming limits was the material flow along the circumferential direction. However, the material flowed along both axial and circumferential directions simultaneously due to the free ends of the overlapping blank. In order to study the material flow along the circumferential direction, the influence of the axial direction should be avoided during the hydroforming of overlapping blanks.

In this paper, the circumferential material flow was investigated using overlapping blanks with axial constraints in the hydroforming of a variable-diameter part [19]. The shape of the variable-diameter part was cylindrical to guarantee that the expansion rates were equal at the bulging area. AISI 304 stainless steel blanks were selected for numerical simulation and experimental research. The circumferential material flow distribution was reflected in the profile of the blank edge. The effect of the overlap angle on the variation of the overlap and thickness distribution was studied.

2. Materials and Methods

2.1. Samples and Material

The sample was a variable-diameter part, as shown in Figure 1. The maximum diameter was 100 mm at the middle region and the diameter was 76 mm at each end. The diameter of the middle region was 31.6% larger than the end. The length of the middle region was 56 mm. The angle of the transition area was 20°. The radius of the fillet was 15 mm in the transitional area.

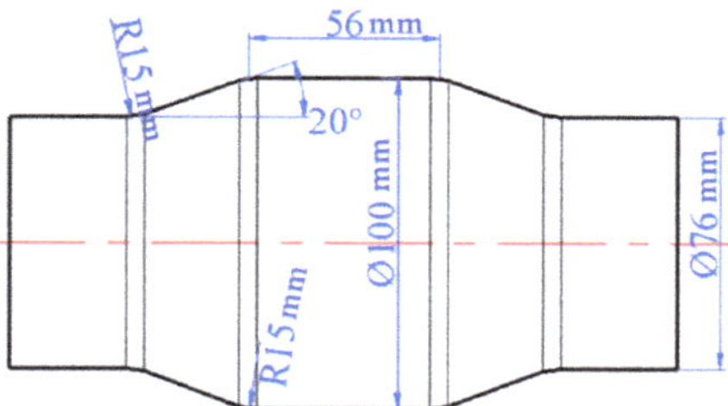

Figure 1. Shape and dimension of the sample.

Overlapping tubular blanks were obtained from sheet blanks after bending and used for the hydroforming of the variable-diameter part. In comparison with the closed cross-sectional tube which was shown in Figure 2a, the cross-section of the overlapping blank was open. The overlap consisted of the inner and the outer layers, as shown in Figure 2b.

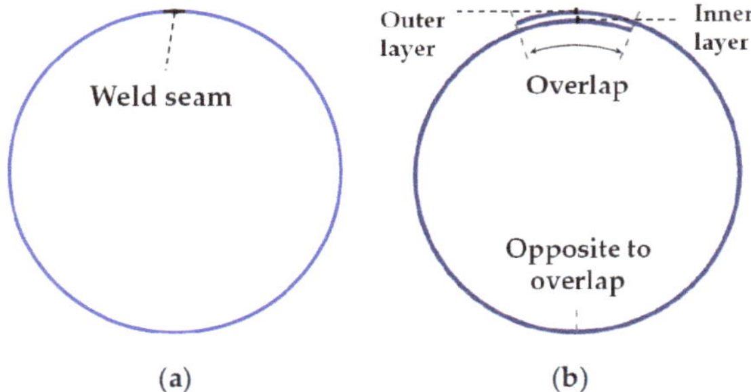

Figure 2. Shape of cross-sections: (**a**) tube blank; (**b**) overlapping blank.

Figure 3a shows the shape and dimension of the overlapping blank. The outer diameter and the length of the tubular blank were 76 and 250 mm, respectively. The level of the overlap was determined by the initial overlap angle α that was defined as the angle between the edges of the inner layer and the outer layer.

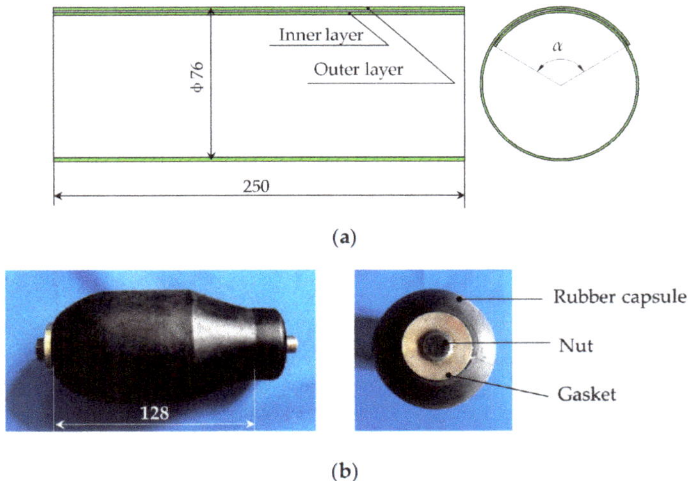

Figure 3. Shape and dimension of: (**a**) the overlapping blank; (**b**) the elastic body.

An elastic body was used for the expansion of the overlapping blank. The function of the elastic body was achieved by rubber capsules that were available on the market. The diameter of the rubber capsule ranged from 50 to 72 mm. The total length of the rubber was 150 mm and the length of the bulging area of the rubber was 128 mm, as shown in Figure 3b. The maximum elastic elongation of the rubber was 100% at least, which was much higher than the expansion rate of the blank. Therefore, the rubber capsule was capable of the hydroforming of the variable-diameter part. There was an entrance for the liquid medium at the right of the elastic body. At the left, the gaskets and the nut could seal the left end of the rubber capsule.

AISI 304 stainless steel blanks were used in the simulation and experimental research. The initial wall thickness of the blank was 0.5 mm. The uniaxial tensile tests were conducted on an AGX-plus 20KN/5KN (SHIMADZU Corp., Kyoto, Japan) machine along the rolling direction of the sheet blanks. Common tensile properties like yield and ultimate tensile strength were obtained from the true stress-strain curve, as shown in Figure 4. The mechanical properties are given in Table 1.

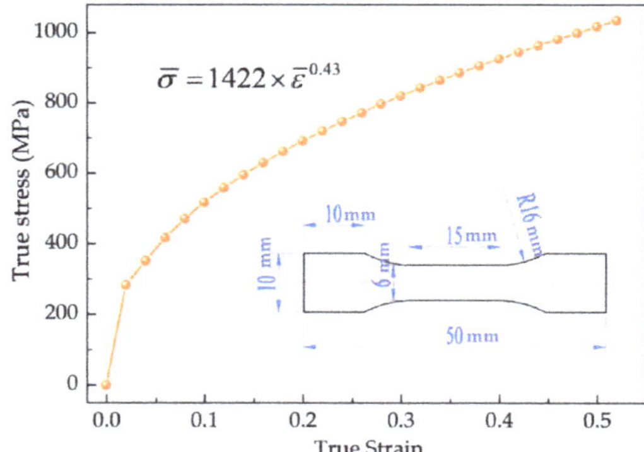

Figure 4. Flow stress curve of AISI 304 stainless steel.

Table 1. Mechanical properties of AISI 304 stainless steel.

Elastic Modulus E (GPa)	Yield Strength σ_s (MPa)	Ultimate Strength σ_b (MPa)	Elongation δ (%)
208	287	903	52.6

2.2. Experimental Setup

The experiments were conducted on the 10 MN hydroforming machine. Figure 5 shows the schematic diagram of the experimental setup for the hydroforming of the variable-diameter part. The hydroforming tools mainly consisted of a lower die, an upper die, two blocks and a self-sealing elastic body.

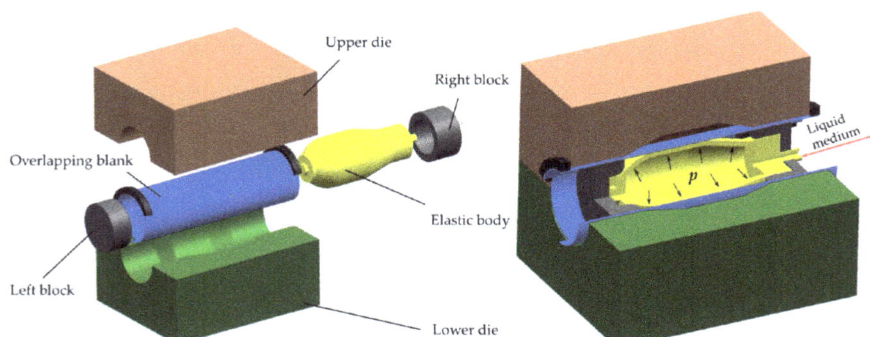

Figure 5. Schematic diagram of the experimental setup.

The liquid medium flowed into the rubber capsule cavity via the right entrance of the elastic body during the hydroforming process. The two blocks were placed against the ends of the elastic body to prevent it from extending along the axial direction, as shown in Figure 6. The pressurized liquid was stored in the expansive elastic body which pushed the overlapping blank to bulge as the internal pressure increased. When the process finished, the internal pressure was unloaded and the capsule could be taken out from the blank due to its high elasticity property.

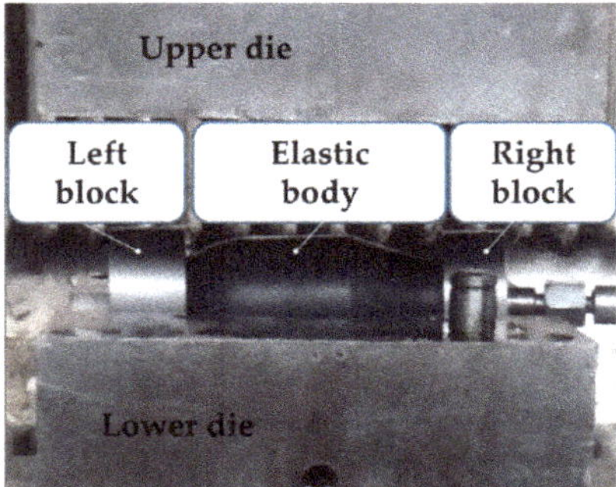

Figure 6. The self-sealing elastic body used for the experiment.

2.3. Finite Element Model

The numerical simulation was conducted with FEM software Abaqus 6.13. (SIMULIA Corp., Providence, RI, USA) Figure 7 shows the finite element model used for the hydroforming of a variable-diameter part with overlapping tubular blanks. The die and the blocks were defined as a rigid body. The overlapping blank was defined as a deformed body using the elastoplastic model. The elastic body was defined as a hyperelastic model and its original shape was simplified into a cylinder in the finite element model. The influence of the elastic body shape was limited to the simulation results due to the high elasticity.

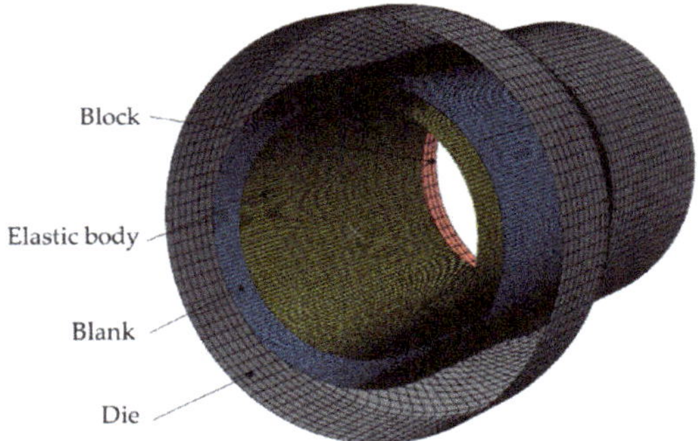

Figure 7. Finite element model.

As for an isotropic and incompressible material, the strain–energy function W can be described as a function of principal elongations [20]:

$$W = W(\lambda_1, \lambda_2, {\lambda_1}^{-1}{\lambda_2}^{-1}) \tag{1}$$

In the Mooney–Rivlin model, W is described as a function of strain invariants (I_1, I_2, I_3) [21]. The relationship between the principal elongations and strain invariants can be written as:

$$\begin{aligned} I_1 &= {\lambda_1}^2 + {\lambda_2}^2 + {\lambda_3}^2 \\ I_2 &= (\lambda_1\lambda_2)^2 + (\lambda_2\lambda_3)^2 + (\lambda_3\lambda_1)^2 \\ I_3 &= (\lambda_1\lambda_2\lambda_3)^2 \end{aligned} \tag{2}$$

In addition, W only depends on the first two invariants I_1 and I_2 because the third invariant I_3 is constant which equals one for incompressible materials [22]. When the level of deformation of an elastic body is relatively low, it is reasonable to apply the original first-order Mooney–Rivlin model, which is expressed as:

$$W = C_{10}(I_1 - 3) + C_{01}(I_2 - 3) \quad (3)$$

where C_{10} and C_{01} represent the material constant of the deviatoric component. In this research, the two coefficients were set at 0.4920 and 0.1752 MPa according to the tensile test result of the rubber material used.

All the meshes of the body were shell elements. The element size of the blank and the elastic body was set at 2.0 mm. The total numbers of elements of the two deformed bodies were 18,960 and 11,280. The friction coefficients were selected as 0.10, 0.10 and 0.18 at the die-blank, blank-blank and elastic body-blank interface, respectively.

2.4. Boundary Conditions

To investigate the deformation process in hydroforming process, the maximum forming pressure should be determined, which can be calculated as follows:

$$p_{\max} = \frac{2t}{d}\sigma_b \quad (4)$$

where t and d are the initial wall thickness and the diameter of the overlapping tubular blank, respectively. σ_b is the ultimate strength, which can be found in Table 1. The maximum forming pressure was 9.03 MPa obtained from Equation (4). In this research, the forming pressure of 9.0 MPa was selected as the maximum forming pressure.

In addition, wrinkling defects at the overlap are likely to occur because of the material flow along the axial direction at each end of the overlapping blank. Therefore, three boundary conditions, free, tied, axial constraint ends, were selected to investigate the effect of boundary constraints on wrinkling defects. For the tied ends, the outer and inner layers were bound together at each end of the blank. The displacements of the inner and outer layers are equal, as shown in Figure 8a. For the axial constraint ends, the material flow along the axial direction was completely restrained at each end. The displacements of the inner and outer layers are equal to zero, as shown in Figure 8b.

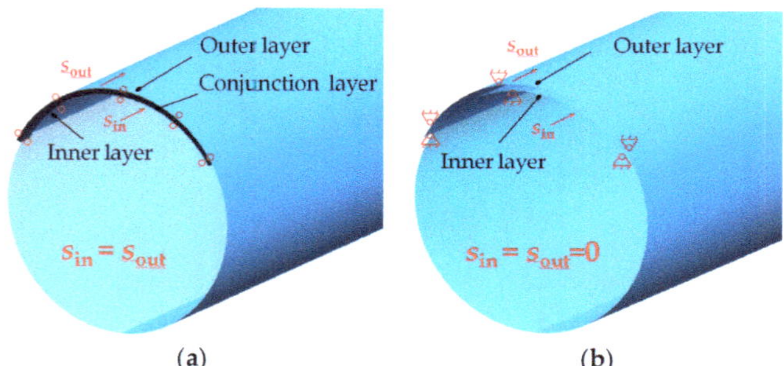

Figure 8. Schematic diagram of boundary conditions: (**a**) tied ends; (**b**) axial constraint ends.

3. Results

3.1. Effect of Boundary Conditions on Wrinkling Defects

Figure 9 shows the deformation process of hydroforming of overlapping blanks with free ends. The blank buckled at the overlap edge of the outer layer at the beginning, as shown in Figure 9a. The wrinkling defects occurred at the outer and inner layer of the overlap as the forming pressure increased, as shown in Figure 9b,c. The material at each end flowed towards the middle area along the axial direction. Meanwhile, the displacement along the axial direction at the edge was larger than that at the interior of the wrinkling zone. The unequal displacement aggregated the wrinkling defects.

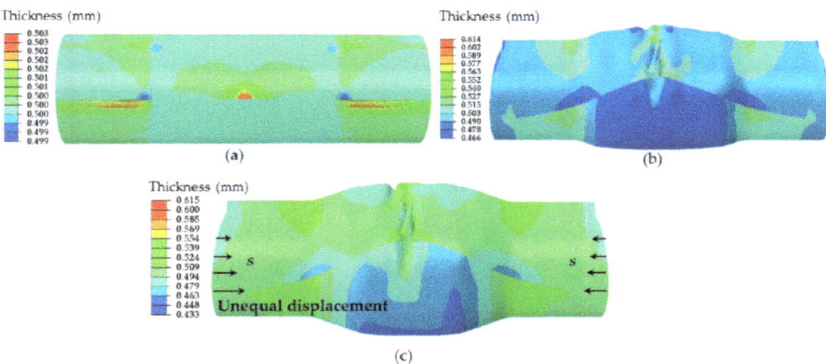

Figure 9. Hydroforming lapping blank with free ends (**a**) p = 2.0 MPa; (**b**) p = 5.0 MPa; (**c**) p = 9.0 MPa.

To decrease the difference in the axial displacement and control the wrinkling defects, the overlaps of the blank were tied together at each end. However, slight wrinkling at the outer layer still existed when the ends were tied, as shown in Figure 10. It is indicated that the occurrence of wrinkling defects cannot be avoided if the axial movement of the material at each end is not restricted.

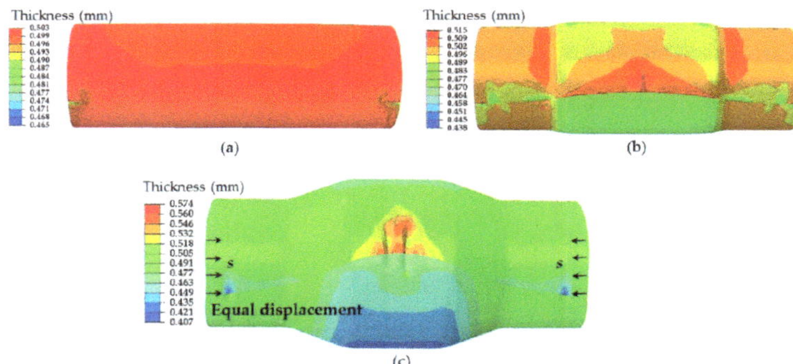

Figure 10. Hydroforming of overlapping blank with tied ends (**a**) p = 2.0 MPa; (**b**) p = 5.0 MPa; (**c**) p = 9.0 MPa.

It is indicated that the occurrence of wrinkling defects cannot be avoided if the axial movement of the material at each end is not restricted. Therefore, a blank with axial constraint ends was used to control the wrinkling defects by restricting the flow of the material towards the middle region. The level of wrinkling defects was eased further, as shown in Figure 11. Hence, the axial constraints had a beneficial effect on the hydroforming of overlapping tubular blanks.

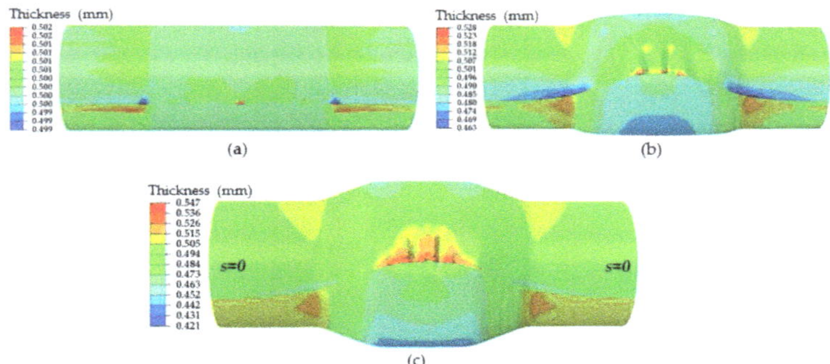

Figure 11. Hydroforming of overlapping blank with axial constraint ends (**a**) p = 2.0 MPa; (**b**) p = 5.0 MPa; (**c**) p = 9.0 MPa.

3.2. Material Flow Along the Circumferential Direction

The most important difference between the overlapping tubular blank and the tube involves whether the material can flow along the circumferential direction during the hydroforming process. Figure 12 shows the distribution of the circumferential material flow at the edge of the inner layer when the initial overlap angle is 120°. The expansion rates were equal at the middle area of the hydroformed part because the shape of the variable-diameter part was cylindrical. Therefore, the circumferential material flow, or the circumferential supplement, could be reflected in the profile of the blank edge directly. The profile of the edge of the inner layer was concave. That the maximum value was located at the middle cross-section could indicate that the maximum level of the circumferential material flow occurred at the middle cross-section for a plane-symmetric part. Additionally, it could be inferred that the nonuniform circumferential material flow led to a plane-bending deformation.

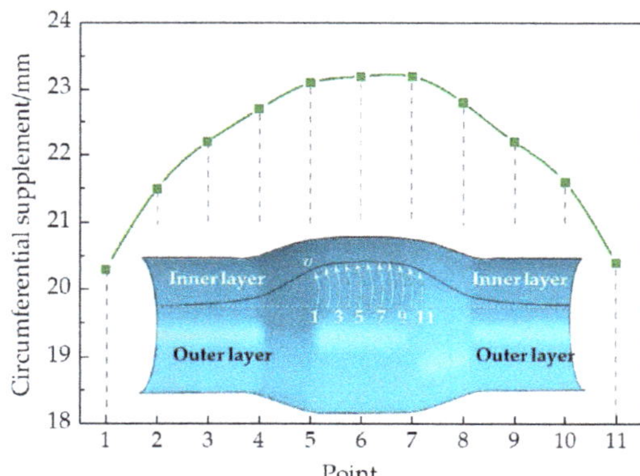

Figure 12. Distribution of circumferential material flow at the edge of the inner layer.

3.3. Thickness Distribution

To achieve the axial constraints, the arc rings were welded onto the ends of the overlapping blank, as shown in Figure 13a. The perimeter of the blank was 320 mm and the initial overlap angle was 120°. Figure 13b shows the hydroformed part under the forming pressure of 9.0 MPa. The wrinkling defects were scattered at the outer layer of the blank and the wrinkling defects of the inner layer were avoided.

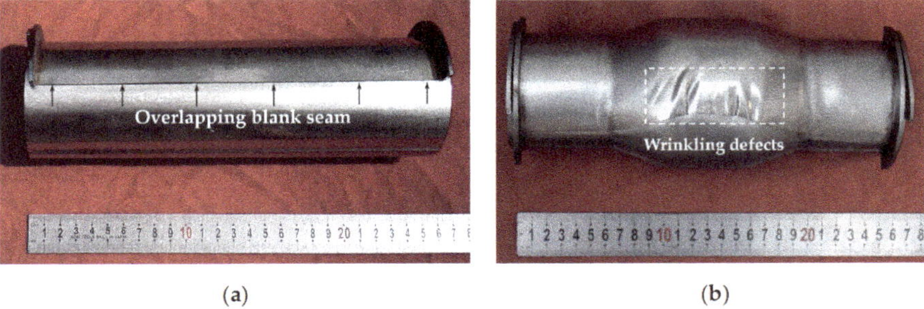

(a) (b)

Figure 13. Hydroforming of overlapping blank under axial constraints in experiment results. (**a**) Initial blank; (**b**) hydroformed part.

Then the redundant overlap and the axial constraints were removed. Then the reserved part was hydroformed again in order to smooth its surface. Finally, a sound part without wrinkling defects was manufactured. Figure 14 shows the final variable-diameter part.

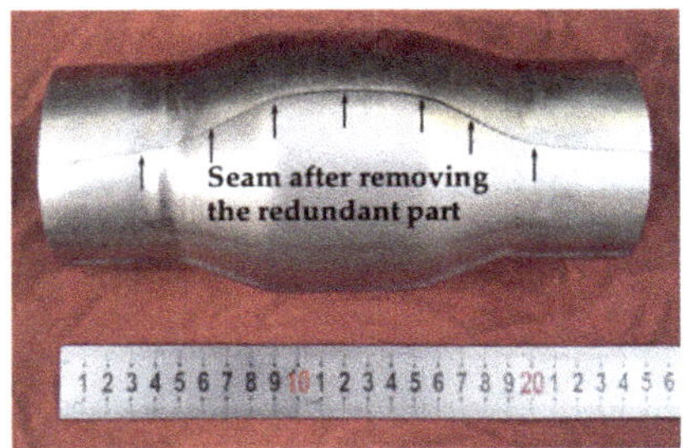

Figure 14. The final variable-diameter part.

The thinning ratio distribution was obtained at typical sections of the final part. Figure 15a shows the thinning ratio distribution along the axial direction. The difference in thinning ratio distribution was obvious between the overlap and the opposite side to the overlap. The maximum thinning ratio was 15.9% opposite to the overlap at the bulging region (Point 11) whereas it was only 4.4% at the transition region of the overlap (Point 8). It is indicated that the thickness was nonuniform at the cross-section of the overlapping blank. Detailed variation of the thickness at the middle cross-section is shown in Figure 15b. There were 25 measuring points marked at the middle cross-section. Points 1 and 25 were at the overlap and Point 13 was opposite to the overlap. It can be seen that the thinning ratio continually increased from the overlap towards the opposite. The maximum thinning ratio was 15.9% close to the opposite side of the overlap (Point 12) whereas the minimum thinning ratio was 2.1% at the outer layer of the overlap (Point 25).

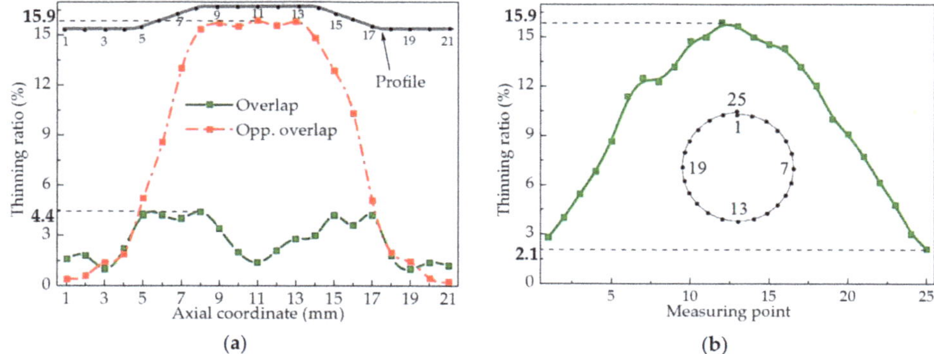

Figure 15. Thinning ratio distribution (**a**) along axial direction; (**b**) at middle cross-section.

4. Discussion

4.1. Stress and Strain Analysis

The simulation and experiment results showed that wrinkling defects are the main forming failure during the hydroforming of overlapping tubular blanks. These wrinkles were usually perpendicular to the axial direction at the overlap. It was implied that the level of compressive axial stress was very high at the region. Figure 16 shows the stress distribution along and opposite to the edge of the outer layer of overlap with both ends of the overlapping blank under axial constraints. The hoop stress and the axial stress increased sharply opposite to the edge, which was analogous to that during the tube hydroforming. On the other hand, the hoop stress was compressive and varied slightly beneath the yield stress and the axial stress was compressive along the edge of the outer layer. Particularly, the compressive axial stress dramatically increased at the axial coordinate of −60 and 60 mm when the internal pressure was 2.0 MPa. With the blank expanding, the maximum compressive axial stress was as many as 589 MPa at the bulging region. Therefore, wrinkling defects occurred at the edge owing to the compressive axial stress during the hydroforming process.

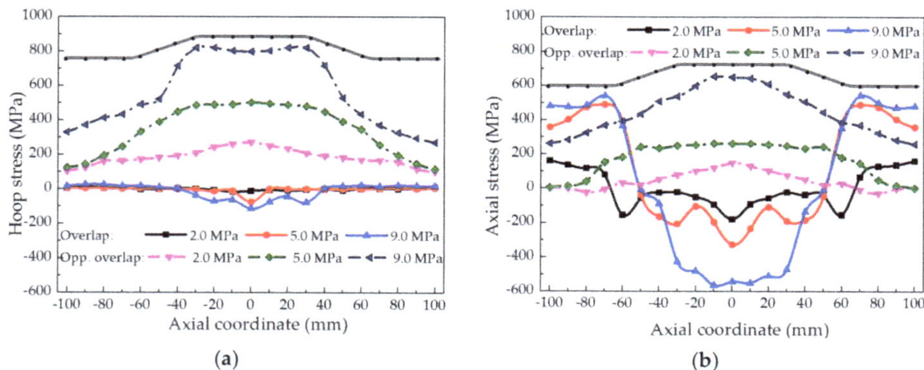

Figure 16. Stress distribution along axial direction. (**a**) Hoop stress; (**b**) axial stress.

Although the variation of axial compressive stress was discussed before, the deformation model at the wrinkling region should be determined. Figure 17 shows the axial stress distribution of the outer overlap at the cross-section where the compressive axial stress sharply increases. The increase of compressive axial stress occurred at section A-A when the internal pressure was 2.0 MPa. The axial stress increased from −152 MPa at the edge (Point 1) to 80 MPa at the interior (Point 11). A bending moment formed at the outer overlap due to the stress distribution at section A-A. This phenomenon was more obvious at section B-B when the internal pressure was 5.0 MPa. The axial stress increased from −325 MPa at the edge (Point 1) to 121 MPa at the interior (Point 11). It is indicated that a bending deformation happens at the edge of the overlap. It is extremely possible that wrinkling defects are caused by a bending moment at the inside radius.

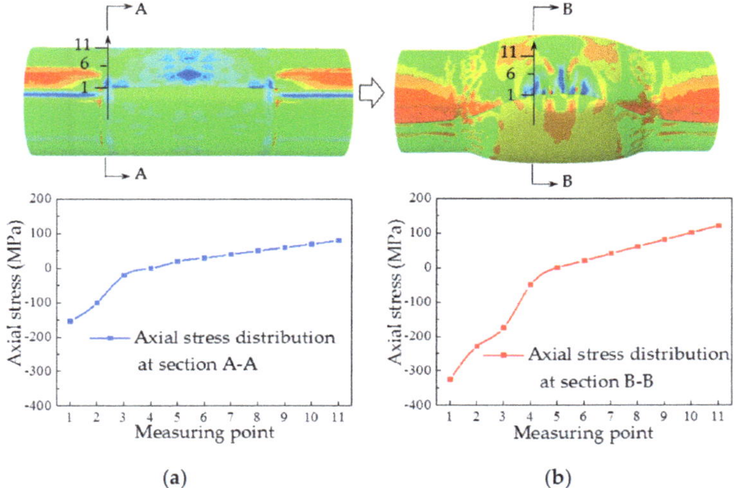

Figure 17. Axial stress distribution at the outer layer of overlap. (**a**) p = 2.0 MPa; (**b**) p = 5.0 MPa.

The strain variation of the hydroformed part was obtained to study the circumferential deformation behavior. Figure 18 shows the circumferential and normal strain distribution at the middle cross-section. The materials of the part elongated along the circumferential direction and thinned along the normal direction in common. The maximum circumferential strain was 0.194 and the minimum normal strain was −0.176 at Point 12. Due to the application of axial constraints, the axial strain of Point 12 was only −0.018, of which the absolute value was much smaller than those of the circumferential and the normal strains. Therefore, the deformation mode was similar to the plane strain state during the hydroforming of an overlapping blank under axial constraints.

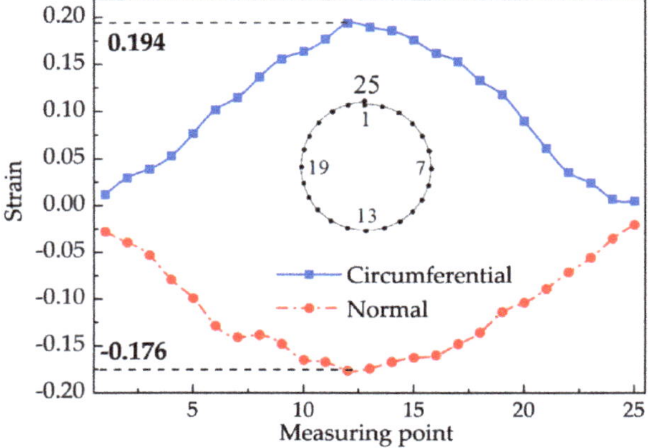

Figure 18. Strain distribution at the middle cross-section.

4.2. Variation of the Overlap Angle

It is difficult for the overlapping blank to deform simultaneously as its geometry is nonaxisymmetric. Figure 19 shows this deformation process of the overlapping tubular blank when the initial overlap angle $\alpha = 140°$. The deformation sequence was the outer layer of overlap at first, then the inner layer of overlap, and the opposite side to the overlap at last. The deformation had already occurred at the outer layer of the overlap when the internal pressure was 2.0 MPa, as shown in Figure 19b. This value of internal pressure was only about half of the yield pressure, in comparison with hydroforming of tubes with closed cross-sections. The yield pressure in tube hydroforming is usually calculated in Equation (5) [23]:

$$p_y = \frac{2t}{d}\sigma_s \tag{5}$$

where σ_s is the yield strength of the material, which can be found in Table 1. Therefore, the yield pressure was 3.78 MPa calculated from Equation (4).

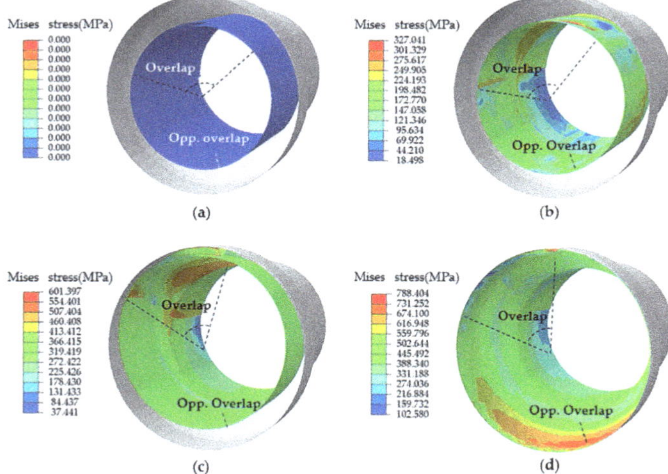

Figure 19. Deformation process of the overlapping tubular blank. (**a**) $p = 0$ (initial state); (**b**) $p = 2.0$ MPa; (**c**) $p = 5.0$ MPa; (**d**) $p = 9.0$ MPa.

Figure 19c shows that the entire overlap contacted the die cavity whereas the opposite side to the overlap did not as the internal pressure reached 5.0 MPa, which was only 1.3 times larger than the yield pressure. The bulging height was higher at the overlap than opposite to the overlap at the same level of forming pressure. Finally, the rest of the blank had already contacted the die cavity completely since the internal pressure reached 9.0 MPa, which can be seen from Figure 19d. The majority region was opposite to the overlap where the value of the Mises stress was relatively high. The final overlap angle β decreased to 43°.

The circumferential material flow can be also reflected in the variation of the overlap angle, which is defined as follows:

$$\Delta = \alpha - \beta \tag{6}$$

where α and β represent the initial and the final overlap angle at the middle cross-section, respectively. Figure 20 shows the variations of the overlap angle under different initial overlap angles. The hydroformed part failed when the initial overlap angle was 80° because the inner and the outer layers were separated. The final overlap angle was negative and its value was −7° in this condition. The value of the final overlap angle was positive when the initial overlap angle was larger than 100°. The variation of the overlap angle kept increasing as the initial overlap angle increased. It meant that more circumferential supplement of the material was achieved if the initial overlap angle became larger. However, the velocity of increase declined gradually and the variation of the overlap angle was constant since the initial overlap angle was larger than 160°. This indicated that there was a limit of the circumferential supplement during the hydroforming using the overlapping blank.

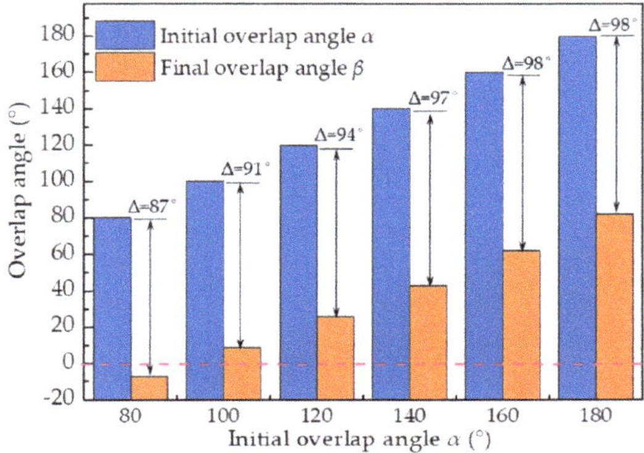

Figure 20. The variations of the overlap angle under different initial overlap angles.

4.3. *Effect of the Overlap Angle on the Thickness Distribution*

Figure 21 shows the effect of the overlap angle on the thickness distribution at the middle cross-section when the initial overlap angle ranges from 100° to 180°. The thinning ratios of $\alpha = 100°$ were much smaller than other conditions near the two edges of the hydroformed part. The maximum thinning ratio of $\alpha = 100°$ was larger than that of $\alpha = 120°$. It seemed that the increase of the variation of the overlap Δ contributed to the improvement of the thickness distribution. However, the maximum thinning ratio became larger with the initial overlap angle increasing since the initial overlap angle was greater than 120°. An optimal thickness distribution was obtained when the initial angle was 120° for the hydroforming of the variable-diameter part with an expansion of 31.6%. To sum up, there was an optimal initial overlap angle to minimize the thinning ratio when taking into account the variation of the overlap angle and the effectiveness of the material flow along the circumferential direction.

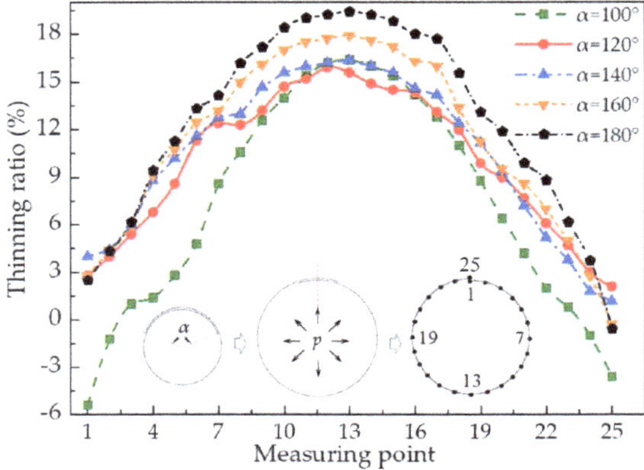

Figure 21. Effect of the overlap angle on the thickness distribution at the middle cross-section.

5. Conclusions

In this paper, the circumferential material flow was investigated using overlapping blanks with axial constraints in the hydroforming of a variable-diameter part. A self-sealing elastic deformation body was used as a loading tool to store the pressurized liquid. Numerical simulation and experimental methods were used to investigate the deformation features. The detailed results were as follows:

(1) The effect of boundary conditions: free ends, tied ends and axial constraint ends, on wrinkling defects at the overlap was studied. Wrinkling defects can be weakened when the blank with axial constraint ends was used.

(2) Wrinkling defects occurred at the edge of the overlap owing to the compressive axial stress. Furthermore, a bending deformation was observed through the stress analysis.

(3) The deformation of the blank at the overlap initiated at a low pressure that was nearly half of the yield pressure for tube bugling. The contact sequence to the die cavity of the blank was the overlap at first, then the inner layer of overlap, and the opposite side to the overlap at last.

(4) The circumferential material flow was obtained when axial constraints were applied to avoid the influence of the axial material flow. The profile was concave at the edge of the overlap due to the circumferential material flow. The peak value is located at the middle cross-section.

(5) The variation of the overlap angle increased with the initial overlap angle increasing but the improvement of the thickness distribution did not. An optimal thickness distribution was obtained when the initial angle was 120° for the hydroforming of the variable-diameter part with an expansion of 31.6%.

Author Contributions: C.H. conceived the experiments and provided all sorts of support during the work; H.F. performed the experiment and wrote the paper. All authors have read and agreed to the published version of the manuscript.

Funding: This work was financially supported by the two National Natural Science Foundations of China (project numbers: 51775136 and U1937205). The authors would like to take this opportunity to express their sincere appreciation.

Acknowledgments: The presentation about this paper was given at the 9th International Conference on tube hydroforming and the paper was recommended to submit to the journal "metals".

Conflicts of Interest: The authors declare no conflict of interest.

References

1. Saboori, M.; Champliaud, H.; Gholipour, J.; Gakwaya, J.; Savoie, A.; Wanjara, P. Evaluating the flow stress of aerospace alloys for tube hydroforming process by free expansion testing. *Int. J. Adv. Manuf. Technol.* **2014**, *72*, 1275–1286. [CrossRef]
2. Cristina, C.; Sánchez-Amaya, J.; Francisco, C.; Juan, V.M.; Javier, B. Springback estimation in the hydroforming process of UNS A92024-T3 aluminum alloy by FEM simulations. *Metals* **2018**, *8*, 404.
3. Hartl, C. Review on advances in metal micro-tube forming. *Metals* **2019**, *9*, 542. [CrossRef]
4. Aue-U-Lan, Y.; Ngaile, G.; Altan, T. Optimizing tube hydroforming using process simulation and experimental verification. *J. Mater. Process. Technol.* **2004**, *146*, 137–143. [CrossRef]
5. Yuan, S.J.; Han, C.; Wang, X.S. Hydroforming of automotive structural components with rectangular-sections. *Int. J. Mach. Tools. Manuf.* **2006**, *46*, 1201–1206. [CrossRef]
6. Kim, C.I.; Yang, S.H.; Kim, Y.S. Analysis of forming limit in tube hydroforming. *J. Mech. Sci. Technol.* **2013**, *27*, 3817–3823. [CrossRef]
7. He, Z.B.; Wang, Z.B.; Lin, Y.L.; Fan, X.B. Hot deformation behavior of a 2024 aluminum alloy sheet and its modeling by Fields-Backofen model considering strain rate evolution. *Metals* **2019**, *9*, 243. [CrossRef]
8. Manabe, K.I.; Amino, M. Effects of process parameters and material properties on deformation process in tube hydroforming. *J. Mater. Process. Technol.* **2002**, *123*, 285–291. [CrossRef]
9. Sadlowska, H.; Kocanda, A. On the problem of material properties in numerical simulation of tube hydroforming. *Arch. Civ. Mech. Eng.* **2010**, *10*, 77–83. [CrossRef]

10. Ngaile, G.; Jaeger, S.; Altan, T. Lubrication in tube hydroforming (THF) Part II. Performance evaluation of lubricants using LDH test and pear-shaped tube expansion test. *J. Mater. Process. Technol.* **2004**, *146*, 116–123. [CrossRef]
11. Fiorentino, A.; Ceretti, E.; Giardini, C. Tube hydroforming compression test for friction estimation-numerical inverse method, application, and analysis. *Int. J. Adv. Manuf. Technol.* **2013**, *64*, 695–705. [CrossRef]
12. Korkolis, Y.P.; Kyriakides, S. Hydroforming of anisotropic aluminum tubes: Part II analysis. *Int. J. Mech. Sci.* **2011**, *53*, 83–90. [CrossRef]
13. Ben Abdessalem, A.; El-Hami, A. Global sensitivity analysis and multi-objective optimization of loading path in tube hydroforming process based on metamodelling techniques. *Int. J. Adv. Manuf. Technol.* **2014**, *71*, 753–773. [CrossRef]
14. Yuan, S.J.; Wang, X.S.; Liu, G.; Wang, Z.R. Control and use of wrinkles in tube hydroforming. *J. Mater. Process. Technol.* **2007**, *182*, 6–11. [CrossRef]
15. Han, C.; Liu, Q.; Lu, H.; Gao, G.L.; Xie, W.C.; Yuan, S.J. Thickness improvement in hydroforming of a variable diameter tubular component by using wrinkles and preforms. *Int. J. Adv. Manuf. Technol.* **2018**, *99*, 2993–3003. [CrossRef]
16. Wang, X.S.; Cui, X.L.; Yuan, S.J. Research on flattening behavior of wrinkled 5a02 aluminum alloy tubes under internal pressure. *Int. J. Adv. Manuf. Technol.* **2016**, *87*, 1159–1167. [CrossRef]
17. Feng, H.; Han, C. Study on wrinkling behavior in hydroforming of large diameter thin-walled tube through local constraints. *Int. J. Adv. Manuf. Technol.* **2018**, *99*, 1329–1340. [CrossRef]
18. Han, C.; Feng, H. A new method for hydroforming of thin-walled spherical parts using overlapping tubular blanks. *Int. J. Adv. Manuf. Technol.* **2020**, *106*, 1543–1552. [CrossRef]
19. Han, C.; Feng, H. A new method for hydroforming of a variable-diameter part with partially overlapping tubular blanks. In Proceedings of the 9th International Conference on Tube Hydroforming, Kaohsuing, Taiwan, 18–21 November 2019; pp. 146–152.
20. Sasso, M.; Palmieri, G.; Chiappini, G.; Amodio, D. Characterization of hyperelastic rubber-like materials by biaxial and uniaxial stretching tests based on optical methods. *Polym. Test.* **2008**, *27*, 995–1004. [CrossRef]
21. Mooney, M. A theory of large elastic deformation. *J. Appl. Phys.* **1940**, *11*, 582–592. [CrossRef]
22. Koubaa, S.; Belhassen, L.; Wali, M.; Dammak, F. Numerical investigation of the forming capability of bulge process by using rubber as a forming medium. *Int. J. Adv. Manuf. Technol.* **2017**, *92*, 1839–1848. [CrossRef]
23. Koc, M.; Altan, T. Prediction of forming limits and parameter in the tube hydroforming process. *Int. J. Mach. Tools. Manuf.* **2002**, *42*, 123–138. [CrossRef]

Article

Oblique/Curved Tube Necking Formed by Synchronous Multipass Spinning

Hirohiko Arai [1,2,*] and Shiori Gondo [1]

1 Advanced Manufacturing Research Institute, National Institute of Advanced Industrial Science and Technology (AIST), Tsukuba 305-8564, Japan; shiori-gondo@aist.go.jp
2 Daitoh Spinning Co. Ltd., Oura-Town 370-0603, Japan
* Correspondence: h.arai@aist.go.jp; Tel.: +81-29-861-7088

Received: 28 April 2020; Accepted: 28 May 2020; Published: 2 June 2020

Abstract: In this paper, we propose a method of forming a tube into an oblique/curved shape by synchronous multipass spinning, in which the forming roller reciprocates in the radial direction in synchrony with the rotation angle of the spindle while the roller moves back and forth along the workpiece in the axial direction to gradually deform a blank tube into a target shape. The target oblique/curved shape is expressed as a series of inclined circular cross sections. The contact position of the roller and the workpiece is calculated from the inclination angle, center coordinates, and diameter of the cross sections, considering the geometrical shape of the roller. The blank shape and the target shape are interpolated along normalized tool paths to generate the numerical control command of the roller. By this method, we experimentally formed aluminum tubes into curved shapes with various radii of curvature, and the forming accuracy, thickness distribution, and strain distribution are examined. We verified that the curved shapes with the target radii of curvature can be accurately realized.

Keywords: metal spinning; tube forming; incremental forming; numerical control

1. Introduction

Metal spinning is a metal forming process in which a metal tube or sheet is rotated by a motor and pushed by a roller tool to form a target shape. Since the tooling cost of metal spinning is lower than that of press forming, it is effective for small-lot production of many kinds of custom-made products and for product prototyping. Moreover, this process is also applied to mass-produced parts and products such as automobile exhaust parts and stainless-steel bottles. It is also advantageous for manufacturing large metal shells requiring a large die. In addition, there is an advantage that the forming force is small because only the portion where the roller is in contact with is locally deformed; hence, the apparatus size, noise, and vibration are small.

To form products with higher added value, research on noncircular spinning, which can produce noncircular cross-sectional shapes such as an ellipse, eccentricity, and polygon, has advanced in recent years [1,2]. In conventional metal spinning, only products with a circular cross section can be formed, since the forming is performed while rotating the workpiece. If metal spinning can easily form noncircular shapes, which have been manufactured by press forming or sheet metal welding so far, the application of metal spinning will be further expanded.

Formation of exhaust system parts is a typical application of metal spinning in the automotive industry. There is a large demand for a manufacturing method by which the end of a tube is formed into an oblique/curved axis. Irie et al. [3] developed a spinning machine for necking a stainless steel tube into an oblique/curved shape by shifting and tilting the central axis of a blank tube relative to the spindle axis of the planetary forming rollers, which revolve around the workpiece while necking it. Xia and coworkers [4,5] built a finite element model (FEM) simulation of the same process and evaluated the forming force, stress, strain, and thickness distribution. They also spun an oblique shape in experiments and compared the forming force and the thickness distribution with the simulation results.

Another way to form an oblique/curved shape by metal spinning is synchronous spinning, in which the roller moves in synchrony with the workpiece rotation. Amano and Tamura [6] first succeeded in synchronous spinning of an elliptic cone from a steel sheet. They used a mechanical cam, gears, and a lever to drive a roller synchronized with the rotation of a mandrel. Xia et al. [7] proposed a profile driving method in which a three-dimensional cam pushes a forming roller to follow the surface of the product shape and formed various types of noncircular cones. The drawback of mechanical synchronization is lack of flexibility, because it requires a dedicated cam as well as a mandrel for each shape. Numerical control of the forming roller and spindle can realize synchronous spinning more easily without such complicated mechanisms. Shimizu [8] used pulse motors to drive the roller and mandrel to form elliptic and square cones by synchronous spinning. Arai et al. [9] built a CNC spinning machine with the spindle axis driven by a servo motor for synchronous spinning and conducted noncircular tube necking of aluminum tubes.

Sugita and Arai [10] proposed synchronous multipass spinning, which deforms the workpiece gradually in multiple steps from a flat blank to the final shape. This method can form noncircular shapes with vertical walls such as a square cup because the sine law for the wall thickness in shear spinning is not valid in multipass spinning. Härtel and Laue [11] formed a tripod shape with two roller passes while avoiding forming failures by optimization of the forming parameters based on FEM simulation. Music and Allwood [12] developed a spinning machine for noncircular shapes in which the internal rollers were controlled to support the workpiece instead of the mandrels. Russo et al. [13] used the spinning machine of [12] for synchronous multipass spinning of square and elliptic cups without using mandrels. They investigated how the degree of asymmetry of the cross-section shape influences the difficulty of forming and the properties of the formed products.

Regarding the synchronous spinning of an oblique/curved shape, Sekiguchi and Arai [14] proposed a method of controlling the tool position in the radial and axial directions in synchrony with the spindle rotation and experimentally formed oblique/curved shapes from aluminum sheets. In contrast, Han et al. [15] geometrically calculated the contact position of the roller with an oblique cone and moved the roller in the radial direction for synchronous spinning. Sekiguchi and Arai [16] used force control of the roller to push the material onto the inclined mandrel and formed an inclined cone by reciprocating the roller in the axial direction. As the processes described in [14–16] involve forming with only one pass and are equivalent to shear spinning, the wall thickness basically follows the sine law. The target shape should be limited to prevent fracture, and vertical walls cannot be formed. Arai and Kanazawa [17] combined this method with multipass spinning, in which the roller reciprocates between the mandrel and the outer edge of the workpiece to deform the workpiece progressively, thereby forming circular and square cups with inclined bottoms from aluminum disks. The position command of the roller is generated by combining the data of a blank disk, the target shape, and the normalized tool paths. Han et al. [18] spun a circular cup with an oblique open end from an offset blank by synchronous multipass spinning. Xiao et al. [19] also used a multipass spinning method to form an angled-flange cylinder from a blank disk.

When the synchronous spinning is performed without a mandrel, the shape of the formed product is completely determined by the roller motion. Therefore, it is necessary to transform the geometrical target shape into the numerical control (NC) program of the roller. In particular, the point of contact between the roller and the target product to form the target shape must be obtained to calculate the tool trajectory. However, the importance of this calculation stage has not been properly recognized so far except in some previous works. Arai [20] applied the synchronous multipass spinning method described in [17] for necking a tube into an oblique/curved shape. In the forming experiment, oblique/curved shapes that were almost the same as the target shape were obtained. On the other hand, the surface of the tube tends to be peeled off owing to friction caused by the side slip of the roller as the roller slides over the workpiece in the axial direction. Since this problem becomes more serious as the tilt angle increases, it is difficult to form a shape with a large inclination angle. Arai [21] utilized a 3D computer aided design (CAD) model in noncircular tube spinning to represent the target shape and calculated the roller contact position using a searching algorithm considering the roller shape geometrically. Although this method can also handle oblique/curved shapes and the spinning of inclined shapes was actually demonstrated in [21], it requires a 3D CAD system to build a target shape model.

In this paper, a synchronous multipass spinning method for oblique/curved tube necking is proposed. The target shape is represented by interpolation of a series of inclined circular cross sections with offsets without using a 3D CAD system. The position where the roller and the product surface come in contact at only one point to form the target shape is calculated by a searching algorithm considering the geometrical shape of the roller. The NC command of the roller is obtained by linear interpolation between the blank shape and the target shape along multiple tool paths, which are normalized in a two-dimensional virtual plane. Since the roller mainly moves in the radial direction in synchrony with the spindle rotation, peeling off of the workpiece surface described in [20] can be prevented. The workpiece is fixed to the spindle and not tilted as in [3–5]. It allows for a larger inclination angle and a smaller radius of curvature of the target shapes. Curved shapes with various radii of curvature are formed from aluminum tubes in the experiments by this method. The formed shapes are measured using a laser range sensor to verify the forming accuracy. The thickness distribution and strain distribution of the products are also evaluated.

2. Calculation of Roller Position

The roller motion in the synchronous spinning of a noncircular shape is expressed as three-dimensional data, consisting of the radial position and the axial position of the roller, and the spindle rotation angle. In the case of multipass spinning, the roller trajectory deforms the blank tube into the target shape step by step. The roller should move along the surface of the target shape in the final pass and the roller should move along the blank tube in the initial pass. Thus, the NC commands of the roller trajectory are calculated as follows.

(1) Calculation of the roller contact position with the product with the target shape.
(2) Interpolation of the blank shape and the target shape along the normalized paths.

2.1. *Calculation of Roller–Product Contact Position*

Since a nonspherical roller is commonly used for a forming tool in metal spinning, it is difficult to analytically solve the contact position between a general oblique/curved product surface and the roller. Therefore, the position where the roller and the target product come in contact at a single point is obtained by a searching algorithm. The target shape is expressed as a series of inclined circles with offset, whose radii, inclination angles, and center coordinates are given individually. Then, the coordinates of a point on the surface of the target shape can be represented by two parameters, the normalized axial distance along the workpiece, λ $(0 \leq \lambda < 1)$, and the circumferential central angle along the cross section, ω $(0 \leq \omega < 2\pi)$. Here, the roller axial position and the spindle angle are assumed to be fixed, which means that the roller only moves in the radial direction. By varying the parameters λ and ω, we obtain the points on the target shape corresponding to each set of parameters. For each of these points, the radial position of the roller when this point is also on the surface of the roller is calculated. Among these positions, the position where the roller is most distant from the spindle axis is the position where the roller and the target product come into contact at only one point.

2.1.1. Expression of Target Shape and Roller Shape

First, a point on the surface of the target shape is expressed by the x, y, z coordinates in the Cartesian coordinate system (Figure 1). The spindle axis of the spinning machine coincides with the z axis, and the radial direction of the roller motion is parallel to the x axis. An oblique/curved shape of a tube with circular cross sections whose centers are on the xz plane is the target shape. In a cross section of the tube with the target shape, the radius of the circle is r, the inclination angle relative to the z axis is α, and the center coordinates are $(x_C, 0, z_C)$. Here, r, α, x_C, and z_C are the functions $r(\lambda)$, $\alpha(\lambda)$, $x_C(\lambda)$, and $z_C(\lambda)$ $(0 \leq \lambda < 1)$ of the parameter λ. $\lambda = 0$ and $\lambda = 1$ represent both ends of the target shape.

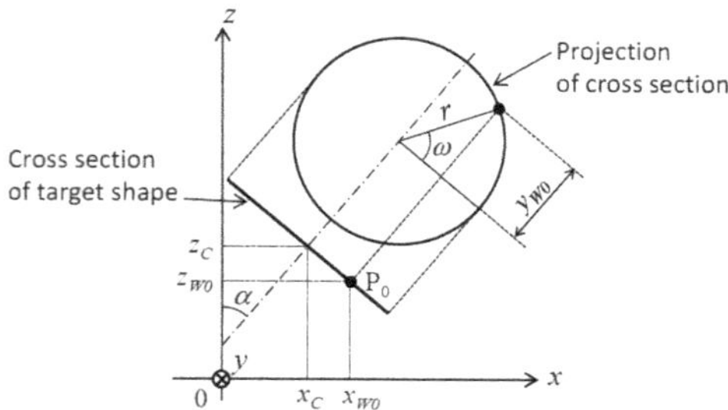

Figure 1. xyz coordinates of point P_0 on a circular cross section of the target shape. The y axis is in the direction into the page.

In practice, the target shape is represented by a series of discrete circular cross sections, and r, α, x_C, and z_C can be obtained by interpolation at intermediate positions. When λ_k and λ_{k+1} are respectively assigned to the kth and k + 1th cross sections, the radius r, the inclination angle α, and the center coordinates $(x_C, 0, z_C)$ in the intermediate cross section are represented as

$$\begin{cases} r = \frac{(\lambda_{k+1}-\lambda)r_k+(\lambda-\lambda_k)r_{k+1}}{\lambda_{k+1}-\lambda_k} \\ \alpha = \frac{(\lambda_{k+1}-\lambda)\alpha_k+(\lambda-\lambda_k)\alpha_{k+1}}{\lambda_{k+1}-\lambda_k} \\ (x_C,\ 0,\ z_C) = \frac{\lambda_{k+1}-\lambda}{\lambda_{k+1}-\lambda_k}\left(x_{C_k},\ 0,\ z_{C_k}\right) + \frac{\lambda-\lambda_k}{\lambda_{k+1}-\lambda_k}\left(x_{C_{k+1}},\ 0,\ z_{C_{k+1}}\right) \end{cases} \tag{1}$$

The point P_0 (x_{W0}, y_{W0}, z_{W0}) on the surface of the tube with the target shape corresponding to the central angle ω is represented as

$$\begin{cases} x_{W0} = x_C + r\cos\alpha\cos\omega \\ y_{W0} = r\sin\omega \\ z_{W0} = z_C - r\sin\alpha\cos\omega \end{cases} \tag{2}$$

When the tube with the target shape is rotated around the z axis by an angle θ, the point P_0 moves to P (x_W, y_W, z_W).

$$\begin{pmatrix} x_W \\ y_W \\ z_W \end{pmatrix} = \begin{pmatrix} \cos\theta & -\sin\theta & 0 \\ \sin\theta & \cos\theta & 0 \\ 0 & 0 & 1 \end{pmatrix} \begin{pmatrix} x_{W0} \\ y_{W0} \\ z_{W0} \end{pmatrix} \tag{3}$$

The shape of the roller is the outer periphery of a donut-shaped torus of diameter D, with a roundness radius of ρ (Figure 2). The radius of the roundness center R is expressed as $R = \frac{D}{2} - \rho$. The center of the roller is on the xz plane and the roller axis is parallel to the z axis. Let the coordinates of the roller center Q be $(x_R, 0, z_R)$. When the point P (x_W, y_W, z_W) on the surface of the tube with the target shape is in contact with the roller, the radius R_W of the roller surface at the contact point is

$$R_W = R + \sqrt{\rho^2 - (z_W - z_R)^2} \tag{4}$$

At this time, the x coordinate of the roller center Q is expressed as

$$x_R = x_W + \sqrt{{R_W}^2 - {y_W}^2} \tag{5}$$

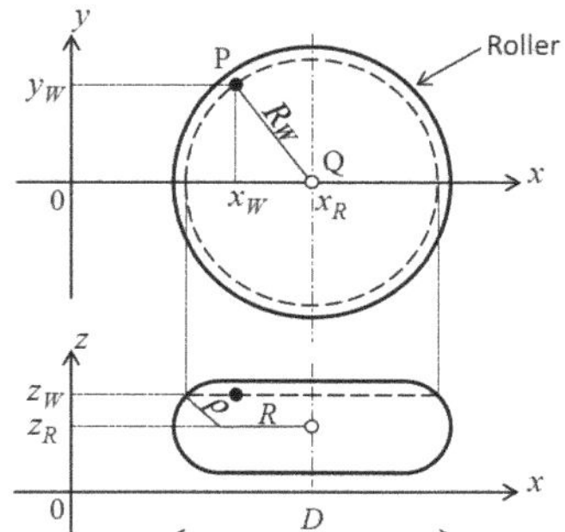

Figure 2. Position of the center of the roller, Q. Point P on the tube with the target shape is also on the surface of the roller.

2.1.2. Search of Contact Position

Then the point of contact between the roller and the target product to form the target shape is searched by the following procedure (Figure 3). The rotation angle θ of the target shape around the z axis and the position z_R of the roller in the axial direction are given. The parameters λ and ω are varied within the range of $0 \leq \lambda < 1$ and $0 \leq \omega < 2\pi$, and the corresponding points (x_W, y_W, z_W) on the tube with the target shape are obtained from Equations (2) and (3). Then, the radial position x_R of the roller is calculated from Equations (4) and (5). The position where x_R becomes maximum is the position where the target shape and the roller come in contact at a single point. The above search calculation for (λ,ω) is repeated while changing the angle θ of the target shape and the axial position z_R of the roller, and the corresponding radial position $x_R(z_R, \theta)$ of the roller is obtained over the entire surface of the tube with the target shape.

The contact point between the roller and the target shape only exists in $z_R - \rho < z_w < z_R + \rho$ considering the thickness of the roller. Moreover, it is not necessary to consider the contact position on the opposite side where the target shape does not face the roller. These limitations can narrow the search range of λ and ω. In practice, the incremental step sizes $\Delta\lambda$ and $\Delta\omega$ are assigned to λ and ω, respectively. After the maximum value of x_R is obtained in the discretized space of (λ,ω) with coarse step sizes, a finer search using smaller step sizes is conducted around the neighborhood of (λ,ω) for the former maximum x_R. This algorithm can reduce the total computation time.

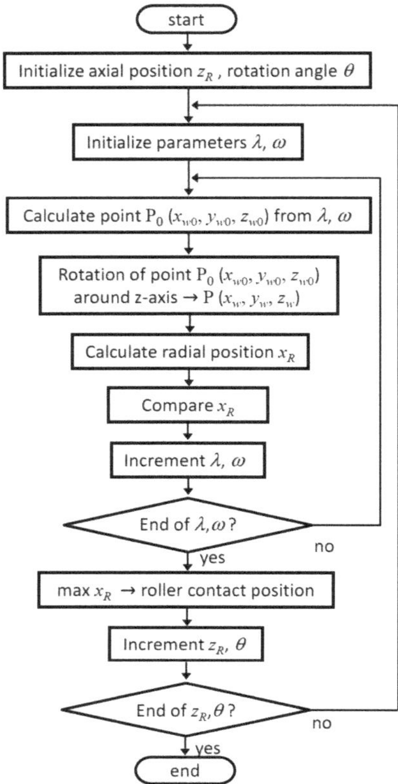

Figure 3. Computation algorithm for searching the roller contact position with the target shape.

2.2. Interpolation of Roller Position

Finally, the roller position during multipass spinning is calculated by interpolation. The roller moves back and forth in the axial direction in synchrony with the spindle rotation to approach the oblique/curved target shape. The computation method in [21] to calculate the NC command from the positions where the roller comes in contact with the product and the blank tube is reviewed here.

The multiple tool paths composed of curved and straight path elements are defined on a virtual two-dimensional plane (Figure 4) to cope with the asymmetry of a noncircular target shape. The virtual plane is normalized from 0 to 1 in the axial direction s_z and radial direction s_x. The components s_x and s_z are used as interpolation coefficients between the target shape and the blank shape. $s_z = 0$ means the tip end of the product and $s_z = 1$ means the base end. When $s_x = 0$, the roller is on the product surface while the roller is on the blank surface when $s_x = 1$.

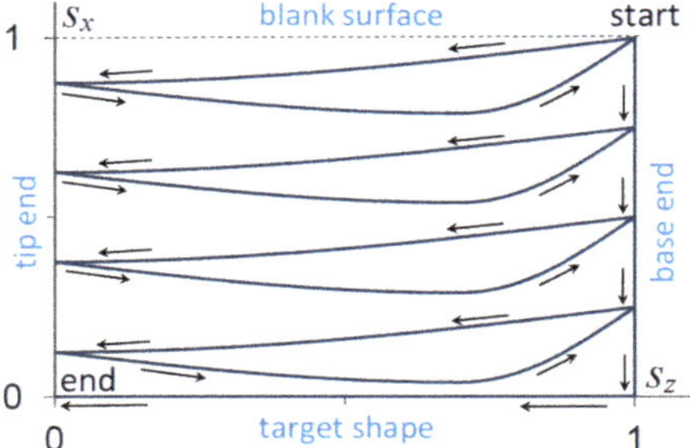

Figure 4. Example of normalized tool paths. The roller starts from the blank surface of the base end and proceeds to the tip end increasing the radial feed. Then, the roller returns to the base end and the radial feed is increased. This is repeated until $s_x = 0$ and the roller moves along the target shape to the tip end.

The axial and radial positions where the roller touches the product with the target shape are defined as $z_t(s_z)$ and $x_p(s_z, \theta)$, respectively, where θ is the spindle angle. $z_t(s_z)$ is obtained from s_z as

$$z_t = z_0 + (z_1 - z_0)s_z \tag{6}$$

where z_0 is the axial position of the tip end and z_1 is that of the base end. The radial position for the blank tube at s_z is also defined as $x_b(s_z, \theta)$. $x_b(s_z, \theta)$ is constant if the blank is a straight circular tube.

The radial positions, $x_p(s_z, \theta)$ and $x_b(s_z, \theta)$, are interpolated using s_x as an interpolation coefficient to obtain the intermediate radial position x_t as follows.

$$x_t = s_x x_b(s_z, \theta) + (1 - s_x)x_p(s_z, \theta) \tag{7}$$

Thus, the roller position (x_t,z_t) is sequentially calculated along the normalized paths to obtain the entire trajectory from the start to the end of the spinning process. The series of roller positions and spindle angles (x_t,z_t, θ) are executed by NC commands for linear interpolation.

3. Forming Experiments

Using the NC programs generated by the method in Section 2, aluminum tubes were experimentally spun into curved shapes with various radii of curvature. The formed products were scanned by a laser range sensor to verify the dimensional accuracy. The thickness and strain distribution of the products were also examined.

3.1. Experimental Setup

The spinning experiments were conducted using a five-axis CNC spinning machine with two rollers (Figure 5). Each roller was driven by two AC servo motors of 1.3 kW and ball screws of 10 mm lead. The spindle was driven by an AC servo motor of 1.3 kW with a planetary reduction gear. The rollers were made from tool alloy steel (SKD 11) with 88 mm diameter and 4 mm roundness radius. The blank tube of pure aluminum (A 1050 TD-H) had 50 mm diameter and 1.55 mm thickness. The surface of the workpiece and the roller were lubricated using compressor oil (ISO VG 68).

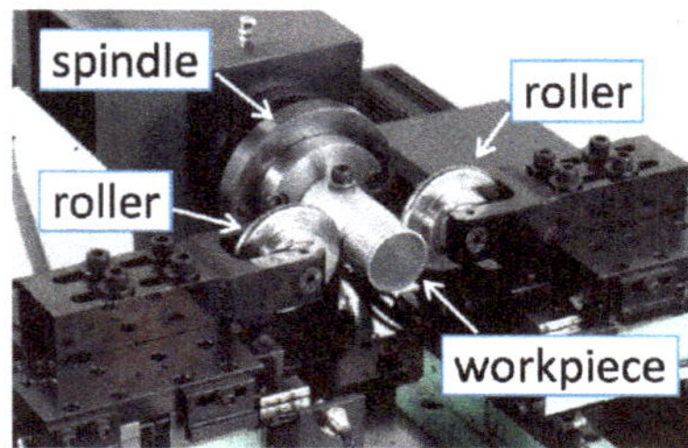

Figure 5. Five-axis two-roller CNC spinning machine.

3.2. Target Shape

Four types of curved shape with different radii of curvature, samples A, B, C, and D, were formed. The diameter of the necked portion was 30 mm. The radii of curvature of the outer side were (A) 40 mm, (B) 50 mm, (C) 60 mm, and (D) 70 mm, therefore, the radii of curvature of the inner side were (A) 10 mm, (B) 20 mm, (C) 30 mm, and (D) 40 mm, respectively. A cross section in the middle of the curved portion was perpendicular to the spindle axis ($\alpha = 0$) and its center had an offset x_C of 7 mm from the axis. As described in Section 2.1.1, the curved portion was composed of discrete circular cross sections. The step of inclination angle was 1° and the position of each center was calculated according to the radius of curvature. The maximum inclination angle, $|\alpha_{MAX}|$, was (A) 64°, (B) 55°, (C) 45°, and (D) 41°.

The calculation algorithm of the contact position of the roller in Section 2.1 was programmed in C language. The contact positions were calculated at 120 points per rotation of the workpiece and every 1 mm in the axial position of the roller. In the case of sample D, 8760 contact positions were calculated in total. The calculations took 12 s using a Windows personal computer (Intel Core i7 CPU, 3.6 GHz).

3.3. Multiple Tool Paths

In the experiments reported in [21], normalized multiple tool paths that were nearly parallel to the axial direction similar to those in Figure 4 were used. Regarding spinning eccentric and oblique shapes, [21] showed that the asymmetry of the radial feed resulted in uneven axial elongation, which depends on the circumferential location. In the case of a small number of paths, the bias of elongation bended the workpiece during the spinning process and it deteriorated the dimensional accuracy.

To solve this problem, a different pattern of normalized paths as shown in Figure 6 was used. Each path element connected between the base side of the target shape and the tip side of the blank. The dimensional error in [21] occurred because the intermediate workpiece was bent beyond the surface of the target shape. The new normalized paths can provide the intermediate workpiece a larger margin to the target shape as the roller approaches the tip end.

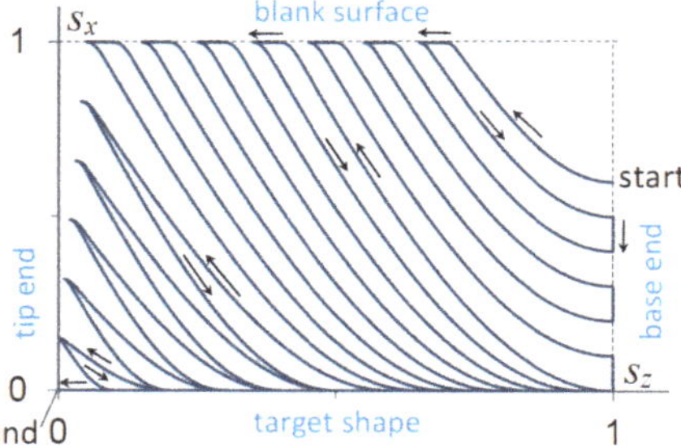

Figure 6. Normalized tool paths used in the experiments. The roller moves along the curved path elements between the base side of the target shape and the tip side of the blank surface. The radial feed to the target shape increases from the base side to the tip side.

The number of roller passes was 12 roundtrips. The axial roller feed was 2 mm/rev. The final pass along the whole target shape had a 1 mm/rev roller feed. The spindle speed was 30 rpm.

3.4. Mearsurement of Products

Three samples were formed for each type of sample (A, B, C, and D). Two samples were used for shape measurement by the laser sensor. White paint was sprayed on the surface of the spun samples to obtain diffused reflection. A laser range sensor of 0.5 μm resolution was used to measure the shape of each product. To obtain the profiles of the outer curve and inner curve, the surface of the sample was measured every 1 mm in the axial direction. The average values of the two samples were plotted. The circular cross section in the middle of the curved portion, which was perpendicular to the spindle axis, was also measured circumferentially using the laser sensor every 9° of workpiece rotation. From these 40 measurement positions, the offset of the center of the necked portion from the axis was calculated. Also, the distance between the center and each measurement point was calculated, and the average radius and the out of roundness, i.e., the difference between the maximum radius and the minimum radius, were calculated.

The wall thickness and the strain distribution along the outer and inner sides of the curvature were measured using another sample. For the measurement of the axial and circumferential strains, grid lines were scribed on the inner surface of the blank tube before forming. The CNC spinning machine was used to control the position of the L-shaped scriber and the rotation of the blank tube to scribe the marks precisely on the inside of the tube. The line marks parallel to the spindle axis were scribed every 7.5° of rotation of the blank tube. The circumferential marks perpendicular to the axis were scribed every 2.5 mm in the axial direction for samples A and B and every 5 mm for samples C and D. The change in the distance between the marks after the spinning process was measured to calculate the axial and circumferential strains. The samples were cut in the axial direction away from the center of the curve. Then pencil graphite was rubbed into the surface grooves. Transparent adhesive tape was placed on the grid surface, removed, and transferred to paper to obtain a high-resolution image of the grid marks for scanning. The positions of the grid points were identified by the image viewer software to calculate the distance between the marks. The wall thickness of the same cut sample was also measured using a digital micrometer with a dual ball anvil along the outer and inner curves every 5 mm in the axial direction.

4. Results and Discussion

4.1. Dimensional Accuracy

Figure 7 shows a photograph of the formed products of samples A, B, C, and D with different radii of curvature. See Video S1 of the supplementary files. The peeling off of the workpiece surface reported in [20] did not occur. Figure 8a,b show the profiles of the outer and inner sides of the curved portion measured using the laser range sensor. The profiles mostly coincide with those of the target shape (thin solid lines) except that the inner sides near the tip end of samples C and D tend to deviate from the target shape. The root mean square of the deviation from the target shape is plotted in Figure 9. The deviations are 0.08–0.12 mm on the outer side and 0.17–0.32 mm on the inner side.

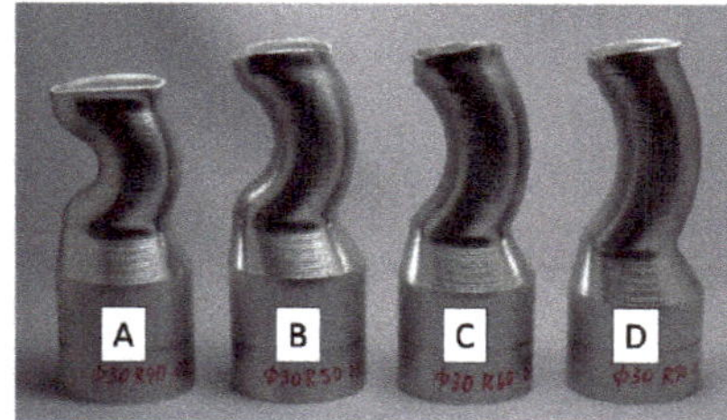

Figure 7. Formed products of samples A, B, C, and D.

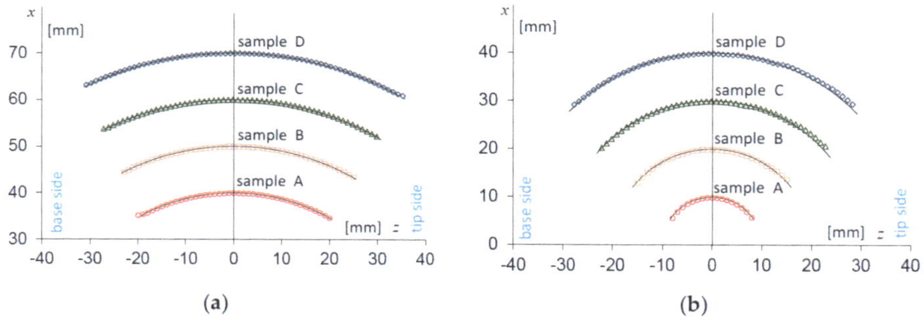

Figure 8. Profiles of the curved portion of samples A, B, C, and D. (**a**) Outer and (**b**) inner sides of the curve. The thin solid lines correspond to target shapes. The horizontal axis represents the axial distance from the middle cross section perpendicular to the spindle axis, in which the base side is negative and the tip side is positive. The origin (0,0) is at the center of each target curvature.

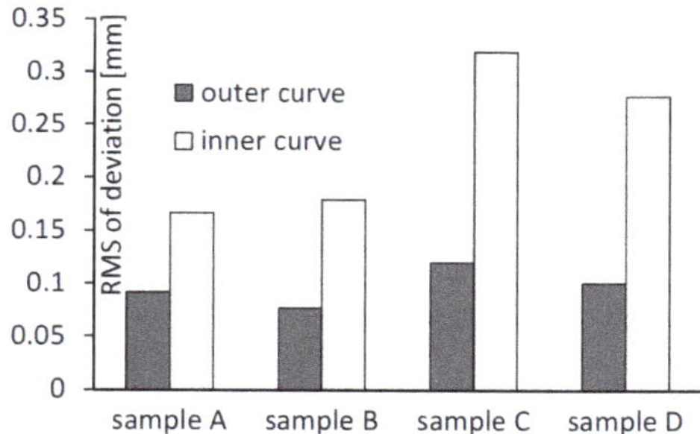

Figure 9. Root mean square of the deviation from the target shape.

The circular cross sections in the middle of the curved portion, which are perpendicular to the spindle axis, were also measured circumferentially using the laser sensor. Table 1 shows the average radius, the offset from the axis, and the out-of-roundness of each sample. The cross sections have the radius and the offset close to the target values, 15 and 7 mm.

Table 1. Average radius, offset from the axis, and out-of-roundness of the cross sections perpendicular to the spindle axis.

Sample Type	Average Radius (mm)	Offset from Axis (mm)	Out of Roundness (mm)
A	15.046	6.893	0.096
B	15.063	6.945	0.082
C	15.047	6.859	0.129
D	15.045	6.917	0.068

4.2. Strain Distribution

Figure 10a,b show the distribution of thickness strain along the outer and inner sides of the curved portion of samples A, B, C, and D. The thickness strain was calculated from the measured thickness of the formed products and the blank tube. The thickness strain of the outer side shows a similar tendency for all samples. The wall thickness increases around the middle part and decreases towards the base end. The thickness also decreases towards the tip end to a lesser extent. On the other hand, the thickness of the inner side decreases around the middle of the curved portion. The peak of thinning deviates from the zero position towards the base end. As the radius of the curvature increases, the thinning becomes smaller and the thickness surpasses the blank thickness on the opposite side. The thinning peak moves towards the base end as the curvature radius increases. The actual wall thicknesses were as follows: sample A, 0.90–2.08 mm; sample B, 1.21–2.11 mm; sample C, 1.49–2.09 mm; sample D, 1.65–2.57 mm.

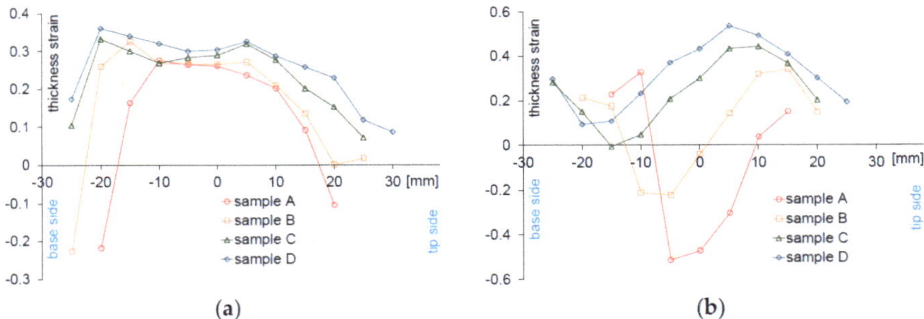

Figure 10. Distribution of thickness strain in the axial direction. (**a**) Outer and (**b**) inner sides of the curve. The horizontal axis represents the distance in the axial direction. The origin of the horizontal axis is at the middle cross section perpendicular to the spindle axis.

Figure 11a,b are plots of the axial strain and the circumferential strain along the outer side of the curved portion. The surface of the outer side shrinks circumferentially and stretches axially. The absolute value of these strains increases towards the tip end. Figure 12a,b show the axial and circumferential strains along the inner side. Although the inner side also shrinks circumferentially and stretches axially, the peak of strain is around the middle cross section. The axial strain at the middle cross section increases as the radius of the curve decreases. This tendency agrees with that of the change in thickness in Figure 10b.

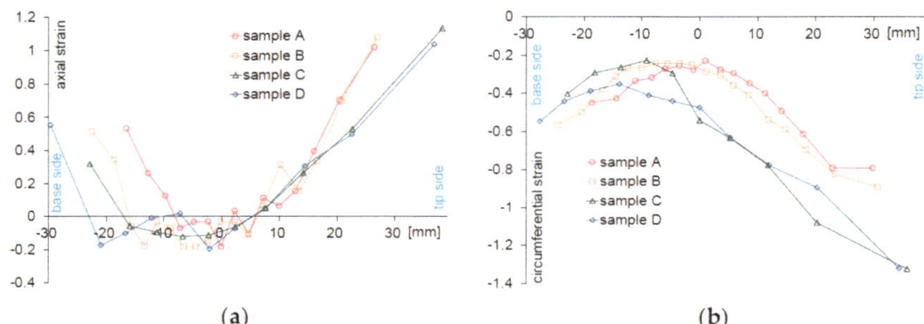

Figure 11. Distributions of (**a**) axial strain and (**b**) circumferential strain in the axial direction of the outer curve. The horizontal axis represents the distance along the curved surface in the axial direction. The origin is at the middle cross section perpendicular to the spindle axis.

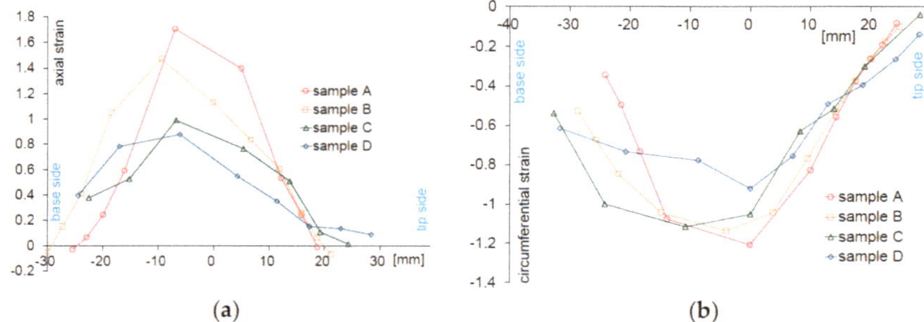

Figure 12. Distributions of (**a**) axial strain and (**b**) circumferential strain in the axial direction of the inner curve. The horizontal axis represents the distance along the curved surface in the axial direction. The origin is at the middle cross section perpendicular to the spindle axis.

4.3. Effect of Normalized Paths

To verify the effect of the new normalized paths in Figure 6, they were compared with simple normalized paths parallel to the spindle axis. Sample B′, which had the same target shape as sample B, was spun using the parallel paths and the dimensional accuracy was measured. Figure 13 shows the actual roller paths for the inner side of the curve after the interpolation described in Section 2.2. Figure 13a shows the actual paths using the parallel normalized paths and Figure 13b shows the actual paths using the new normalized paths. The same axial roller feed and spindle speed were used for both paths.

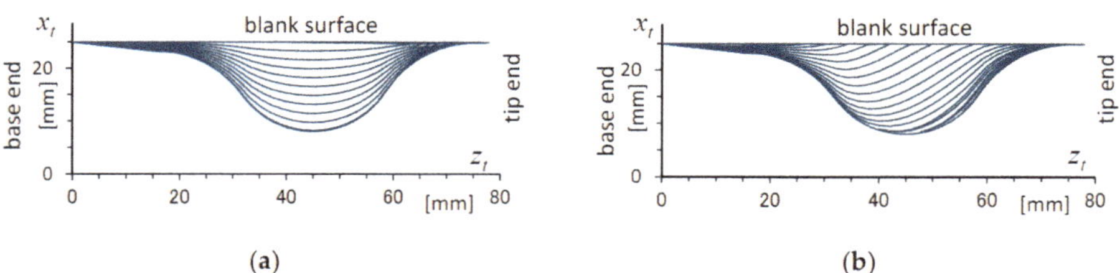

Figure 13. Actual roller paths for samples (**a**) B′ and (**b**) B. x_t and z_t are the positions of the roller in the radial and axial directions. The origin of the horizontal axes is at the base end of the product.

The shapes of the inner curve measured using the laser sensor are plotted in Figure 14. The profile of sample B′ deviates from the target shape near the tip end of the curve. In addition, the roller deviated from the workpiece surface of the inner curve near the tip end while sample B′ was spun. The radial feed of the inner side of the curve was larger than that of the outer side. Then, the inner side elongated more than the outer side, and the tip end of the workpiece was bent towards the outer side beyond the target shape. In contrast, the final roller paths of Figure 13b show less radial feed at the base side and a larger margin to the target shape at the tip side than Figure 13a. Hence, the dimensional error near the tip end was smaller in sample B.

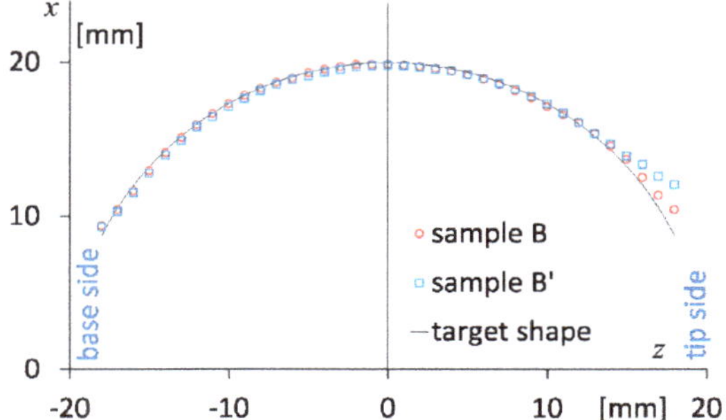

Figure 14. Profiles of the inner curves of samples B (red) and B′ (blue). The thin solid line corresponds to the target shape. The horizontal axis represents the axial distance from the middle cross section perpendicular to the spindle axis. The origin (0,0) is at the center of the target curvature.

5. Conclusions

The proposed searching algorithm can find the contact position between the roller and the oblique/curved shapes expressed as a series of inclined circular cross sections with offset. The experiments focused on relatively simple curved shapes with cross sections of a constant radius and with a constant curvature. However, more complicated curved shapes of changing radius and curvature can be handled by the same algorithm. It is also easy to calculate the roller contact position with the oblique target shapes of the straight-line profile.

In this study, the position of a point on the surface of the target shape is calculated from the two parameters λ and ω using Equations (2) and (3). Here, it is assumed that the center of the circular cross section is on the xz plane, and the cross sections are rotated around the y axis. These equations can be naturally extended to handle more general cases where the center of the cross section has offset in the x and y directions, and the cross section is inclined around the x and y axes. Furthermore, the framework of this method is theoretically applicable to an arbitrary tubular surface that can be parametrically expressed as functions of two parameters, e.g., ellipsoids, even if the cross-sectional shape is noncircular.

The experimental results demonstrated that the proposed method can form the curved shapes according to the target shape parameters. A small radius of curvature (10 mm) and a large inclination angle (64°) were achieved in sample A. It is difficult for the spinning machine reported in [3–5] to form such a shape because of the mechanical interference between the rollers and the workpiece. Reference [20] showed peeling off of the workpiece surface owing to the side slip of the roller in the normal direction. In contrast, the motion of the roller relative to the workpiece in the proposed method is in the rolling direction of the roller and does not have the side slip. The proposed method prevents the peeling off and hence allows a large inclination angle. Comparison of these two methods can be seen in Videos S2 and S3 of the supplementary files.

Reference [21] suggests that the asymmetry of the radial roller feed resulted in the partial elongation of the workpiece in the axial direction and bending of the workpiece. The improvement of the normalized tool paths as shown in Figure 6 gave the tip side of the intermediate workpiece a larger margin with respect to the target shape. Even when the workpiece is bent, its surface does not go beyond the target profile. The experimental results presented in Figure 14 show the suppression of the dimensional error in the proposed method compared with the previous method. Nevertheless, the error near the tip end is still observed for the new normalized paths of sample B. The deviation from the target profile is also found in samples C and D in Figure 8b. The deviation of the inner curve is larger than that of the outer curve in Figure 9. These dimensional errors may also be attributed to the bending due to the asymmetric radial feed. Further investigation on methods to deal with this problem more generally would be necessary in future work.

Supplementary Materials: The following are available online at http://www.mdpi.com/2075-4701/10/6/733/s1, Video S1: forming process of sample type B by the proposed method. Video S2: peeling off of workpiece surface by the method of Reference [20]. Video S3: spinning of the same shape without peeling off by the proposed method.

Author Contributions: Conceptualization, H.A.; Data curation, S.G.; Formal analysis, S.G.; Investigation, H.A. and S.G.; Methodology, H.A.; Software, H.A.; Validation, S.G.; Writing—original draft, H.A.; Writing—review and editing, H.A. and S.G. All authors have read and agreed to the published version of the manuscript.

Funding: This research received no external funding.

Acknowledgments: The authors wish to thank Shuhei Murakami, Yamagata Research Institute of Technology, for measurement of experimental results.

Conflicts of Interest: The authors declare no conflict of interest.

References

1. Music, O.; Allwood, J.M.; Kawai, K. A review of the mechanics of metal spinning. *J. Mater. Process. Technol.* **2010**, *219*, 3–23. [CrossRef]
2. Xia, Q.; Xiao, G.; Long, H.; Cheng, X.; Sheng, X. A review of process advancement of novel metal spinning. *Int. J. Mach. Tools Manuf.* **2014**, *85*, 100–121. [CrossRef]
3. Irie, T. Method and Apparatus for Forming a Processed Portion of a Workpiece. U.S. Patent US6223993B1, 22 May 2001.
4. Xia, Q.; Cheng, X.; Hu, Y.; Ruan, F. Finite element simulation and experimental investigation on the forming forces of 3D non-axisymmetrical tubes spinning. *Int. J. Mech. Sci.* **2006**, *48*, 726–735. [CrossRef]
5. Xia, Q.; Cheng, X.; Long, H.; Ruan, F. Finite element analysis and experimental investigation on deformation mechanism of non-axisymmetric tube spinning. *Int. J. Adv. Manuf. Technol.* **2012**, *59*, 263–272. [CrossRef]
6. Amano, T.; Tamura, K. The study of an elliptical cone spinning by the trial equipment. In Proceedings of the 3rd International Conference on Rotary Metalworking Processes, Kyoto, Japan, 8–10 September 1984; pp. 213–224.
7. Xia, Q.; Lai, Z.; Zhan, X.; Cheng, X. Research on Spinning Method of Hollow Part with Triangle Arc-Type Cross Section Based on Profiling Driving. *Steel Res. Int.* **2010**, *81*, 994–998. [CrossRef]
8. Shimizu, I. Asymmetric forming of aluminum sheets by synchronous spinning. *J. Mater. Process. Technol.* **2010**, *210*, 585–592. [CrossRef]
9. Arai, H.; Okazaki, I.; Fujimura, S. Synchronized metal spinning of non-axisymmetric tubes. In Proceedings of the 56th Japanese Joint Conference for the Technology of Plasticity, Naha, Japan, 18–20 November 2005; pp. 687–688. (In Japanese)
10. Härtel, S.; Laue, R. An optimization approach in non-circular spinning. *J. Mater. Process. Technol.* **2016**, *229*, 417–430. [CrossRef]
11. Sugita, Y.; Arai, H. Formability in synchronous multipass spinning using simple pass set. *J. Mater. Process. Technol.* **2015**, *217*, 336–344. [CrossRef]
12. Music, O.; Allwood, J.M. Flexible asymmetric spinning. *CIRP Ann. Manuf. Technol.* **2011**, *60*, 319–322. [CrossRef]
13. Russo, I.; Cleaver, C.; Allwood, J.M.; Loukaides, E. The influence of part asymmetry on the achievable forming height in multi-pass spinning. *J. Mater. Process. Technol.* **2020**, *275*, 116350. [CrossRef]
14. Sekiguchi, A.; Arai, H. Synchronous die-less spinning of curved products. *Steel Res. Int.* **2010**, *81*, 1010–1013. [CrossRef]

15. Han, Z.; Xu, Q.; Jia, Z.; Li, X. Experimental research on oblique cone die-less shear spinning. *Proc. Inst. Mech. Eng. Part B J. Eng. Manuf.* **2017**, *231*, 1182–1189. [CrossRef]
16. Sekiguchi, A.; Arai, H. Development of oblique metal spinning with force control. In Proceedings of the 2010 International Symposium on Flexible Automation, Tokyo, Japan, 12–14 July 2010; pp. 1863–1869.
17. Arai, H.; Kanazawa, T. Synchronous multipass spinning of oblique-bottom shape. *J. Mater. Process. Technol.* **2018**, *260*, 66–76. [CrossRef]
18. Xiao, Y.; Han, Z.; Zhou, S.; Jia, Z. Asymmetric spinning for offset blanks. *Int. J. Adv. Manuf. Technol.* **2020**, *107*, 2433–2448. [CrossRef]
19. Xiao, Y.; Han, Z.; Fan, Z.; Jia, Z. A study of asymmetric multi-pass spinning for angled-flange cylinder. *J. Mater. Process. Technol.* **2018**, *256*, 202–215. [CrossRef]
20. Arai, H. NC programming for oblique/curved tube necking using synchronous multipass spinning. In Proceedings of the 2015 Japanese Spring Conference for the Technology of Plasticity, Yokohama, Japan, 29–31 May 2015; pp. 171–172. (In Japanese)
21. Arai, H. Noncircular tube spinning based on three-dimensional CAD model. *Int. J. Mach. Tools Manuf.* **2019**, *144*. [CrossRef]

Article

A Vision-Based Fuzzy Control to Adjust Compression Speed for a Semi-Dieless Bellows-Forming

Sugeng Supriadi [1,*], Tsuyoshi Furushima [2] and Ken-ichi Manabe [3]

1 Department of Mechanical Engineering, Universitas Indonesia, Jakarta 16424, Indonesia
2 Department Mechanical and Bio-functional Systems, Tokyo University, Tokyo 153-8505, Japan; tsuyoful@iis.u-tokyo.ac.jp
3 Department of Mechanical Systems Engineering, Tokyo Metropolitan University, Tokyo 192-0397, Japan; manabe@tmu.ac.jp
* Correspondence: sugeng@eng.ui.ac.id

Received: 9 April 2020; Accepted: 29 April 2020; Published: 28 May 2020

Abstract: A novel semi-dieless bellows forming process with a local heating technique and axial compression has been initiated for the past years. However, this technique requires a high difficulty in maintaining the output quality due to its sensitivity to the processing conditions. The product quality mainly depends on not only the temperature distribution in the radial and axial direction but also the compression ratio during the semi-dieless bellows process. A finite element model has clarified that a variety of temperature produced by unstable heating or cooling will promote an unstable bellows formation. An adjustment to the compression speed is adequate to compensate for the effect of the variety of temperatures in the bellows formation. Therefore, it is necessary to apply a real-time process for this process to obtain accurate and precise bellows. In this paper, we are proposing a vision-based fuzzy control to control bellows formation. Since semi-dieless bellows forming is an unsteady and complex deformation process, the application of image processing technology is suitable for sensing the process because of the possible wide analysis area afforded by applying the multi-sectional measuring. A vision sensing algorithm is developed to monitor the bellows height from the captured images. An adaptive fuzzy has been verified to control bellows formation from 5 mm stainless steel tube in to bellows profile up to 7 mm bellows height, processing speed up to 0.66 mm/s. The adaptive fuzzy control system is capable of appropriately adjusting the compression speed by evaluating the bellows formation progress. Appropriate compression speed paths guide bellows formation following deformation references. The results show that the bellows shape accuracy between target and experiment increase become 99.5% under given processing ranges.

Keywords: bellows forming; vision-based sensor; fuzzy control; semi-dieless forming; local heating

1. Introduction

Metal bellows are convoluted metal tube that provide high flexibility on various direction. It is widely applied in the flexible joining of piping systems for water, oil and gas provisions. Metal bellows are usually produced through a hydroforming or a gas-forming process. When we increase the internal pressure of the tube, the bellows shape will be formed following the die shape. A new process of manufacturing expansion joint bellows from Ti-6Al-4V alloys using a gas pressure [1]. A fluid pressure instead of a gas pressure in the metal semi-dieless bellows forming process [2]. A single-step tube hydroforming process for producing metal bellows has been conducted with specific dies to fabricate a rectangular, circular, and triangular bellows profile [3]. However, not only do those methods require expensive dies and a complex machine but they also are inefficient in manufacturing a small quantity or various sizes of bellows. I the previous work, a novel technique had been proposed to fabricate metal bellows without employing internal fluid pressure or dies. The new technique only employed a compression-assisted local heating using a high-frequency induction heating source [4]. An observation on deformation mechanism of semi-dieless bellows forming using finite element analysis has been verified that bellows formation was caused by localized heating on top of the convolution due to the effect of various heating techniques that could be employed using induction heating [5,6]. Since the deformation is not guided using dies, the forming phenomena is very vulnerable to the processing conditions such as temperature, compression ratio, and cooling process. As a result, the produced bellows have low dimensional/shape accuracies of convolutions. Low efficiency is one of the technical issues in this process. Therefore, in this study, a vision-based fuzzy control system for semi-dieless bellows forming was developed to improve bellows accuracy and repeatability. Finite element analysis was also observed to establish appropriate control for this process.

2. The Principle of Semi-Dieless Bellows Forming through Local Heating

Semi-dieless bellows forming with a local heating technique is a novel process of producing a bellows shape without the uses of internal pressure and dies. This process is borrowed from previous technology of dieless drawing process where using force and localized heat to produce incrementally metal deformation such as stainless steel, magnesium and ceramic [7–12]. However, the forming principle of semi-dieless belows forming is entirely different from all other applied loading and deformation phenomena. Figure 1 shows the schematic of semi-dieless bellows forming with a local heating technique [5]. A combination of local heating and compression moving promotes continuous local bucking deformation of a tubular workpiece with a mandrel inner tube. We start the process of the semi-dieless bellows forming by thickening the tube under a heating coil. Afterwards, the local buckling is induced and moved to the cooling region as the first convolution. Simultaneously, a new local buckling occurs behind the first convolution leading to the second convolution. This process is repeated until a series of convolutions produces bellows. The local buckling results from the low flow stress of the material at an elevated temperature. We obtain the compressive load by applying the compression speed v_1 faster than the feeding speed v_2 ($v_1 > v_2$). To measure the convolution formation progress, the compression ratio (C) is used, as in $C = v_1/v_2$. After the first convolution formed by the buckling passes through the cooling area, the next convolution will be initiated behind the first convolution located inside the heating coil. When the convolutions are continuously produced in the tube, bellows are produced. We can control the bellows height by selecting the compression ratio [4,5]. The increasing compression ratio increases the bellows height to a specific value, while the bellows pitch remains constant. The mechanism has been applied for non-ferrous material such aluminum with similar behavior [13]. Effect when using bigger tube size has been verified that able to produce bellows using semi dieless bellows forming with similar characteristic [14].

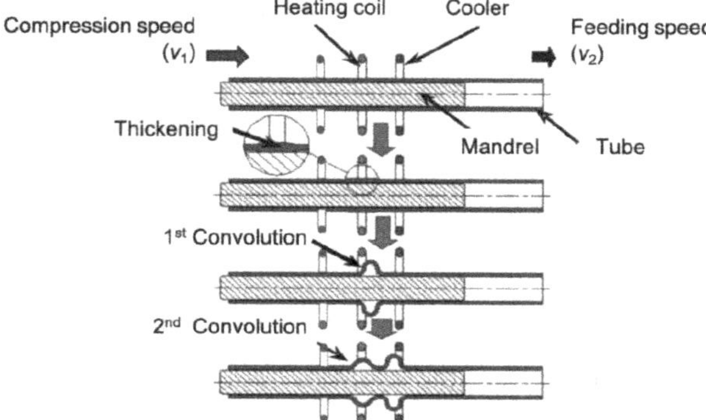

Figure 1. Schematic of the semi-dieless bellows forming with a local heating technique.

Deformation mechanism of the semi-dieless forming process is under a free deformation. Therefore, this technique is very sensitive to the forming conditions and disturbances. As a result, the bellows height and pitch are not constant and do not vary as shown in Figure 2a,b. Another problem that occurs is asymmetric bellows as shown in Figure 2c. These phenomena also happened in aluminum tube [13].

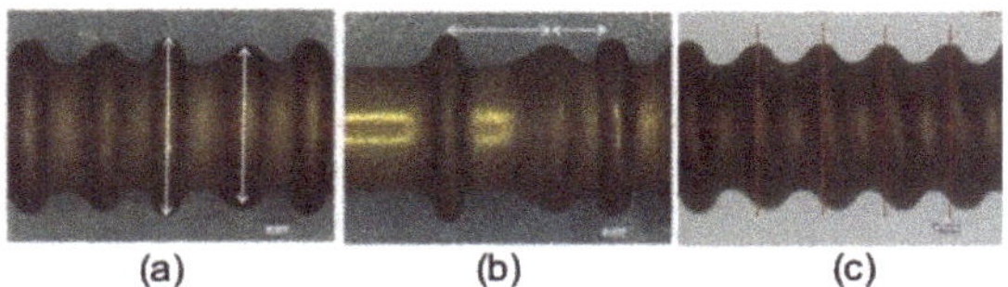

Figure 2. The problem of semi-dieless bellows forming on SUS 304 (**a**) Variety of bellows height, (**b**) Variety of bellows pitch, and (**c**) Asymmetric bellows shape.

3. Finite Element Analysis (FEA)

To evaluate the dieless bellows forming process and to verify problems in the dieless bellows forming process, we have developed the finite element model to simulate the dieless bellows forming process by using a finite element method. Figure 3 shows the schematic of an axisymmetric bellows forming model on the FEM carried out on an MSC Marc Mentat version 2005 commercial software from MSC Software USA. A deformable tube and a rigid mandrel are comprised of 5-node, asymmetrically quadrilateral shell elements. We simulate local heating by applying heat flux on the tube surface. The magnitude of heat flux in the induction heating is inversely proportional to the distance from the coil to the workpiece surface. Therefore, Equation (1) is utilized to model various heating when taking into consideration the bellows formation progress already verified in our previous work [5], where q is heating quantity, a is a constant, and y is the position of a node in the y-axis in a global position. qs and a are set to achieve the processing temperature up to 1200 °C. Just before the convolution occurs, the heating quantity is qs; after convolution is formed, the heat flux quantity increases. We model a double cooling system by using film subroutines to reduce the temperature by increasing the heat transfer capacity. Figure 3 shows the geometry of heating and cooling system model referring to the experimental conditions with a 3.5-mm cooling distance (Cd), a 5-mm cooling length (Cl), and a 4-mm heating length (Hl).

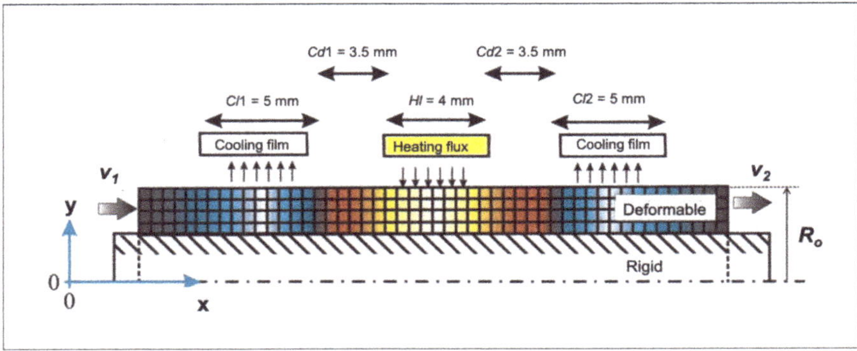

Figure 3. Schematic of the finite element model for the asymmetrical semi-dieless bellows forming process.

Taking into consideration strain hardening (n), strain rate sensitivity (m), and strength coefficient (K), we utilize a tube material made of SUS 304 Stainless steel. It was clarified that n, m, and K (Mpa) are function constants of temperature as shown in Equations (2)–(5), where, σ is flow stress (Mpa), ε is strain, and $\dot{\varepsilon}$ is strain rate [15]. We assume that the material is isotropic in mechanical and thermal properties. v_1 (t) is set at 0.6 mm/s, while v_2 (t) is set constantly at 0.3 mm/s. While boundary condition for the FEA is shown in Table 1.

$$q = q_0 + \left(a \cdot (4 - y)^2\right) \tag{1}$$

$$\sigma = K \cdot \varepsilon^n \cdot \dot{\varepsilon}^m \tag{2}$$

$$K(T) = \left(7.10^{-4} \cdot T^2\right) - (2.14 \cdot T) + 1788.3 \tag{3}$$

$$n(T) = \left(-2.472 \cdot 10^{10} \cdot T^3\right) + \left(4.645 \cdot 10^2 \cdot T^2\right) + 0.4367 \tag{4}$$

$$m(T) = (4.374 \cdot T) + 1.449 \cdot 10^{-3} \tag{5}$$

Table 1. Material model and Boundary conditions of SUS 304 used in Finite Element Model (FEM).

Boundary Conditions	
Heating quantity, q ($W \cdot mm^{-2}$)	Equation (1)
Heating temperature (°C)	1100
Heating length, Hl (mm)	5
Cooling Length, Cl (mm)	5
Heat transfer coefficient of cooling, h_c ($W \cdot mm^{-2} \cdot K^{-1}$)	1000
Heat transfer coefficient of radiation to air, h_a ($W \cdot mm^{-2} \cdot K^{-1}$)	30
Thermal conductivity, λ ($W \cdot mm^{-1} \cdot K^{-1}$)	0.0163
Specific heat ($J \cdot g^{-1} \cdot K^{-1}$)	0.502
Mass density, ρ ($g \cdot mm^{-3}$)	0.008

We promote deformation in the semi-dieless bellows forming by adding the compression speed and localized elevated temperature. By assuming that the compression speed driven by the servo motor is stable, we suspect that the cause of deformation instability is variations in the processing temperature. In the real semi-dieless bellows forming process, heating system and cooling system effect to temperature stability. The unstable heating system results from a variety of distances between the heating coil and the workpiece. On the other hand, the unstable cooling is produced by the unstable water cooling. To verify these effects on the semi-dieless forming process, we model a variety of temperatures by varying the quantity of heat flux (q_{sv} (W/mm^2)) and heat transfer in the cooling area (h_{cv} ($W \cdot mm^{-2} \cdot K^{-1}$)) as shown in Equations (6) and (7) respectively. Figure 4 shows the schematic of variety of heating and cooling in time series.

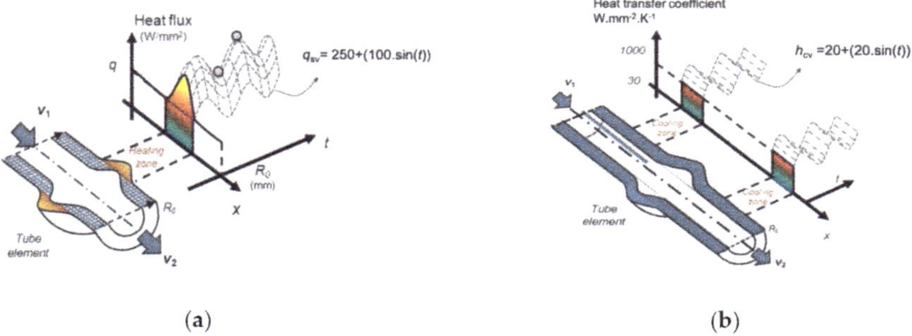

(a) (b)

Figure 4. (**a**) Schematic of a variety of heating (q_s) and (**b**) a variety of cooling system (h_c).

$$q_{sv} = 250 + (100 \cdot \sin t) \tag{6}$$

$$h_{cv} = 20 + (20 \cdot \sin t) \quad (7)$$

Various heating and cooling are modeled using a sinusoidal function as shown in Equations (6) and (7), while the visualization of the heating and cooling models in time series is shown in Figure 5.

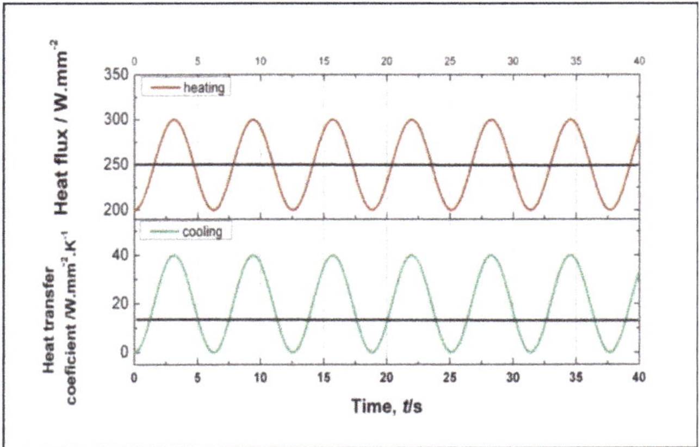

Figure 5. Various models of heating and cooling in time series.

The results of the simulation using the variation of heating and cooling condition produce different bellows profile as shown in Figure 6. Changes in temperature distributions promoted by a variety of heating and cooling affect the bellows profile stability of the bellows pitch and height. We obtain the bellows height by measuring the outer diameter of the bellows profiles. Compared to the constant heating and cooling, various heating and cooling produce various bellows heights and pitches. When a temperature difference occurs in the radial direction, an asymmetric bellows shape is created.

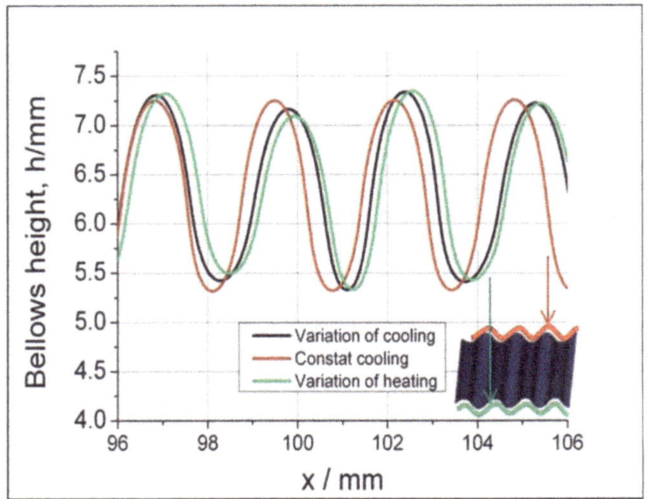

Figure 6. Bellows profiles under different conditions.

4. Selecting the Compensation Method for the Unstable Deformation Promoted by a Variety of Temperatures

We have clarified the unstable deformation resulting from a variety of temperatures by using a Finite element model (FEM). We have also conducted a FEM to find appropriate compensations for the unstable deformation. We model this unstable deformation by varying the temperatures obtained from various heat transfers. We propose two compensation methods of adjusting the heating flux and adjusting the compression ratio to compensate for those various temperatures. Since the feeding speed (v_2) affects the processing speed and stability, v_2 is kept constant. Therefore, we adjust the compression ratio by adjusting the compression speed (v_1).

The results of the two-compensation methods using the heating power and compression speed are shown in Figure 7. The results indicate that adjustment to the heating power cannot achieve the desired effect, as shown in Figure 8 since not only does the bellows temperature depend on the heating and heating condition, but it also depends on the bellows height. We promote the deformation of the semi-dieless bellows forming by localizing the heating on top of the convolution while applying a compression stroke. Therefore, when compensating the temperatures compensated from the adjustment to the heat flux, we have difficulties in adjusting to the bellows height due to the fact that the temperature during the process also performs a non-linear delay. Therefore, it is difficult to implement and regulate the temperature; it will require further studies on temperature behaviors.

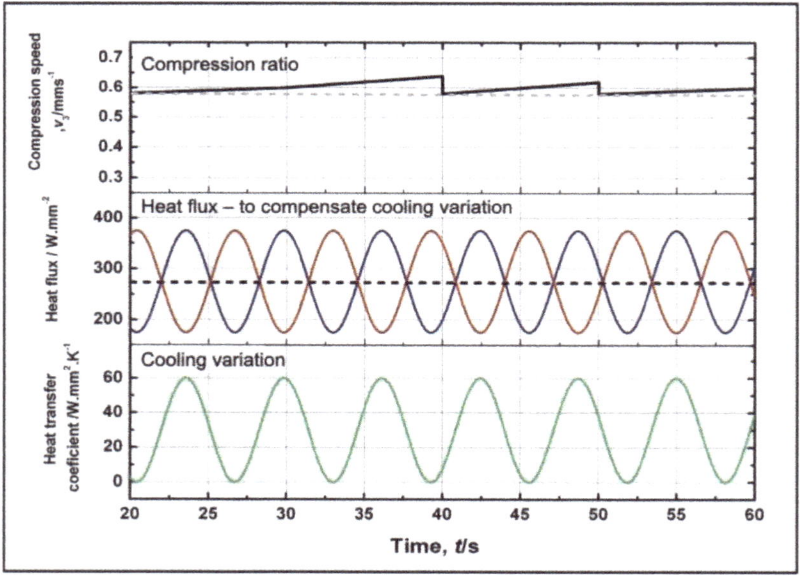

Figure 7. Compensations under various, different temperatures through applying changes in the heat flux and the compression speed.

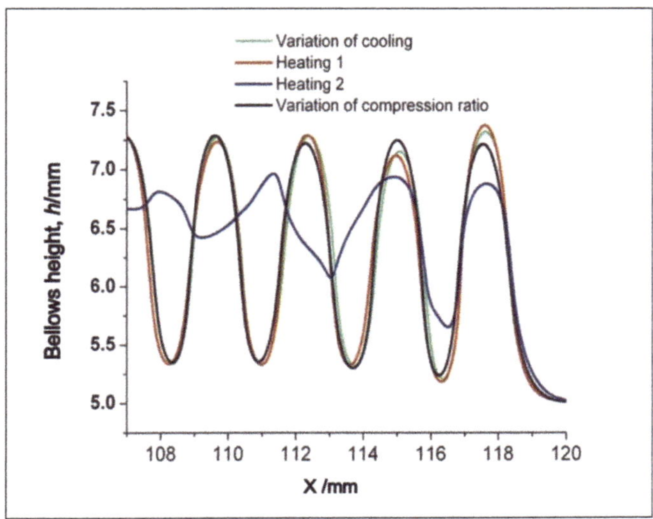

Figure 8. Results of the heating and the compensation of the compression speed to various cooling.

On the other hand, adjustment to the compression speed can effectively and efficiently modify the bellows height. When we use the compression speed to adjust the bellows height, we can obtain a homogeneous bellows height. Figure 9 shows the effects of the compression stroke (since the compression speed is higher than the feeding speed, it significantly affects the bellows height). Therefore, we should adjust the bellows height by adjusting compression stroke (v_1 or v_2).

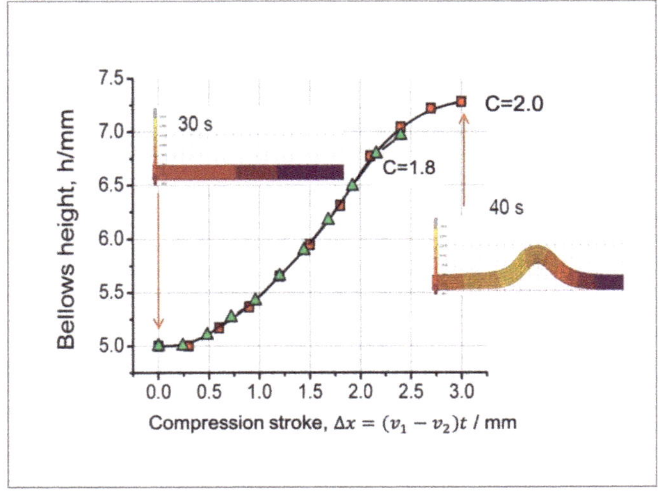

Figure 9. Effects of the compression stroke on the bellows heights.

5. Improving the Dieless Bellows Forming Process with the Application of a Feedback Control System

Since the deformation of semi-dieless bellows forming is a free and complex deformation, a feedback control system is a reasonable choice to improve the process and eliminate problems during the process. We have to establish components for the feedback control system such as the deformation reference, the sensing system, and the controller as shown in Figure 11. In the previous work, we had developed a feedback control system for a dieless drawing process by using an adaptive fuzzy logic control [16]. However, in this work, the difference lies in deciding on the deformation target of the bellows profile and selecting the fuzzy control parameters to obtain the best results. Since the deformation characteristic of semi-dieless bellows forming is different from that of the previous work, we have to decide on another control strategy.

5.1. Deformation Reference

In the real bellows geometry, the outer diameter of the bellows target shape is a function of the x domain (D (x)). In the real-time monitoring, the control program is run under a time series. Therefore, it is necessary to construct a diameter reference in the time domain. According to the sequence of bellows formation that we had obtained by employing a FEM from our previous work, bellows formation is a cyclic process. A series of bellows height has a specific period as shown in Figure 10.

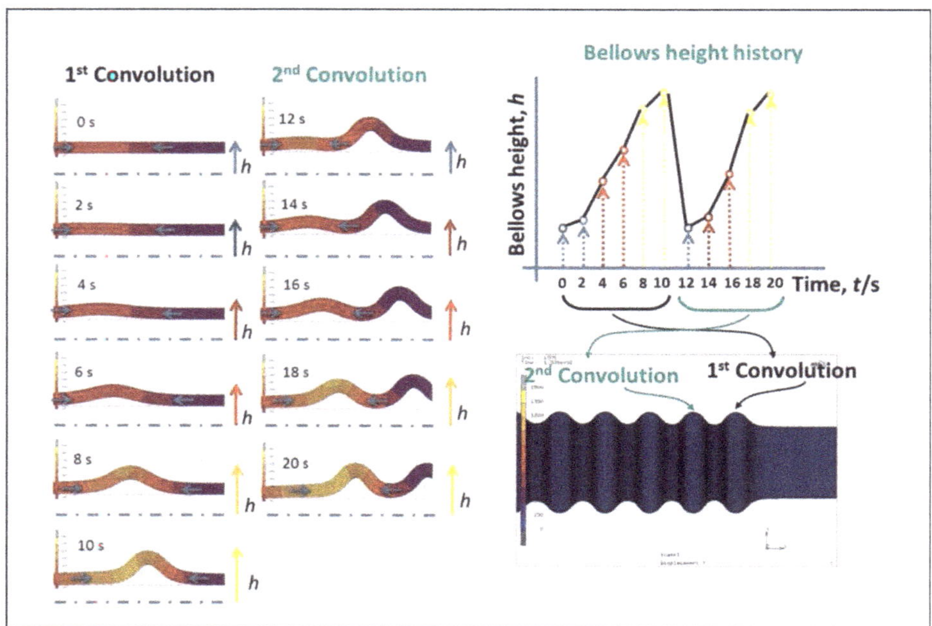

Figure 10. FEM results showing bellows formation history.

In this paper, the reference of bellows height progress uses a sinusoidal function. The bellows height reference in a time series is expressed in Equation (8), where D is the bellows height, Amp is the amplitude of bellows, ω is the frequency of bellows, ϕ is the phase shift, D_0 is the initial diameter, and t is time (s).

$$D\,(t) = Amp\,\sin\,(\omega t - \phi) + D_0 \tag{8}$$

5.2. A Vision-Based Feedback Control

A Continuous Semi-dieless bellows forming has complex mechanism. Therefore, deformation is free deformation due to elevated temperature and compression force. The deformation area of Continuous semi dieless bellows forming is wide and move. Therefore, it requires a contactless measurement technique to monitor deformation progress in continuous semi-dieless bellow forming.

An imaging-based sensing system appropriate sensing technique that compromise measuring condition in continuous semi-dieless bellows forming [17]. An image is captured with an infrared filter to improve focused on the object at elevated temperature [18,19]. An infrared-pass filter removes visible light passing through the image sensor that improves detection on the edge of bellows. An Algorithm has been adopted from dieless drawing process. Therefore, semi dieless bellows forming need additional rule to measure dynamic bellows height formation [20].

The components in the vision-based feedback control are bellows image capturing, image and data processing, calibration of the machine vision. Since difficulties in capturing deformation zone inside heating coil, the camera is tilted 10° from the perpendicular direction. In this study, the picture is captured using an infrared-pass filter. Therefore, a high contrast image of the workpiece by can be obtained for image and data processing.

5.3. Establishing a Control System

Fuzzy inference is an artificial intelligent computation method in which linguistic calculation is converted into a mathematical calculation. Fuzzy sets improve the handling of uncertainty by reducing it and developing a correct conclusion although the real world is not precise or specific. The fuzzy inference is selected since this method is suitable for controlling a non-linear system with limited knowledge of the system. The fuzzy control system provides transparency, and the intuitive nature of the rule base and input variables adopted to make it relatively easy to be developed, tested and modified [21]. To establish the appropriate fuzzy control system of the semi-dieless bellows forming, a conventional fuzzy and adaptive fuzzy logic control is prepared. The conventional fuzzy has fixed parameters during the control process such as membership function and fuzzy rules while the adaptive fuzzy logic control has adjustable parameters such as membership function and fuzzy rules according to the deformation characteristics (bellows progress, error and changing error).

Figure 11 shows the control flowchart of two fuzzy control strategies for semi-dieless bellows forming. The fuzzy input is the diameter error (e), changing rate of error ($\Delta\dot{e}$), and bellows formation progress in percent (h) only for adaptive fuzzy control. The outputs of fuzzy control are the compression speed. The adaptive fuzzy logic control has a supervisory system to adjust the membership span of main fuzzy control by using gain value. The gain is used to modify the membership span of compression speed. The membership functions for error, changing rate of error, Δv_1, and gain are shown in Figure 12. Membership of each category has specific name that used for making fuzzy rule in the controller. Each member ship consist of range of parameter (x axis) relative to level of membership value (from 0 to 1). The membership value 0 means the parameter is not affect to fuzzy inference calculation. The membership value depends on parameter (x axis) of membership function. The rule logic represents an expert of dieless bellows forming. The Fuzzy rule is construct based on previous works and FEA. There are two fuzzy controller that applied in this paper, a conventional fuzzy controller, and an adaptive fuzzy controller.

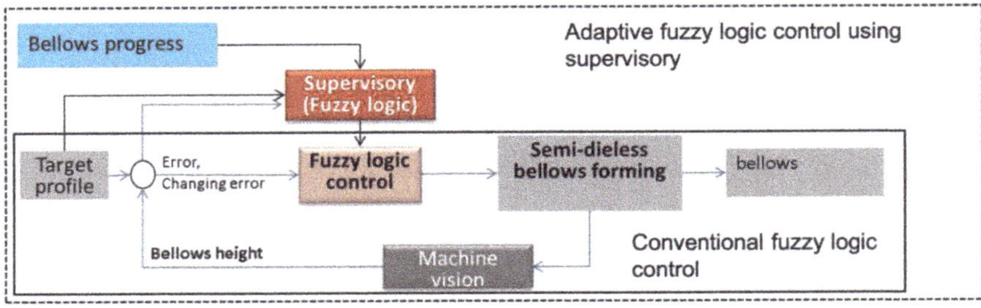

Figure 11. Flowchart of the semi-dieless bellows forming with a vision-based fuzzy control.

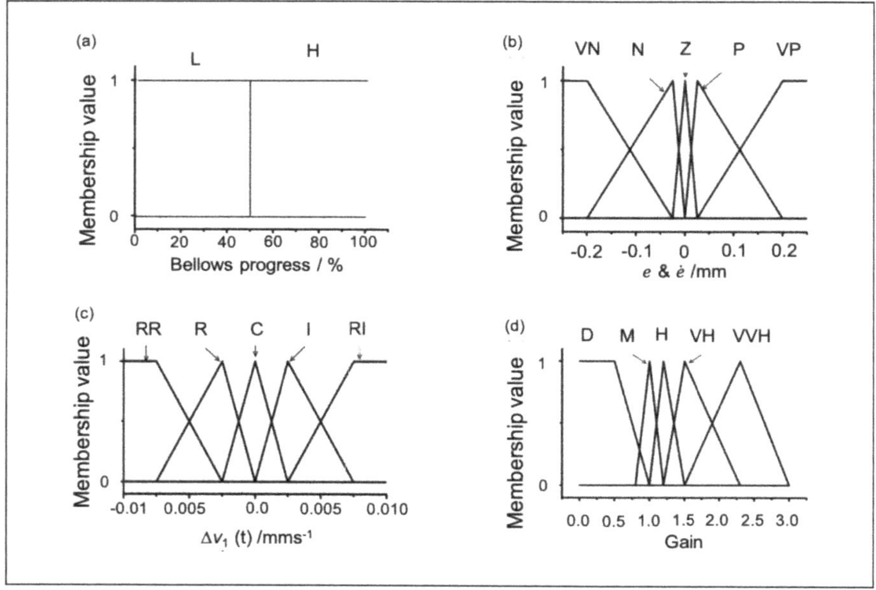

Figure 12. Membership functions of conventional fuzzy controller: (**a**) bellows formation progress consist of two membership function L (Low), H (High), (**b**) bellows height error and changing rate of the error has similar membership function VN (Very Negative), N (Negative), Z (Zero), P (Positive), VP (Very Positive), (**c**) incremental compression speed (Δv_1) consist of membership function RR (Rapid Reduce), R (Reduce), C (constant), I (Increase), RI (Rapid Increase), and (**d**) gain consist of membership function D (Decrease), M (Medium), H (High), VH (Very High), and VVH (Very Very High).

The conventional fuzzy rule is used to adjust compression speed according to the error and changing error value as shown in Table 2. When machine vision measure the belows height, then compare with target profile resulting error (e_h) and changing error (Δe_h). As the input, value of e_h and Δe_h will determine membership function in Figure 12b. According to fuzzy rule in Table 2, membership function of e_h and Δe_h determines membership function of output on incremental compression speed (Δv_1). The fuzzy inference calculates Δv_1 to adjust compression speed (v_1). Therefore compression speed will be adjust according to error and changing error of bellows simuntanously until bellows formation finish. Membership function on the adaptive fuzzy rules are divided into two different parts, below and over 50% of the bellows height. This strategy is adopted because, in the early stage of bellows formation, the error tends to be positive owing to the process characteristic in bellows formation. Thus, the fuzzy rules are not appropriate when the positive error of $v_,$ is reduced which makes the convolution height low since the deformation area moves. In the case of the control action command for below 50% of the convolution progress, the compression speed, v_1, is set as a slow response to the error. If convolution progress is over 50%, the fuzzy rules will be configured as a fast response to the error.

Table 2. Conventional fuzzy rules to determine the compression speed adjustment Δv_1. (*Refer to Figure 12).

Input		Output	Input 2: Changing Error (Δe_h)				
			VN*	N*	Z*	P*	VP*
Input 1: *Bellows height error* (e_h)	VP*	Δv_1	I	I	R	R	RR
	P*	Δv_1	I	C	C	R	R
	Z*	Δv_1	I	C	C	R	R
	N*	Δv_1	I	C	C	C	I
	VN*	Δv_1	RI	I	R	I	I

Tables 3 and 4 show the details of adaptive fuzzy rules to adjust the v_1 and gain in accordance with the fuzzy input of the convolution progress, the diameter error, and the changing rate error. v_2 is kept constant since it affects the reference target. The fuzzy rule is divided into two conditions: below and above 50% of the reduction progress. Fuzzy rules for below 50% are set as the slow response because, in this step, the error is always positive. Therefore, if the fuzzy response is high, v_1 will decrease. Consequently, it is difficult to achieve a maximum bellows height.

Table 3. Adaptive fuzzy rules under 50% of the bellows progress to determine the compression speed adjustment Δv_1 and gain. (*Refer to Figure 12).

Input		Output	Input 2: Changing Error (Δe_h)									
			VN*		N*		Z*		P*		VP*	
Input 1: *Bellows height error* (e_h)	VP*	Δv_1/gain	RI	/H	I	/M	I	/M	C	/M	R	/H
	P*	Δv_1/gain	RI	/M	C	/M	C	/M	C	/M	C	/M
	Z*	Δv_1/gain	C	/M	C	/M	C	/M	C	/M	R	/M
	N*	Δv_1/gain	C	/M	C	/M	C	/M	I	/M	RI	/H
	VN*	Δv_1/gain	R	/H	C	/M	I	/M	I	/M	RI	/VH

Table 4. Adaptive fuzzy rules over 50% of the bellows progress to determine the compression speed adjustment Δv_1 and gain. (*Refer to Figure 12).

Input		Output	Input 2:Changing Error (Δe_h)									
			VN*		N*		Z*		P*		VP*	
Input 1: *Bellows height error* (e_h)	VP*	Δv_1/gain	I	/H	I	/M	R	/M	R	/H	RR	/H
	P*	Δv_1/gain	I	/H	C	/M	C	/M	R	/M	R	/H
	Z*	Δv_1/gain	I	/M	C	/M	C	/M	R	/M	R	/H
	N*	Δv_1/gain	I	/M	C	/M	C	/M	C	/M	I	/M
	VN*	Δv_1/gain	RI	/H	I	/M	R	/M	I	/M	I	/H

6. Experimental Method

6.1. Experimental Equipment

The Vision based fuzzy controller is implemented in the horizontal semi-dieless compression machine, a computer for implementing the fuzzy rule, control speed of two servo motors. The machine is equipped with an induction heating source, as shown in Figure 13. The compression speed v_1 and feeding speed v_2 of the specimen is controlled by rotation of ball screw from servo motor. A narrow heating area is achieved with a high-frequency induction heating apparatus using a 4-mm induction coil width (*Hl*). A water-cooling system is applied to maintain a constant temperature distribution with a 7.5-mm cooling distance (*Cd*). The image acquisition system using a commercial CCD camera to grab the image up to 30 fps. The image of bellows progress is transferred to image processing unit. The fuzzy algorithm is designed to adapt with bellows formation progress. Image processing unit and fuzzy logic controller is construct under platform of LabVIEW software. The output obtained by the fuzzy inference is an input to the servo motors to adjust the compression speed of v_2.

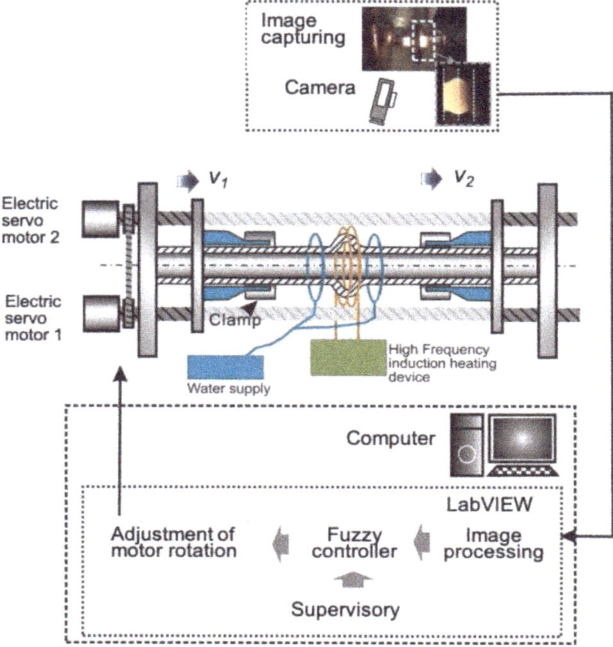

Figure 13. Experimental apparatus for the fuzzy control using a machine vision system.

6.2. Materials and Experimental Conditions

A stainless-steel, SUS304 tube with a 5-mm outer diameter of and a 0.5-mm-thick wall is used in this experiment. The maximum processing temperature is 1100 °C. We use a mandrel with a 3.9-mm outer diameter to ensure a stable axisymmetric deformation of the convolutions by inserting it into the workpiece. The feeding speed is kept constant at 0.3 mm/s.

In the case of the bellows target shape shown in Figure 14, the bellows shape parameters in the x domain must be converted to that of the time domain where the original diameter, D_0 is 5 mm and the maximum bellows diameter is 7 mm. After transformed into Equation (8), the amplitude, A, will amount to 1 mm. Pursuant to a previous study, the pitch distance between the convolutions on these processing conditions are fixed at 2.7 mm [4,5].

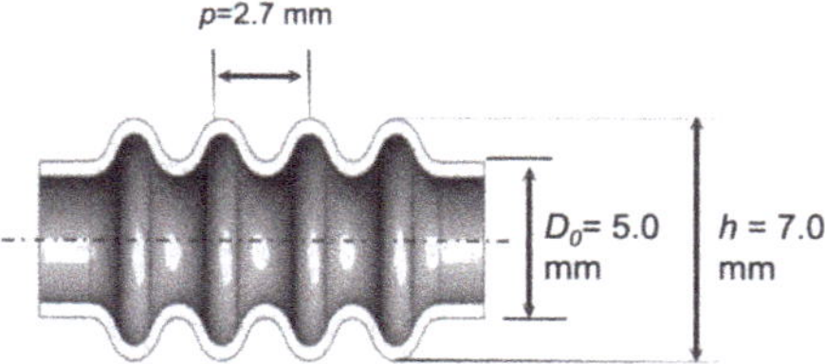

Figure 14. Geometry and dimensions of the metal bellows target with the example of h = 7.0 mm.

$$t_{\text{pitch}} = p/v_2 \tag{9}$$

$$\omega = 2\pi / t_{pitch} \tag{10}$$

With the 0.3-mm/s feeding speed (v_2), the time required to make one convolution is 9.0 s (t_{pitch}). However, for the target reference, only half of the sinusoidal function (the positive slope) is used; it means that a half reference equals one convolution. Therefore, the period of one unit wave is twice as long as the time required to make one convolution. The period of the target reference in the time domain is 18.0 s. Therefore, the reference frequency is 0.348 obtained from Equations (9) and (10). Since the function is sinusoidal and the convolution starts from the lowest value, the function should be shifted 90° to the right ($\phi = 1.57$ rad). The diameter reference converted to the time domain for the given bellows geometry is shown as Equation (11):

$$D\ (t) = \sin\ (0.348t - \phi) + 5. \tag{11}$$

The reference value is set, where the convolution starts from the outer diameter of the tube and evolves with the increasing axial compression. After the diameter reaches the maximum value, the convolution is held owing to the moving heating zone, a new convolution is initiated, and the phenomenon is repeated. When the derivative of the diameter reference is negative ($dD/dt < 0$), the function will be reset. Figure 15 shows the graph of diameter reference converted to the time domain for the profile target in Figure 9. As a result, the metal bellows are measured using a KEYENCE 900 optical microscope to obtain a bellows profile dimension.

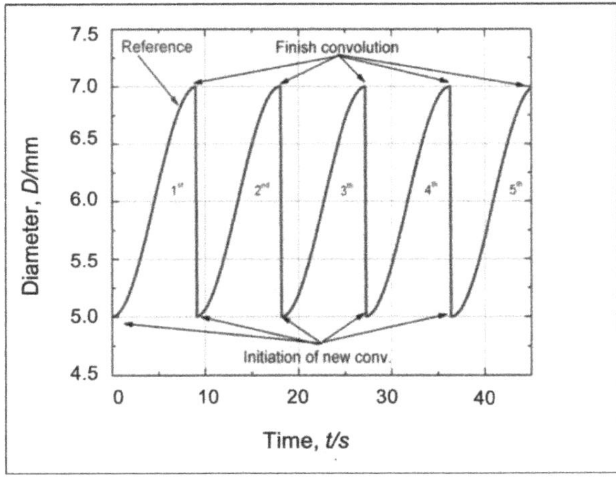

Figure 15. Diameter reference used in the semi-dieless bellows forming with vision-based fuzzy control.

7. Results and Discussion

Figures 16 and 17 show history of the compression speed paths, the bellows error, the adjustment compression speed (Δv_1) and the bellows height obtained from the conventional fuzzy and the adaptive fuzzy control respectively. The speed path using a conventional fuzzy is obtained from the adjustment of the initial compression speed according to the error and the changing error condition following the fuzzy rules. This adjustment does not take into considerations the bellows formation progress. The fuzzy control adjusts the compression speed by only taking into considerations the bellows height error from the reference and the actual bellows height. Adjustment of the compression speed follows the fuzzy rules as shown in Table 1. The history of the bellows height produced by using a conventional fuzzy control shows little agreement with the deformation reference and little agreement with the bellows height. It results from the deformation delay in the semi-dieless bellows forming process. The adjustment results of the compression speed in the beginning of the bellows formation will affect the middle-end bellows formation. In the beginning of the bellows formation, the error is always positive. Therefore, the fuzzy reduces the compression speed. The decreased compression speed reduces the compression stroke that decreases the bellows growth and that approaches the deformation reference. When the history of the bellows height is lower than that in the deformation reference, the fuzzy increases the compression speed. However, due to the delay process and the limited bellows formation timing (10 s), the increased compression speed is not capable of achieving the desired bellows height target before it moves to the cooling zone.

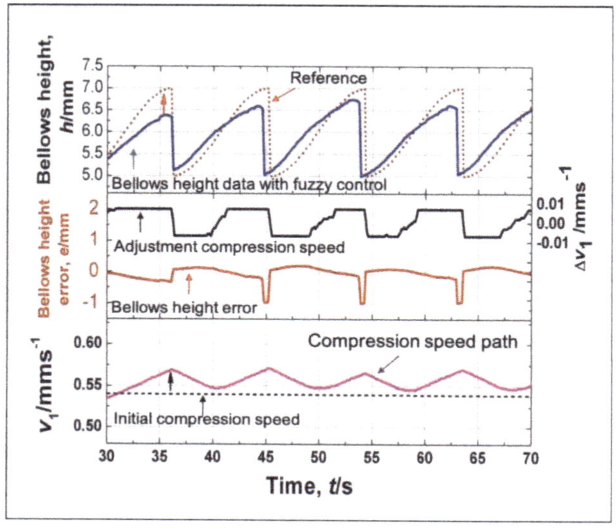

Figure 16. Bellows profile produced using a conventional fuzzy control.

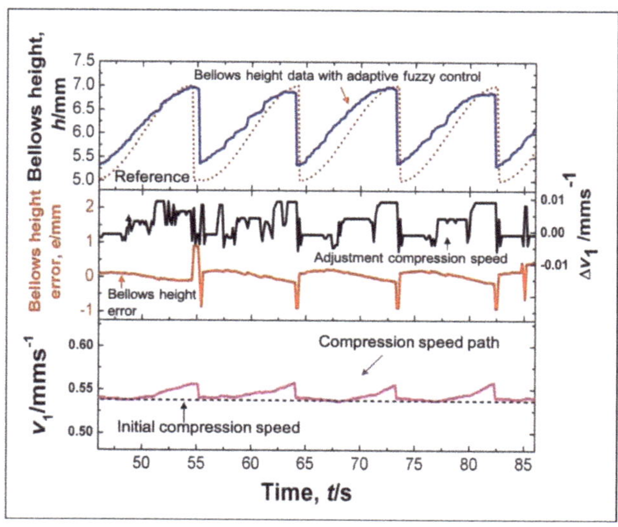

Figure 17. Bellows profile produced using an adaptive fuzzy control.

The speed path from the adaptive fuzzy control adjustment fluctuates according to the error and changing error during the semi-dieless bellows forming process. For every convolution formation, however, the compression speed is slightly adjusted when the bellows progress is over 50%, and the compression speed is significantly changed. We apply this strategy to cope with the deformation delay and the positive error in the beginning of each bellows formation. History of the bellows height shows that homogeny and accuracy of the bellows height produced by the semi-dieless bellows forming increases if we apply an adaptive fuzzy control. It shows that from evaluating the bellows progress, applying appropriate adjustment using an adaptive fuzzy control is adequate to control the deformation of the semi dieless bellows forming process.

The vision-based with adaptive fuzzy control performs good agreement with the bellows height target with low variations as shown in Figure 18. It is shows that the compensation based on progress of bellows formation is effective on adjusting bellows progress with low standard deviation. According to FE analysis, bellows height formation can be adjusted effectively after bellows initiation is produce. Adjustment of the compression speed is effective when it is applied to over half of the bellows progress as it has been proven in the FEA. To verify the reliability of the vision-based adaptive fuzzy control system, several experiments have been conducted with different processing conditions such as various feeding speeds, various bellows height targets, and various initial compression ratios.

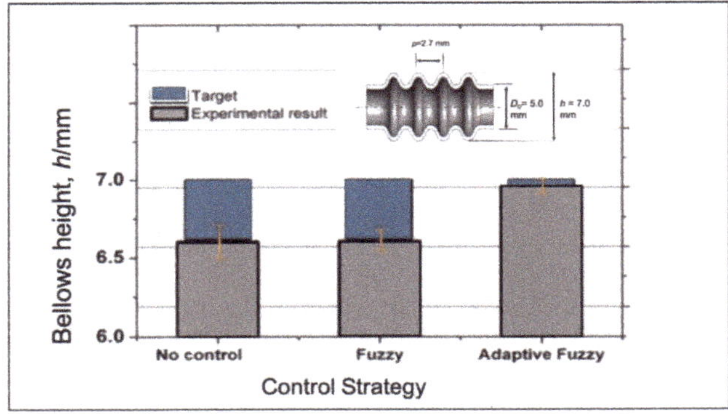

Figure 18. Comparison between the bellows height accuracy and its variation under different control strategies.

7.1. *Verification of the Proposed Control System for the Semi-Dieless Bellows Forming under Various Feeding Speeds*

An increased feeding speed reduces the convolution cycle time since the bellows speed formation increases when the feeding speed increases. One cycle of the bellows formation at the feeding speeds of 0.2, 0.3, and 0.5 mm/s are 12.5, 10, and 5 s respectively. Therefore, the time required to produce bellows becomes shorter. The conventional approach produces a lower bellows height, while the adaptive fuzzy controls can adapt to a decreased bellows formation time. Under a given range of the feeding speeds, the proposed control system can maintain the accuracy and homogeny in accordance with the deformation references. The real bellows geometry is analyzed using an optical microscope as shown in Figure 19. The results show that an accurate bellows profile is produced under various feeding speeds at the given range. It indicates that reproducibility of the proposed control system has been verified under different feeding speeds if we perform a similar bellows height and pitch.

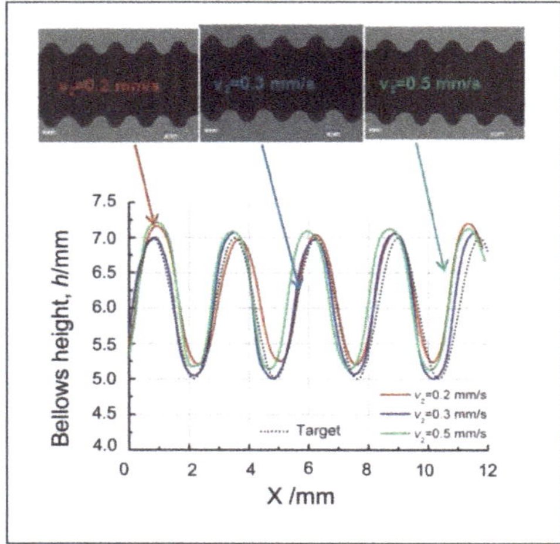

Figure 19. Bellows profile at various feeding speeds.

Performance of vision-based adaptive fuzzy control under different initial feeding speed can be seen in Figure 20. All conditions correspond to the target of bellows height. For low speeds, the bellows height tends to be slightly lower than the target height. For high initial compression speeds, the bellows tend to be more elevated than the target even if we apply the adaptive fuzzy to adjust delta v_1 in accordance with the error and changing error. Since the adaptive fuzzy is only adjusted for each bellows formation, when one bellows formation is finished, it will be back to the initial feeding speed. Therefore, the initial compression speed is essential and can be changed using a fuzzy. If the bellows height is below the target, the fuzzy will increase the initial feeding speed and the other way around. Figure 21 shows the evaluation of bellows pitch at different initial compression speeds. Even though it produces low accuracy, the proposed control system delivers lower variations compared to that of the conventional approach due to the fact that each bellows formation is a guide that makes every bellows formation more homogeneous.

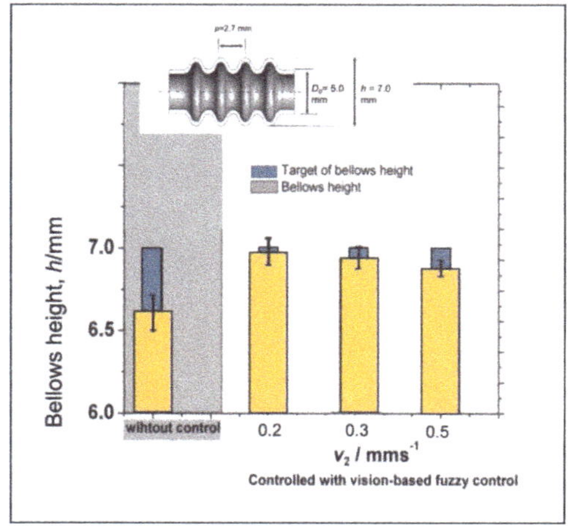

Figure 20. Evaluation of the bellows height under different initial compression speeds.

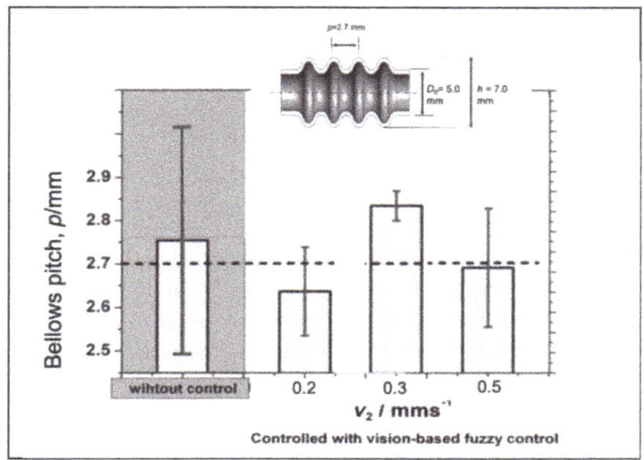

Figure 21. Evaluation of the bellows pitches under different initial compression speeds.

7.2. Verification of the Proposed Control System for the Semi-Dieless Bellows Forming under Various Bellows Targets

Three different bellows targets have been prepared to verify the performance of the proposed control system. The objectives are indicated with the maximum bellows height (h) of the 6.0, 6.5, and 7-mm references. Figure 22 shows that the bellows height data of different targets well correspond to the target. It indicates that the adaptive fuzzy controller with a machine vision sensor has a good performance to enhance the accuracy of the semi-dieless bellows forming. At lower bellows heights, the target shows low accuracy in the beginning of the bellows formation. When we verify this condition with the real bellows profile by using the optical microscope, the bellows show low accuracy in the minimum bellows height and pitch. This behavior results from a thickening process occurring at a low compression ratio as verified in the previous findings. These results indicate that the bellows pitch is not constant [4].

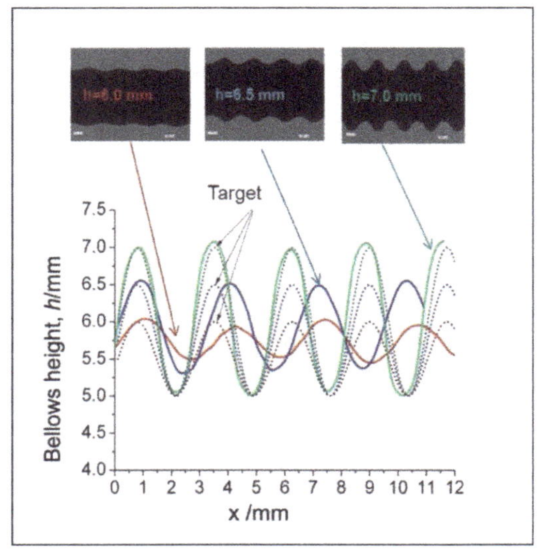

Figure 22. Bellows profile at various targets.

Figure 23 shows the comparison between the proposed control system and the conventional approach to control the bellows height. The proposed control system shows an excellent performance to produce a desired bellows height, while, under a similar condition, the conventional approach using a given chart delivers low accuracy and a higher bellows variation. It indicates that the vision based fuzzy control is sufficient to control the bellows height in accordance with the target. However, the proposed control system has a poor performance to achieve the accurate bellows pitch as indicated in Figure 24. Figure 24 shows that the bellows pitch is not constant at various bellows heights as assumed in the previous work.

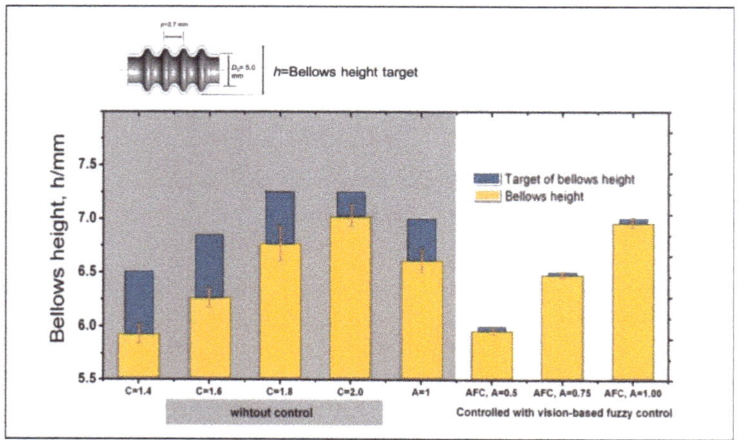

Figure 23. Evaluation of the bellows height under different bellows targets.

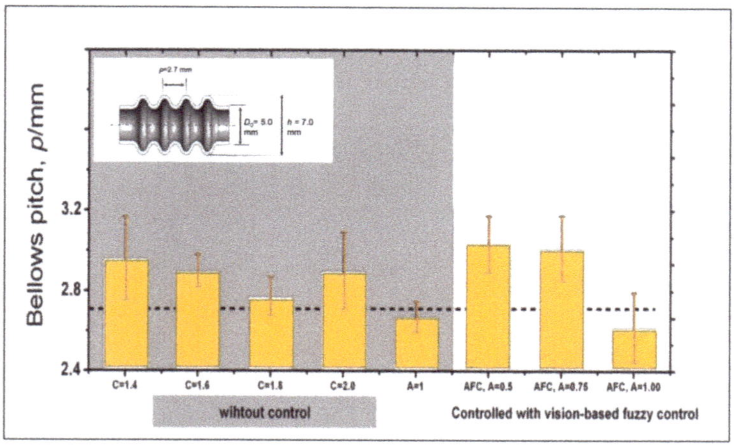

Figure 24. Evaluation of the bellows pitches under different bellows targets.

7.3. Verification of the Proposed Control System for the Semi-Dieless Bellows Forming under Various Initial Compression Speeds

The proposed control system adjusts the compression speed based on the initial value of the compression speed. Therefore, it is necessary to verify the implementation of the proposed control system when we apply it for different initial set ups of the compression speed or the compression ratio. At lower speeds, the initial compression speed tends to be lower than the bellows height; conversely, it tends to be higher at a higher initial compression speed. We have verified these findings using an optical microscope to observe the real bellows profile as shown in Figure 25. Not only does a higher initial compression speed perform a higher bellows height but it also performs a shorter bellows pitch. This condition indicates that the fuzzy supervisor should consider an initial compression speed to obtain the appropriate compensation value.

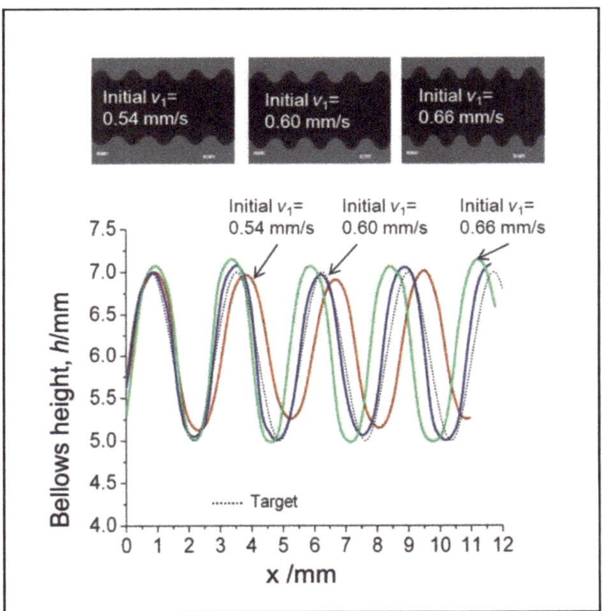

Figure 25. Bellows profile at different initial compression ratios.

Making the evaluation on the larger area, we find out that the accuracy of the bellows height will slightly decrease when the processing speed increases as shown in Figure 26 probably since the time to produce one convolution (5 s) is short, so it makes the effectiveness of the vision-based fuzzy control decrease. We can enhance the effectiveness of the adaptive fuzzy by considering the processing speed to adjust the fuzzy parameters such as modifying a membership function span and the adaptive fuzzy to rule at a higher processing speed. Figure 27 shows the evaluation of the bellows pitch that shows different accuracies. However, for a variety of the bellows pitch, the utilization of the proposed control system has produced a lower variation that increases reproducibility.

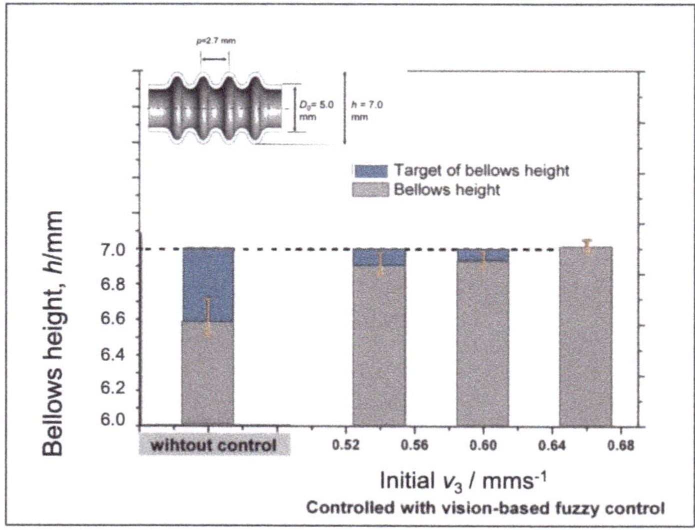

Figure 26. Evaluation of the bellows height under different feeding speeds.

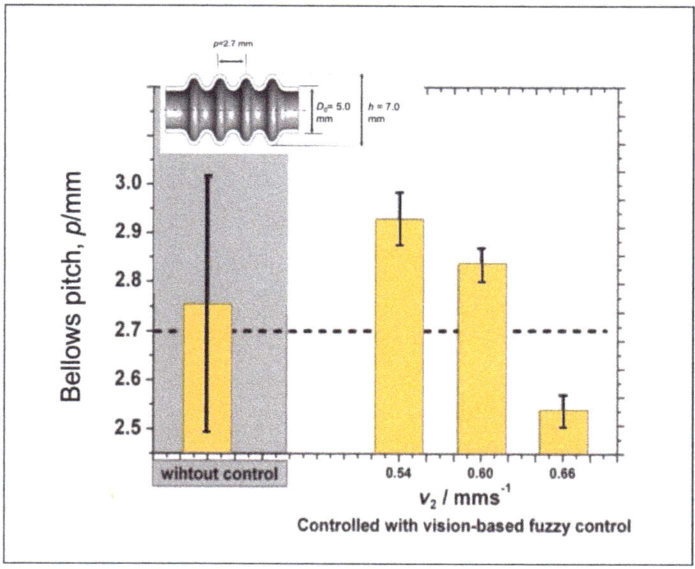

Figure 27. Evaluation of the bellows pitch under different feeding speeds.

8. Conclusions

We have conducted an investigation to low accuracy of the bellows produced by the semi-dieless bellows forming process by using a finite element method. The results show that low accuracy and low reproducibility of the bellows results from temperature disturbances from an unstable heating or cooling system. Adjusting the compression speed is effective and efficient to compensate for a variety of the bellows during the process created by those temperature disturbances. To implement a real-time adjustment, it is necessary to establish a feedback control system. A vision-based fuzzy control for the semi-dieless bellows forming has been developed. It is already confirmed that a machine vision is appropriate for the real-time monitoring of the bellows height and pitch. The camera of the machine vision can flexibly be placed and monitor the bellows formation process inside the heating coil to provide deformation data although other measurement devices cannot be applied. The validity of the machine vision has been verified with an optical microscope. The reference target shape for this bellows forming can be used as a sinusoidal function similar to that of the bellows height formation. The adaptive fuzzy controller shows an excellent performance to guide the deformation and to produce accurate bellows. The performance of the proposed feedback control system using vision sensing and an adaptive fuzzy control to provide accurate bellows have been verified under various processing conditions and bellows targets. A remaining problem that still occurs is low accuracy in the bellows pitch due to an incorrect assumption. Further research might be attended to develop a fuzzy supervisor that will consider the initial set up of the dieless bellows forming process.

Author Contributions: Supervision, T.F.; Writing–original draft, S.S.; Writing–review & editing, K.-i.M. All authors have read and agreed to the published version of the manuscript.

Funding: The APC was funded by Universitas Indonesia: HIbah QQ.

Acknowledgments: The author would like to thank to grant from Universitas Indonesia to support this work and Publication.

Conflicts of Interest: The authors declare no conflict of interest.

Nomenclature

A_0	: Initial cross section area of tube, mm^2
Amp	: Amplitude, mm
a	: Constanta for increasing heat quantity, $W{\cdot}mm^{-2}$
Cd	: Cooling distance, mm
Cl	: Cooling zone, mm
c	: Heat capacity, $J{\cdot}Kg^{-1}{\cdot}K^{-1}$
D	: Diameter, mm
D_0	: Initial diameter, mm
e_h	: Bellows height error, mm
Δe_h	: Increment of bellows height error, mm
$\bar{e}_h$	: Average bellows height error, mm
G	: Gain
h	: Bellows height
h_a	: Heat transfer coefficient of radiation to air, $W.mm^{-2}{\cdot}K^{-1}$
h_c	: Heat transfer coefficient of cooling, $W{\cdot}mm^{-2}{\cdot}K^{-1}$
Hl	: Heating length, mm
K	: Strength coefficient, MPa
n	: Strain hardening index
m	: Strain rate sensitivity
h_{Rev}	: Reference bellows height, mm
h_{Prog}	: Progress of bellows height, %
q	: Heat flux quantity, $W{\cdot}mm^{-2}$
q_0	: Initial heat flux quantity, $W{\cdot}mm^{-2}$
p	: Pitch, mm
t	: Time, s
X	: Elongation, mm
t_{pitch}	: Time to produce one bellows, s
v_1	: Compression speed, $mm.s^{-1}$
v_2	: Feeding speed, $mm{\cdot}s^{-1}$

y	: Position of node in y axis in global position, mm
σ	: Flow stress, MPa
ε	: Strain
$\dot{\varepsilon}$	: Strain rate, s^{-1}
ω	: Frequency, Hz
ϕ	: Phase, rad
λ	: Thermal conductivity, $W \cdot mm^{-1} \cdot K^{-1}$
σ_e	: Standard deviation of error, mm
Δv_1	: Changing of compression speed, $mm \cdot s^{-1}$
Δv_2	: Changing of feeding speed, $mm \cdot s^{-1}$
Δx	: Compression stroke

References

1. Wang, G.; Zhang, K.F.; Wu, D.Z.; Wang, J.Z.; Yu, Y.D. Superplastic forming of bellows expansion joints made of titanium alloys. *J. Mater. Process. Technol.* **2006**, *178*, 24–28. [CrossRef]
2. Faraji, G.; Mashhadi, M.M.; Norouzifard, V. Evaluation of effective parameters in metal bellows forming process. *J. Mater. Process. Technol.* **2009**, *209*, 3431–3437. [CrossRef]
3. Kang, B.H.; Lee, M.Y.; Shon, S.M.; Moon, Y.H. Forming various shapes of tubular bellows using a single-step hydroforming process. *J. Mater. Process. Technol.* **2007**, *194*, 1–6. [CrossRef]
4. Furushima, T.; Hung, N.Q.; Manabe, K.; Sasaki, O. Development of Semi-dieless Metal Bellows Forming Process. *J. Jpn. Soc. Technol. Plast.* **2012**, *53*, 251–255. [CrossRef]
5. Supriadi, S.; Hung, N.Q.; Furushima, T.; Manabe, K. A Novel Dieless Bellows Forming Process Using Local Heating Technique. *Steel Res. Int.* **2011**, 950–955.
6. Furushima, T.; Suzuki, Y.; Manabe, K.; Sasaki, O. Convolution Formation Behavior of Semi-dieless Bellows Forming Process for Metal Tubes. In Proceedings of the 6th International Conference on Tube Hydroforming (TUBEHYDRO 2013), Jeju, Korea, 25–28 August 2013; pp. 268–274.
7. Sekiguchi, H.; Kobatake, K.; Osakada, K.A. Fundamental Study on Dieless Drawing. In Proceedings of the 15th MTDR Conference, Birmingham, UK; 1974; pp. 539–544.
8. Sekiguchi, H.; Kobatake, K.; Osakada, K. Dieless Drawing Process. *J. Jpn. Soc. Technol. Plast.* **1976**, *70*, 67–71.
9. Milenin, A.; Kustra, P.; Furushima, T.; Du, P.; Němeček, J. Design of the laser dieless drawing process of tubes from magnesium alloy using FEM model. *J. Mater. Process. Technol.* **2018**, *262*, 65–74. [CrossRef]
10. Kustra, P.; Milenin, A.; Płonka, B.; Furushima, T. Production Process of Biocompatible Magnesium Alloy Tubes Using Extrusion and Dieless Drawing Processes. *J. Mater. Eng. Perform.* **2016**, *25*, 2528–2535. [CrossRef]
11. Furushima, T.; Manabe, K. A novel superplastic dieless drawing process of ceramic tubes. *CIRP Ann. Manuf. Technol.* **2017**, *66*, 265–268. [CrossRef]
12. Furushima, T.; Manabe, K. Large reduction die-less mandrel drawing of magnesium alloy micro-tubes. *CIRP Ann. Manuf. Technol.* **2018**, *67*, 309–312. [CrossRef]
13. Zhang, Z.; Manabe, K.; Furushima, T.; Tada, K.; Sasaki, O. Deformation behavior of aluminum alloy tube in semi-dieless metal bellows forming process with local heating technique. *Adv. Mater. Res.* **2014**, *936*, 1742–1746. [CrossRef]
14. Zhang, Z.; Furushima, T.; Manabe, K.; Tada, K.; Sasaki, O. Development of dieless metal bellows forming process with local heating technique. *Proc. Inst. Mech. Eng. B J. Eng. Manuf.* **2015**, *229*, 664–669. [CrossRef]
15. Hashizume, S. Resistance to plastic deformation of metals (stainless steel). *J. Jpn. Soc. Technol. Plast.* **1965**, *6*, 71–75.
16. Supriadi, S.; Manabe, K. Enhancement of dimensional accuracy of dieless tube-drawing process with vision-based fuzzy control. *J. Mater. Process. Technol.* **2013**, *213*, 905–912. [CrossRef]
17. Dworkin, S.B.; Nye, T.J. Image processing for machine vision measurement of hot formed parts. *J. Mater. Process. Technol.* **2006**, *174*, 1–6. [CrossRef]
18. Kurada, S.; Bradley, C. A machine vision system for tool wear assessment. *Tribol. Int.* **1997**, *30*, 295–304. [CrossRef]
19. Kurada, S.; Bradley, C. A review of machine vision sensors for tool condition monitoring. *Comput. Ind.* **1997**, *34*, 55–72. [CrossRef]

20. Supriadi, S.; Furushima, T.; Manabe, K. Real-time Monitoring System of Dieless Bellows Forming using Machine Vision. *Adv. Mater. Res.* **2013**, *789*, 429–435. [CrossRef]
21. Rahman, S.M.; Ratrout, N.T. Review of the Fuzzy Logic Based Approach in Traffic Signal Control: Prospects in Saudi Arabia. *J. Transp. Syst. Eng. Inf. Technol.* **2009**, *9*, 58–70. [CrossRef]

Article

Influence of Internal Pressure and Axial Compressive Displacement on the Formability of Small-Diameter ZM21 Magnesium Alloy Tubes in Warm Tube Hydroforming

Hajime Yasui [1,*], Taisuke Miyagawa [1], Shoichiro Yoshihara [2], Tsuyoshi Furushima [3], Ryuichi Yamada [4] and Yasumi Ito [4]

1 Faculty of Engineering, Integrated Graduate School of Medical, Engineering, and Agricultural Sciences, University of Yamanashi, 4-3-11 Takeda Kofu-shi, Yamanashi 400-8511, Japan; taielle39@gmail.com
2 Department of Engineering and Design, Shibaura Institute of Technology, 3-9-14 Minato-ku, Tokyo 108-8548, Japan; yoshi@shibaura-it.ac.jp
3 Institute industrial science, The University of Tokyo, 4-6-1 Komaba, Meguro-ku, Tokyo 153-8505, Japan; tsuyoful@iis.u-tokyo.ac.jp
4 Graduate Faculty of Interdisciplinary Research Faculty of Engineering, Mechanical Engineering (Mechanical Engineering), University of Yamanashi, 4-3-11 Takeda Kofu-shi, Yamanashi 400-8511, Japan; ryamada@yamanashi.ac.jp (R.Y.); yasumii@yamanashi.ac.jp (Y.I.)
* Correspondence: hy.swim.3vf@gmail.com

Received: 31 March 2020; Accepted: 19 May 2020; Published: 21 May 2020

Abstract: In this study, the influence of internal pressure and axial compressive displacement on the formability of small-diameter ZM21 magnesium alloy tubes in warm tube hydroforming (THF) was examined experimentally and numerically. The deformation behavior of ZM21 tubes, with a 2.0 mm outer diameter and 0.2 mm wall thickness, was evaluated in taper-cavity and cylinder-cavity dies. The simulation code used was the dynamic explicit finite element (FE) method (FEM) code, LS-DYNA 3D. The experiments were conducted at 250 °C. This paper elucidated the deformation characteristics, forming defects and forming limit of ZM21 tubes. Their deformation behavior in the taper-cavity die was affected by the axial compressive direction. Additionally, the occurrence of tube buckling could be inferred by changes of the axial compression force, which were measured by the load cell during the processing. In addition, grain with twin boundaries and refined grain were observed at the bended areas of tapered tubes. The hydroformed samples could have a high strength. Moreover, wrinkles, which are caused under a lower internal pressure condition, were employed to avoid tube fractures during the axial feeding. The tube with wrinkles was expanded by a straightening process after the axial feed. It was found that the process of warm THF of the tubes in the cylinder-cavity die was successful.

Keywords: tube hydroforming; small-diameter tube; magnesium alloy; warm working; deformation characteristics; forming defects; forming limit

1. Introduction

In recent years, magnesium and its alloys have been widely applied in the automotive, aircraft and telecommunication industries. The reason for the employment of magnesium alloys is their excellent characteristics, such as light weight and high strength. In addition, magnesium alloy has been expected to be employed as a material in medical devices, owing to its outstanding biocompatibility [1,2]. Some recent researches on bio-absorbable implants, including magnesium alloys, have focused on the technical utilization of magnesium alloys as biomaterials [3,4]. On the other hand, it has become necessary to miniaturize medical devices, and there is a demand for the establishment of deformation methods of small-diameter tubular products to apply them in miniaturized devices [5]. To respond to this demand, tube hydroforming (THF) emerged as a method to manufacture the small-diameter tubular parts [6]. THF is one of the plastic working methods. It deforms tubular parts by loaded internal hydro pressure and compressive force along the axial direction of the tube. Additionally, THF can deform tube products with a complex cross-section shape integrally. This characteristic leads to a weight reduction and high strength of the products. Therefore, it has become a research direction to use THF in the manufacture of small-diameter magnesium alloy tubes for medical devices. However, using THF with magnesium alloy tubes at room temperature (RT) is difficult, because the material has a low formability. This problem is caused by the high critically resolved shear stress of the slip planes, except for the {0001} plane of the hexagonal close-packed structure, in magnesium alloy at RT. Warm working has generally been employed to improve formability. So far, the following investigations have been carried out using warm THF on magnesium alloy tubes.

Chan [7] carried out experiments and a FE analysis using THF at elevated temperatures (220 °C, 250 °C and 280 °C) to deform an AZ31 magnesium alloy tube with a 22 mm outer diameter and 1.5 mm wall thickness. The tube, with an expansion ratio of 1.4 times compared with the initial outer diameter, was hydroformed successfully. Additionally, failure during the process was predicted. Manabe et al. [8] conducted a study on the T-shape forming of an AZ31 tube with an outer diameter of 42.7 mm and a wall thickness of 1.0 mm at 250 °C. The deformation characteristics were evaluated experimentally and numerically. In addition, it was suggested that the temperature distribution created in the die would lead to a more uniformed wall thickness of the deformed T-shaped product [9]. Liewald et al. [10] conducted investigations on the formability of ZM21 magnesium alloy tubes using warm THF at 350 °C. The tubes used had a 42 mm outer diameter and 2.0 mm wall thickness. ZM21 tubes, which have a 1.5 times expansion ratio, were deformed successfully, and an example of the product manufactured by this warm hydroforming process was presented.

However, there have been very few studies employing warm THF with a small-diameter Mg alloy tube, with an outer diameter of less than 10 mm. Using THF with small tubes requires high precision and fine tooling, as well as a suitable forming machine [11]. Additionally, simply scaling down might be impossible because of the size effect [12]. In particular, there is a critical problem that needs to be solved in relation to the consideration of a difference in the ratio of the outer diameter and the wall thickness (t/D) between a large tube and a small tube. The wall thickness of a microtube increases with downscaling, and handling of the microtube becomes difficult [13]. This increase of t/D is caused because of the forming limit in the drawing process and the security of the rigidity of the tube materials. The scales of t/D employed in some previous studies were 10^{-1}, but it is assumed that a manufacturable t/D for a small-diameter tube, which is required for medical device parts, is 10^{-2} [13]. Simply scaling down the processes is not easy. Therefore, the deformation behavior of small-diameter tubes should be elucidated separately. It is necessary to evaluate the formability and the deformation characteristics of small-diameter magnesium alloy tubes using warm THF.

Our previous paper focused on using warm THF with a small magnesium alloy tube, developed a new die-moving-type THF system, and applied warm free bulge forming of a small-diameter ZM21 magnesium alloy tube, with an outer diameter of 2.0 mm and a wall thickness of 0.15 mm, at a maximum of 300 °C [14]. The basic deformation behavior of the tube was evaluated, and the following results were obtained: (1) the tube was most expanded at 250 °C, and the maximum expansion ratio of a hydroformed sample was 1.76; (2) when the loaded pressure was lower, wrinkles occurred on the tube, and deformation behavior was observed; and (3) the thinning of the wall thickness was exceeded by the increase of the axial feeding amount ΔL. Additionally, it was clarified that the axial feeding limit L_{max} in comparison with the outer diameter D of the tube was found to be around $L_{max}/D = 0.5$.

However, the deformation characteristics and the hydroformability of a small-diameter ZM21 tube, when it is deformed to a die shape, has not been studied yet. In addition, the demand for integrated complex tubular micro-components, without welding and mechanical joining, has been increasing in recent years [15], and its shape is not only symmetrical, but also asymmetrical [16]. Accordingly, it is necessary to clarify the deformation characteristics of the tube during warm THF using the asymmetrical-shape die.

In this study, warm THF was carried out with small-diameter ZM21 magnesium alloy. The ZM21 tubes used in this study have a 2.0 mm outer diameter and 0.20 mm wall thickness. The purpose of the current study is to evaluate the influences of internal pressure and axial compressive displacement on formability and deformation behaviors, such as expansion, buckling and wrinkling. In addition, to deform the tubes into a die shape, straightening was applied to the tube with wrinkles, which were caused during the axial feeding, and the influence of the straightening on the hydroformability of the tube was clarified.

2. Materials and Methods

2.1. Overview of the THF System

Figure 1 shows the appearance of the THF machine, and Figure 2 shows a schematic of the THF system used for the experiment in this study. The machine consisted of an oil pump, servo-motors, load cells, cartridge heaters (250 W) and a heater controller. The pump loaded the internal oil pressure, and the motors applied the axial feeding for both tube ends. The processing was carried out at 250 °C. The heaters could heat the tube to 250 °C in approximately 140 s. The heat resistance of the oil was 350 °C. During the heating, the machine was covered by a case made of aluminum, and the case was filled by Ar gas to avoid fire.

Figure 3a shows the oil sealing mechanism. The oil was sealed by two O-rings. The deformation of each tube end inside the oil pool was restrained by a restraining ring made of steel to avoid expansion of the tube in areas other than the die cavity. Figure 3b shows the loading mechanism of the axial feeding and the internal pressure. The axial feeding was carried out by closing the die, in which process the distance between the left and the right die was changed. Therefore, the optional axial feeding amount was set as the distance between the left and the right die. These mechanisms could resolve problems, such as the preparation of a high-precision axial compressive punch and oil-sealing part, which were required in using THF with small tubes.

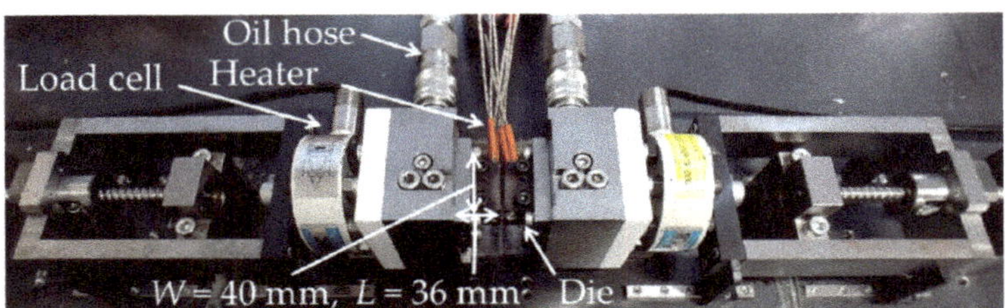

Figure 1. Appearance of the small warm hydroforming system used for the experiments.

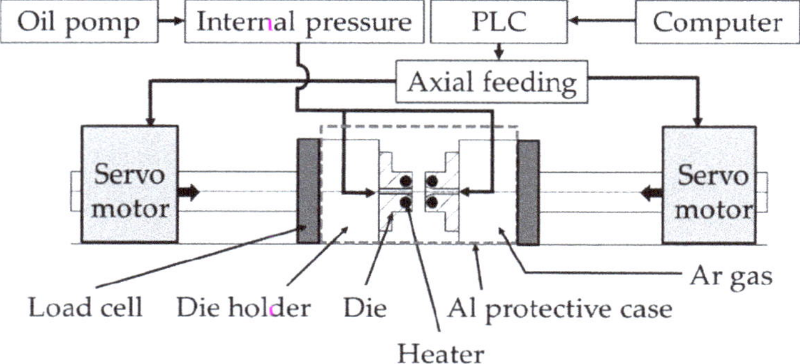

Figure 2. Schematic of the small warm hydroforming system.

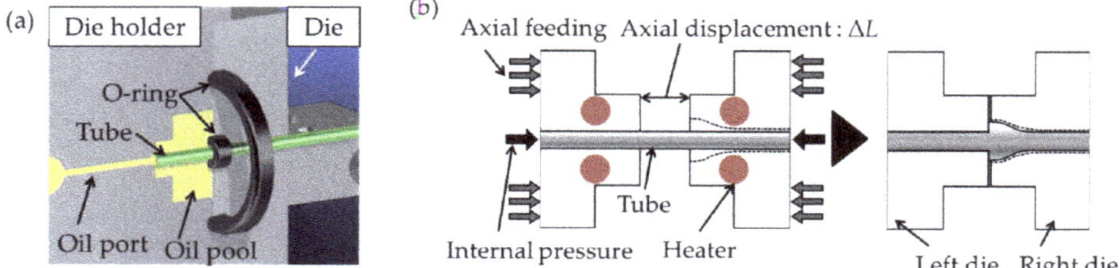

Figure 3. Geometry diagram of the die for the warm tube hydroforming (THF): (**a**) sealing mechanism; (**b**) loading mechanisms of the axial feeding and internal oil pressure in the tube for the experiment.

2.2. Material Characteristic and Results of the Tensile Test and Bursting Test

The mechanical properties of ZM21 magnesium alloy were obtained by tensile and bursting tests. Each specimen undergoing these tests were annealed at 300 °C for 24 h. First, to understand the rough tensile characteristics of ZM21, a tensile test using ZM21 magnesium alloy rods was carried out. Table 1 shows the chemical composition of the ZM21 rods used for the tensile test. The rods, with an outer diameter of 6.0 mm, were manufactured by hot extrusion. The tensile speed was 0.6 mm/min, and the test temperature in the warm region was raised by increments of 50 °C. Figure 4 shows the nominal stress–nominal strain diagram of the specimens, obtained from the tensile test. As a result, the decrease of the deformation resistance and the improvement of the ductility were confirmed at elevated temperatures. The ductility of the ZM21 rod at 250 °C is 220% higher than the test result at RT. It is considered that using warm THF with ZM21 magnesium alloy at 250 °C is appropriate, because of its excellent ductility and low resistance to deformation.

Then, the material characteristic of the small-diameter ZM21 tube, which was used for THF, was evaluated by the tensile test at RT and 250 °C. The tube material manufactured by hot drawing [16] has a similar chemical composition to the ZM21 rod, as shown in Table 1. The tube has a 2.0 mm outer diameter and 0.2 mm wall thickness. The tubular test specimen was chucked between the upper and lower chucks, with a distance between each chuck of approximately 10 mm, which is the same as the gauge length. The tensile speed was set to 0.6 mm/min. Figure 5 shows the nominal stress–nominal strain diagram of the ZM21 tube. The ductility of the ZM21 tube at 250 °C is 210% higher than the test result at RT. Comparing the tensile test results, shown in Figure 4, with the test results for ZM21 rods, shown in Figure 5, correlations are observed between the tube and the rod materials in relation to changes of the deformation resistance and the ductility.

Moreover, the bursting pressure p_b of the ZM21 tube was measured by the bursting test at RT and 250 °C, in order to define the fracture limit of the tube in THF. The internal pressure was loaded linearly by 1.0 MPa/s, until the tube burst. Figure 6 shows the result of the test. The bursting limit at 250 °C is 23% of the limit at RT.

Furthermore, to evaluate the expansion characteristics of the tube in the THF process, warm free bulge forming was conducted at RT, 150 °C, 250 °C and 300 °C [14]. The tubes used had a 2.0 mm outer diameter and 0.15 mm wall thickness. Figure 7 shows a comparison of the expansion characteristics of the hydroformed tubes. In this process, the internal pressure was set at 20% lower than the bursting pressure p_b of the initial tube. From this experiment, it was confirmed that the warm-hydroformed ZM21 tube at 250 °C (Figure 7b) could be deformed sufficiently, and expanded most in these temperature conditions. The expanded tube had a 3.52 mm outer diameter. Thus, under this temperature condition, the highest expanding height could be obtained.

Table 1. Composition of specimens for experiments (mass%).

Material	Al	Zn	Mn	Si	Fe	Cu	Ni	Mg
ZM21	0.004	1.81	0.68	0.01	0.028	0.001	0.001	Bal.

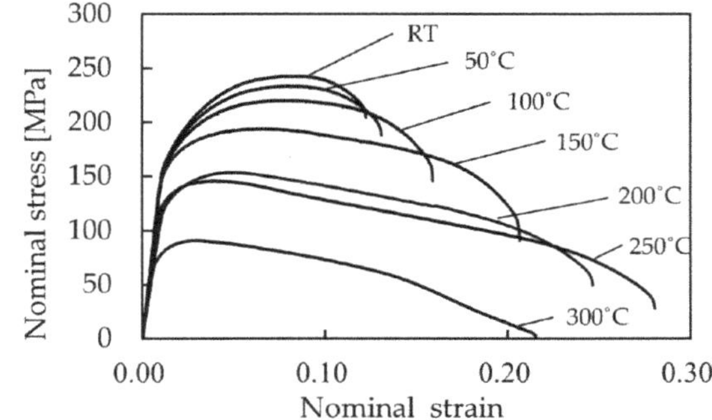

Figure 4. Stress–strain curves of the ZM21 rod obtained from the tensile test result under various temperature conditions.

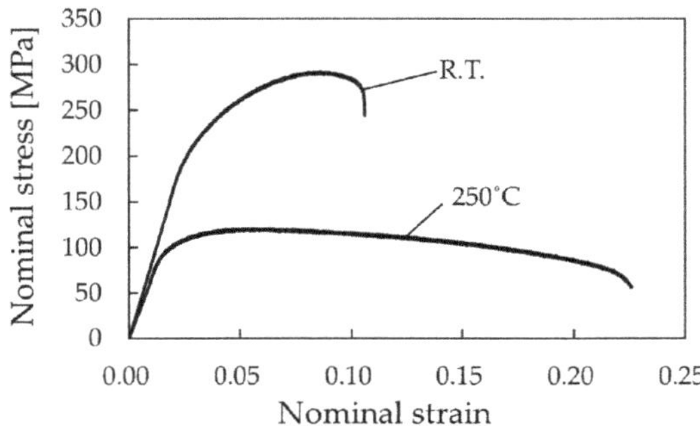

Figure 5. Stress–strain curves of the small-diameter ZM21 tube obtained from the tensile test result at R.T and 250 °C.

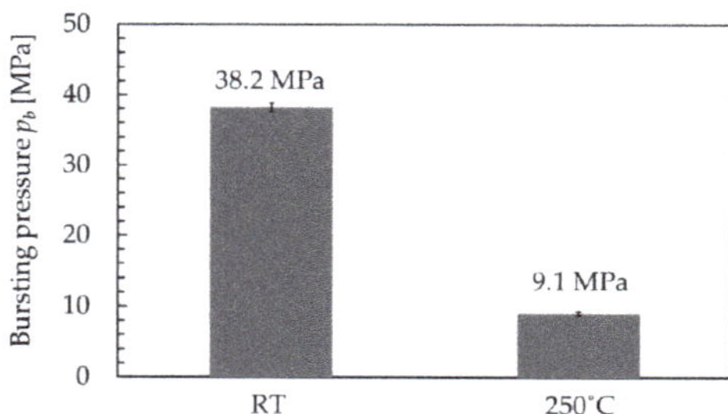

Figure 6. Bursting pressures of the small diameter ZM21 tube obtained from the bursting test result at R.T and 250 °C.

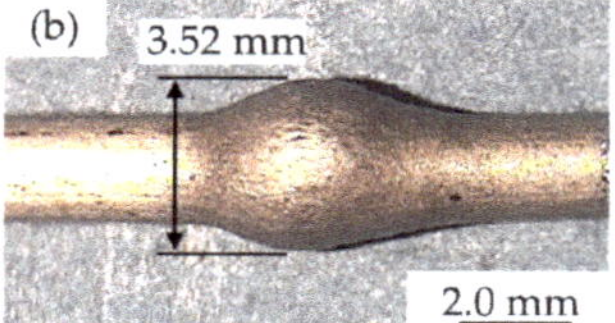

Figure 7. Expansion behavior of the tube at R.T and 250 °C: (**a**) fractured tube at R.T; (**b**) expanded tube at 250 °C.

2.3. Experimental Methods

Warm THF was conducted on the small-diameter ZM21 tube with a 2.0 mm outer diameter and 0.2 mm wall thickness under an elevated temperature condition (250 °C). Silicone oil was used as the hydraulic pressure medium, and molybdenum disulfide (MoS_2) paste was employed to provide lubrication between the tube and the die. Figure 8 shows the loading path of the internal pressure p_h and the axial feeding displacement ΔL adopted in the experiment. In the loading path, Stage (1) is the hydraulic pressure stage. The internal oil pressure is loaded on the predetermined internal pressure p_h. The pressure is maintained after the pressure medium is filled, and the air in the tube is vented. The value of *ph* was defined based on α. It is defined by a percentage of the loaded internal pressure *ph* to the bursting pressure *pb*. After that, in Stage (2), axial feeding is carried out to deform the tube, which produces the axial feeding displacement ΔL = a maximum of 1.5 mm, while maintaining the internal pressure p_h. In Stage (2), the axial feeding is applied at a speed of 0.025 mm/s.

Figure 9 shows the die used for the warm THF experiment. The deformation characteristics of the tubes, which are deformed in the taper-cavity die (Figure 9a) and the cylinder-cavity die (Figure 9b), were evaluated. The divided asymmetrical dies with A-side and B-side parts, as shown in the figures, were employed. Each side has a different cavity: the A-side has a straight cavity, without bulging, and the B-side has a taper- or a cylinder-cavity. The taper- and the cylinder-cavity have an expansion ratio that is 1.5 times the maximum, in comparison with the initial outer diameter of the ZM21 tube. Before processing, every tube was annealed at 300 °C for 24 h.

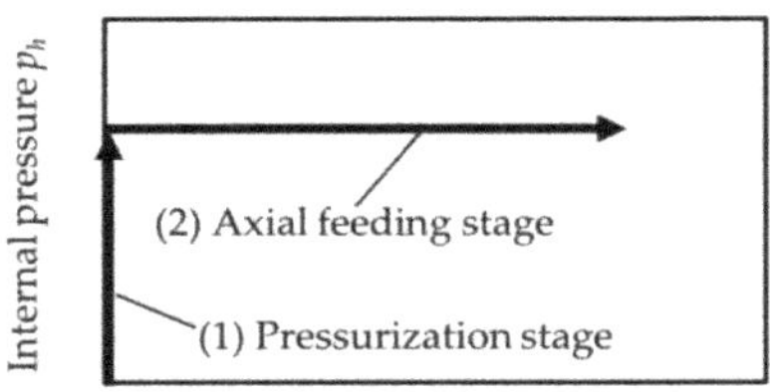

Figure 8. Schematic of the loading path used in the experiment and finite element (FE) method (FEM).

Figure 9. Geometry and dimensions of the tube hydroforming dies used in this study: (**a**) a taper-cavity die; (**b**) a cylinder-cavity die.

2.4. FEM Model and FE Analysis Conditions

The warm THF process was analyzed by the FE method (FEM) to evaluate the deformation behavior of the tube in the processing. The FEM code used was the dynamic explicit LS-DYNA 3D (ver. 19. 2) (ANSYS Inc. Canonsburg, PA, USA). The FE models were constructed for one-quarter of the entire part, considering the geometrical symmetry of the THF model, as shown in Figure 10. Figure 10a is an FEM model that includes a taper-cavity die, and Figure 10b includes a cylinder-cavity die. In these models, the tube has a length of 17 mm, an outer diameter of 2.0 mm and a wall thickness of 0.2 mm. Solid elements were used for the tube and the dies. Rigid bodies were employed for the die parts. The isotropic elastoplastic material was applied for the material model of the ZM21 tube. The model was meshed, and the element was a ten-node element. The tube has 6 elements in the thickness direction. The minimum element size of the tube was 33 μm × 33 μm × 39 μm. The number of total elements for the tube blank and the dies are 192,000 and 1666, respectively.

The mechanical properties of the ZM21 magnesium alloy tube at 250 °C used in the simulations are shown in Table 2. The material characteristics of the tube were determined using a true stress–true strain diagram, which was created based on the nominal stress–nominal strain diagram shown in Figure 5. The stress–strain data were inputted as the values approximated by a multiline approximation approach. In the simulation, the Coulomb friction model was used, and the friction coefficient was 0.1. The kinetic friction coefficient was defined as the same value as static. In the simulation, the internal pressure and the axial feeding were linearly loaded along the loading path shown in Figure 8.

To define the internal pressure conditions in the simulation, the bursting pressure was calculated by the bursting test in FEM. In this test, the internal pressure p_h was loaded linearly, without loading of the axial feeding. Figure 11 shows the maximum equivalent von-Mises stress of the tube obtained from the test. The tube reached around the maximum value of the stress at p_h = 12.5 MPa. After that, the simulation stopped due to the occurrence of a negative volume of the elements in the tube. Therefore, it is considered that the bursting pressure p_b of the tube in FEM is 12.5 MPa.

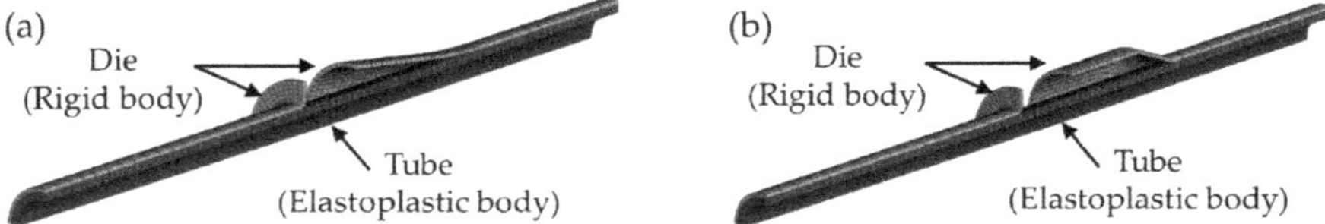

Figure 10. Schematic of the finite element (FE) model: (**a**) FE model for the taper-shape warm THF; (**b**) FE model for the cylinder-shape warm THF.

Table 2. Mechanical properties of the small-diameter ZM21 magnesium alloy tube used in FEM.

Material	ZM21 (250 °C)
Material model	Elastoplastic body (Multilinear plasticity)
Mass density (kg/mm^3)	1800
Elastic modulus (GPa)	8.20
Yield stress (MPa)	90.76
Poisson's ratio	0.35

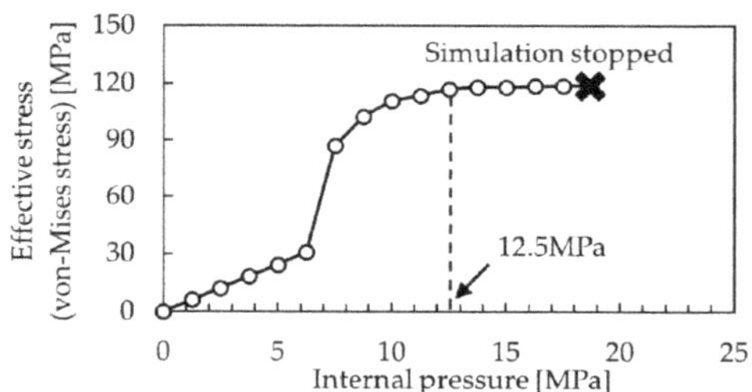

Figure 11. Definition of the bursting pressures of the small-diameter ZM21 tube in FEM at 250 °C.

3. Results and Discussion

3.1. Fracture and Buckling Limits of the Tube in Relation to the Loaded Internal Pressure

To determine the value of the internal pressure that is loaded during the axial feeding in Stage (2) shown in Figure 8, the fracture and the buckling limits of the small-diameter ZM21 tube were evaluated by warm free bulge forming at 250 °C. The die used for this evaluation is shown in Figure 12. The employed internal pressure conditions α were 90%, 80%, 70% and 50%, and the axial feeding was performed at ΔL = 1.5 mm. Figure 13 shows the hydroformed tubes obtained under each condition. When the loaded pressure was high (Figure 13a), the tube fractured immediately after the axial feeding was applied (ΔL = 0.27 mm), and the tube could not be expanded. On the other hand, when the loaded pressure was low (Figure 13c), wrinkles appeared on the tube. Additionally, the tube was fractured by severe folding deformation when the maintained internal pressure α was 50%, as shown in Figure 13d. Therefore, it is considered that the buckling limit of the tube α = 50%, and the fracture limit α = 90%. However, the tube successfully expanded when α was 80%. Figure 13b shows the successfully hydroformed tube, without wrinkles, folds or fractures. The outer diameter of the hydroformed tube was 3.23 mm, in this case.

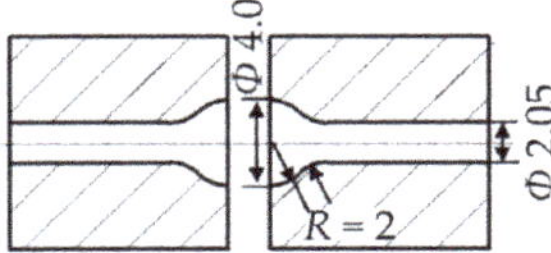

Figure 12. Schematic of the die cavity shape used for the evaluation of the fracture and bursting limit of the tube.

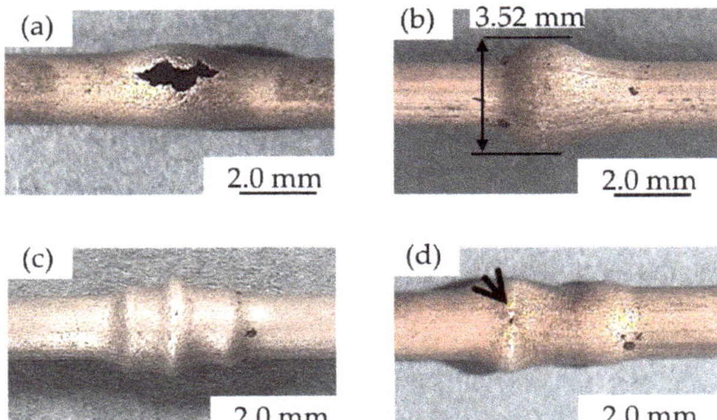

Figure 13. Deformation characteristics of the small ZM21 tube deformed under the various internal pressure conditions: (**a**) a fracture (α = 90%); (**b**) bulge without fractures or wrinkles (α = 80%); (**c**) a wrinkle (α = 70%); (**d**) a wrinkle and folding (α = 50%).

3.2. *Influence of the Axial Feeding Direction on the Deformation Characteristic in Warm Tube Hydroforming*

Since asymmetrical-shape dies would be used in this experiment, as mentioned in Section 2.3, it was considered that the difference in the axial feeding direction may influence the deformation of the tube [17]. Therefore, two conditions regarding the applied axial feeding direction for the tube were established, and the deformation characteristics of the small-diameter ZM21 tube was evaluated under these conditions at 250 °C. The conditions were named CASE A and CASE B. The axial feeding was applied only from the A-side in CASE A, and in CASE B, it was applied from the B-side only. In these conditions, the loaded axial feeding displacement ΔL was 1.5 mm, and the employed internal pressure condition α was 80%. Figure 14 shows the hydroformed tubes and cross-section of the deformed tube under both conditions using the taper-cavity die. The taper-shaped tube was successfully hydroformed into the die cavity shape in CASE A (Figure 14a). However, a wrinkle appeared on the surface of the tube in CASE B (Figure 14b). The cause of the occurrence of the wrinkle is considered to be the influence of the difference of the shoulder shape in the moved die on the deformation of the tube. When the axial feeding was applied to the die with a straight-cavity (A-side), the process led to an excellent formability of the tube.

Figure 15 shows the axial compressive force measured by the load cell mounted on each die end during the processes. In these results, the increase of the axial compressive force from around ΔL = 0.5 to 1.0 mm was caused by a rise in the hardness of the tube material, which was affected by the occurrence of the twin boundaries during the process [15,18]. Additionally, it is considered that the increase from around ΔL = 1.2 to 1.5 mm was caused by the bending deformation added to the expanding deformation in filling the tube material in the die corner. By comparing the measured axial compressive force in CASE A and CASE B, a difference was found in the value of the measured axial compressive force. The incline of changes showed that it clearly differed. Therefore, it seems that the difference between the deformation behavior of the tubes was caused at around ΔL = 1.0 mm, and a wrinkle appears at that time.

In addition, Figure 16a shows the microstructure in the taper-shaped tube obtained in the CASE A condition. Grain with twin boundaries and grain that is finer than the grain in the annealed tube were confirmed in the taper-shaped tube. Figure 16b shows the equivalent von-Mises stress, and Figure 16c shows the equivalent plastic strain in the deformed area of the tube. These plots show that higher compressive stresses and strain appeared conspicuously on the inside of the bended part of the hydroformed tube. There is a correlation between the area where the grain with twin boundaries and the refined grain were observed and that where the higher compressive stresses and strain appeared. It is considered that the deformed tube has advantages, such as a high strength and high-stress corrosion resistance, because of the refinement of the grain. However, their occurrence behavior during the process is yet to be observed and noted in detail. More specific observations of the changes of the microstructure in the tube during the deformation is necessary in evaluating the refinement behavior of the grain in the hydroformed tube.

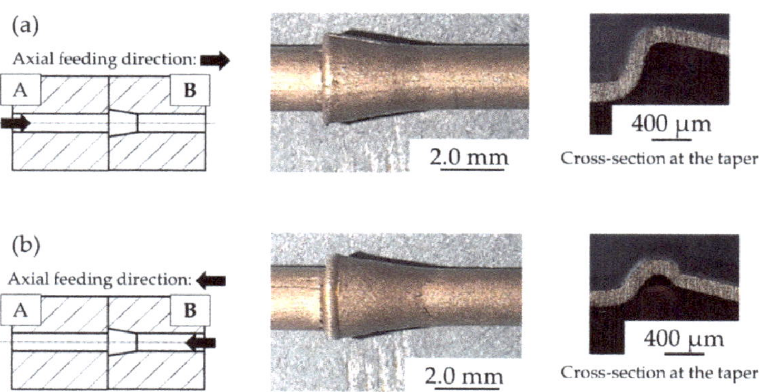

Figure 14. Effect of the axial feeding direction on the deformation characteristics of the small ZM21 tube: (**a**) taper-shaped tube; (**b**) wrinkle.

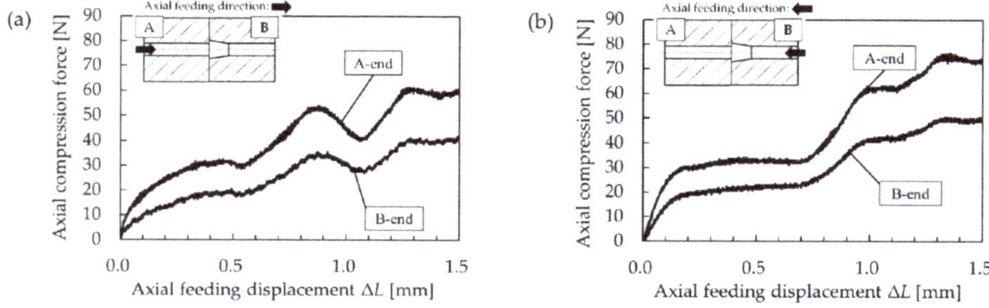

Figure 15. Effect of the axial feeding direction on the axial compressive force during the process: (**a**) no wrinkle; (**b**) wrinkle.

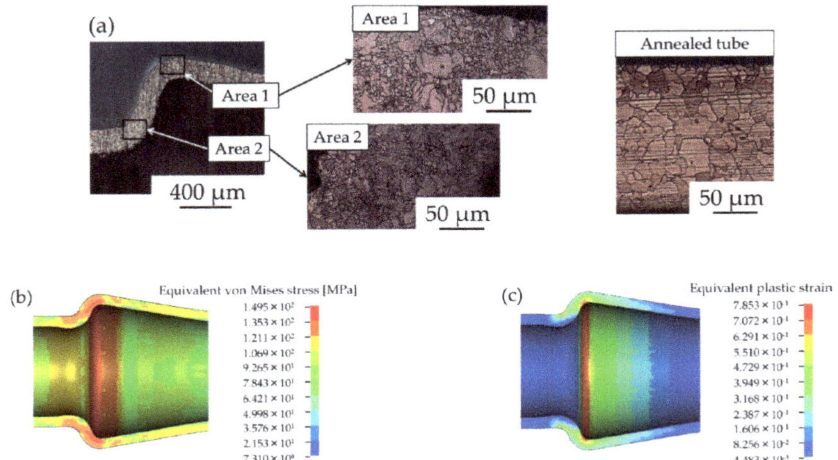

Figure 16. Microstructure at different positions of the taper-shaped tube and the FEM results of the taper-shape forming: (**a**) Microstructure of the taper-shaped tube; (**b**) distribution of the equivalent von-Mises stress at the taper; (**c**) distribution of the equivalent plastic strain.

3.3. Forming Defects

When the cylinder-cavity die was employed, warm THF was impossible under both axial feeding conditions, because the tube was fractured during the processing. Figure 17 shows the deformation behavior of the tube, which was deformed in the cylinder-cavity die obtained by FEM. In addition, Figure 18 shows the thinning behavior of the wall of the tube, in the area where the maximum reduction of the wall thickness was observed in FEM. As shown in Figure 17, the tube appeared to have been deformed successfully at first glance. However, it is clear that a severe decrease of the thickness occurred at ΔL = 0.525 mm. The maximum thickness strain in that area was −31.5% at ΔL = 0.9 mm, and the wall thickness of the tube was decreased from 0.2 mm to 0.137 mm. It is considered that the tube fractured during the axial feeding, due to the severe thinning deformation of the wall in the experiment. Therefore, the suppression of the thinning of the wall is necessary to avoid fracture when the cylinder-cavity die is employed for warm THF.

On the other hand, Figure 19 shows a deformation defect obtained when warm THF was employed with the taper-shape die. This defect was caused because the closing die pinched the tube, which expanded to an extreme extent during the axial feeding. Excessive expansion of the tube is caused by uncontrollable fluctuation of the internal pressure during the processing. In the experiment, the internal pressure fluctuates within a maximum error of approximately 5%, because of the change of the tube volume with the expansion, the thinning of the wall, and a leak of the pressure medium.

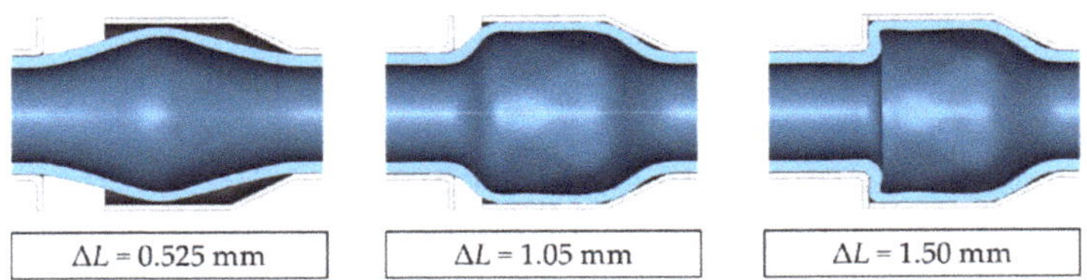

Figure 17. Deformation behavior of the small ZM21 tube in the cylinder-shape warm THF obtained by FEM.

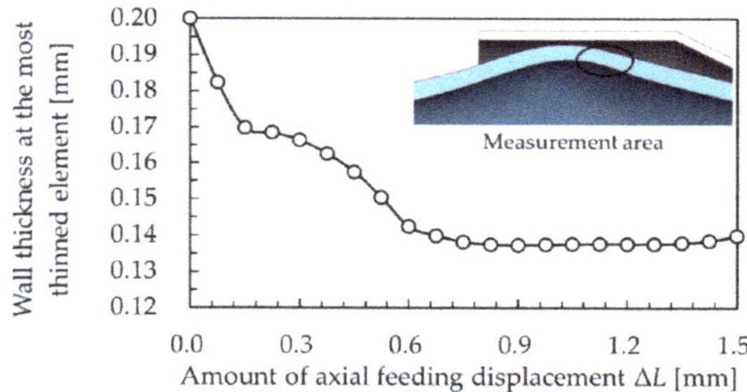

Figure 18. Thinning behavior of the wall thickness in the cylinder-shape warm THF obtained by FEM.

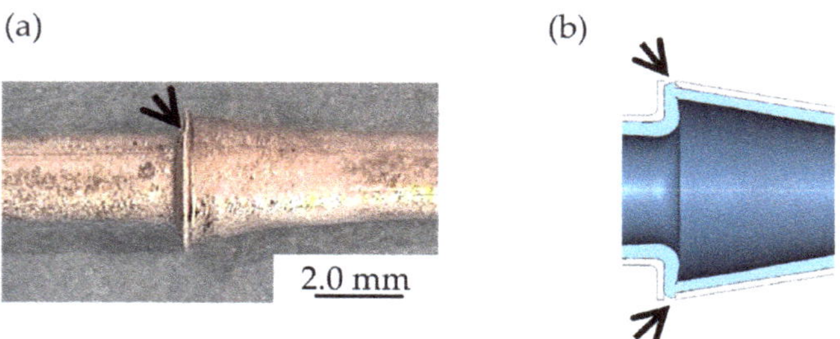

Figure 19. Forming defect of the small ZM21 tube in the taper-shape warm THF: (**a**) experiment; (**b**) FEM.

3.4. Influence of Deference in the Loading Path on the Deformation Characteristics in Warm Tube Hydrofoming

Figure 20 shows the modified loading path. To avoid forming defects, as mentioned in Section 3.3, the internal pressure p_h at Path (2) was maintained with a lower value, which can suppress the occurrence of forming defects. In this experiment, the internal pressure was loaded, and it was α = 60% in Stage (1) and (2). In Stage (2), the axial feeding was applied while maintaining the internal pressure. The occurrence of wrinkling is allowed in this stage. After that, the straightening stage was installed, as Stage (3) in the loading path, to expand the wrinkled tube. The internal pressure was raised from α = 60% to 80% in the straightening stage. The experiments were carried out at 250 °C, and the axial feeding condition used was CASE A, with ΔL = 1.5 mm. The hydroformed samples were observed in Stage (2) and (3).

Figure 21 shows the hydroformed small-diameter ZM21 tubes, which were obtained by warm THF using the modified loading path. Every picture of the hydroformed tube shown in the figure was deformed from other materials. The experiments were not carried out using the same tube. The tubes with wrinkles were expanded into a taper shape (Figure 21a) and cylinder shape (Figure 21b) successfully in Stage (3). Figure 22 shows the deformation behavior of the tube during warm THF using the cylinder-cavity die in FEM. There is no area where the wall of the tube is very thin. Additionally, Figure 23 shows the thickness strain of the tube deformed using the cylinder-cavity die in the experiment. The thinning amount of the wall of the wrinkled tube was suppressed, compared with the tube that was deformed using the initial loading path shown in Figure 8. The maximum thickness strain was −2.5%. Moreover, the maximum thickness strain of the tube at the end of Stage (3) was −16.8%. From these results, it is considered that the warm THF using the cylinder-cavity die was successful because of the suppression of the thinning of the wall. It is clear that the value of the loaded internal pressure during axial feeding is an important factor that influences the deformation of the tube.

On the other hand, the wall thickness of the hydroformed tubes was not uniform. Figure 24 shows the expansion distribution of the hydroformed tubes. Remarkably, the expansion of the tube occurs in the valley of the wrinkle. Compared with the deformed tube shape, the position of the notably thickened area was the shoulder and the valley part between the left wrinkle and right wrinkle in Stage (2), as shown in Figure 23. After the straightening (Stage (3)), the thinning remarkably proceeded in the valley. This thinning is caused by a local expansion in the valley of the wrinkles. The deformation of the tube is constrained due to the top of the wrinkles touching the die cavity at the end of Stage (2). Therefore, the tube is expanded in the valley of the wrinkles only in Stage (3), and this area shows severe thinning deformation of the wall.

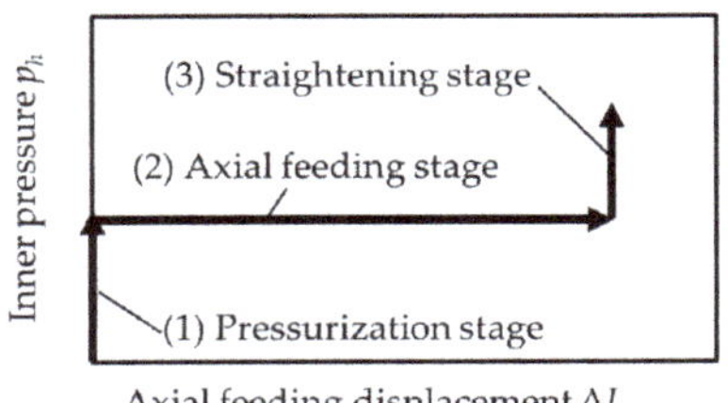

Figure 20. Schematic of the modified loading path used for the experiment and FEM.

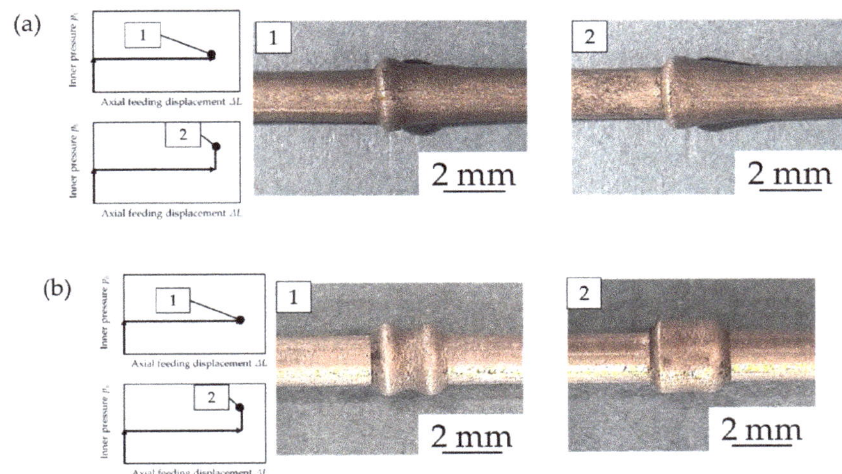

Figure 21. Deformation behavior of the small ZM21 tubes, which were deformed based on the modified loading path at 250 °C: (**a**) taper-shaped tube; (**b**) cylinder-shaped tube.

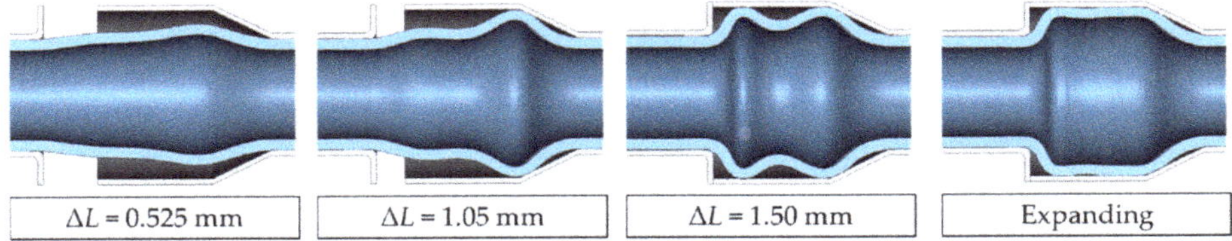

Figure 22. Deformation behavior of the small ZM21 tube in the cylinder-shape warm THF obtained by FEM at 250 °C using the modified loading path.

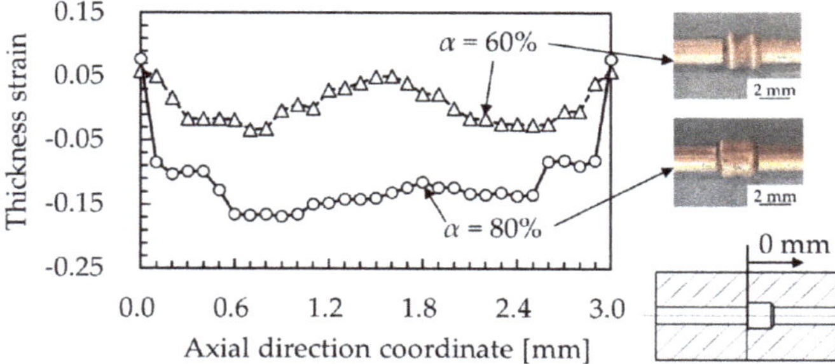

Figure 23. Thickness strain of the deformed tubes in the cylinder-shape warm THF at 250 °C.

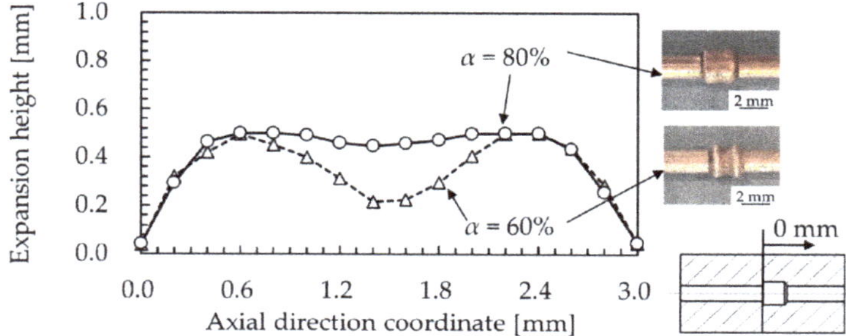

Figure 24. Expansion distribution of the deformed tubes in the cylinder-shape warm THF at 250 °C.

3.5. Forming Limit

The fracture limit of the small-diameter ZM21 tube in Stage (3) in the modified loading path was investigated in the experiment. The cylinder-cavity die was employed for this evaluation. In Stage (3), the internal pressure was loaded at a speed of 1.0 MPa/s, until fracturing. Figure 25 shows the deformed tubes in Stage (3). The wrinkled tubes could withstand a higher internal pressure than the bursting pressure p_b of the initial tube obtained from the bursting test. The warm THF was successful when α = 200%, as shown in Figure 24. However, when the loaded pressure α was 200% or more, the tubes tended to fracture, as shown in Figure 25b,c. Bursting of the tube under the higher-pressure region occurred in the area around the right-angle die corner. It is considered that the tubes could withstand a higher pressure than the bursting pressure, owing to the die cavity and the thickening of the wall in the expanded valley of wrinkle constraining the expansion of the tube, as mentioned in Section 3.4.

In addition, the expansion distribution of the hydroformed tube with the loaded pressure condition α = 200% in Stage (3) is shown in Figure 26. The results of the warm THF, without the straightening (α = 60%) and with the straightening (α = 80%), are shown in that figure for comparison. The tube expanded sufficiently from the wrinkled to the cylinder-shape when α = 200%. Conversely, the expansion of the tube, which was straightened with raised pressure α = 80%, was incomplete, compared with the hydroformed tube with α = 200%. The cause of the insufficient expansion when α was 80% in the straightening stage is considered to be the high strength of the tube material, owing to grain refinement, as mentioned in Section 3.2. Figure 27 shows the microstructure of the tube that was wrinkled in Stage (2). At the top of the wrinkle, there were approximately 20 fine-grained layers in the wall thickness direction, as shown in Figure 27b, and the average grain size was approximately 9.3 μm (Figure 27a). On the other hand, the grain size in the non-deformed area was 21.5 μm. Therefore, warm THF of the tube, from the wrinkle to the die cavity shape, requires the loading of a higher internal pressure to straighten the wrinkle.

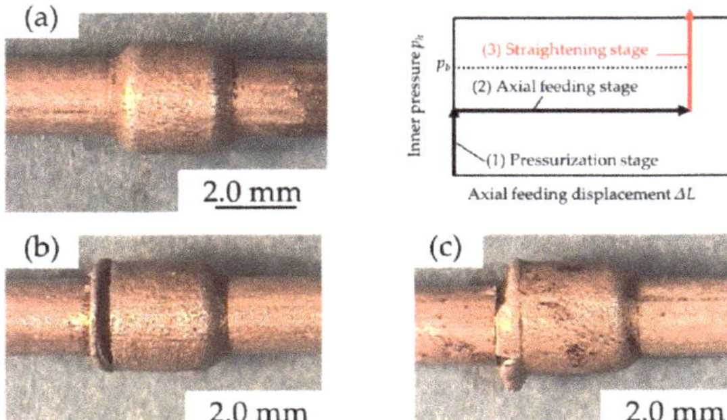

Figure 25. Forming limit of the small ZM21 tube in the cylinder-shape warm THF at 250 °C: (**a**) no fracture (α = 200%); (**b**) fracture (α = 230%); (**c**) fracture (α = 250%).

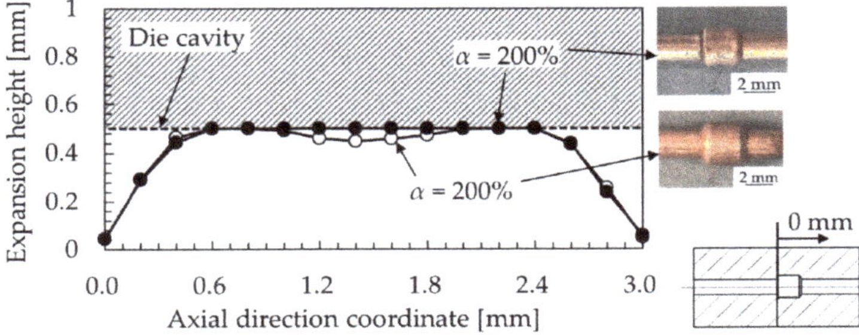

Figure 26. Expansion distribution of the deformed tube, which was loaded with a higher internal pressure for the straightening of the wrinkle in the cylinder-shape warm THF at 250 °C.

Figure 27. Microstructure of the wrinkled tube obtained from the cylinder-shape warm THF in Stage (2) at 250 °C: (**a**) microstructure in the non-deformed area; (**b**) microstructure at the top of the wrinkle.

Moreover, the forming limit for the die corner shape in the right-angle die corner was evaluated. Figure 28 shows the axial cross-sections of the taper-shaped tube (Figure 28a) and the cylinder-shaped tube (Figure 28b) under the conditions of ΔL = 1.5 mm and α = 80% in Stage (3). In each processing stage, filling to the die corner was insufficient. The obtained radii of the corner-shape R were 0.318 mm in Figure 28a and 0.384 mm in Figure 28b. From this result, it is considered that the forming limit of the tube in the die corner is approximately R = 0.3 to 0.4 mm. For the filling of the tube material to the sharper die corner, it seems that the material flow into the die is necessary. In this study, the axial feeding was carried out, without an axial compressive paunch. Thus, the tube material could barely flow into the die cavity. Additionally, there is a disagreement in the bending shape of the hydroformed tubes between the taper- and the cylinder-shaped tube. It is considered that this disagreement is caused by the difference in the ratio of the axial feeding displacement to the die-cavity volume between these dies. The cylinder-cavity die has a lager die cavity volume than the volume of the taper-cavity die. Therefore, optimization of the amount of the axial feeding displacement that is applied to the tube in relation to the die cavity volume is required.

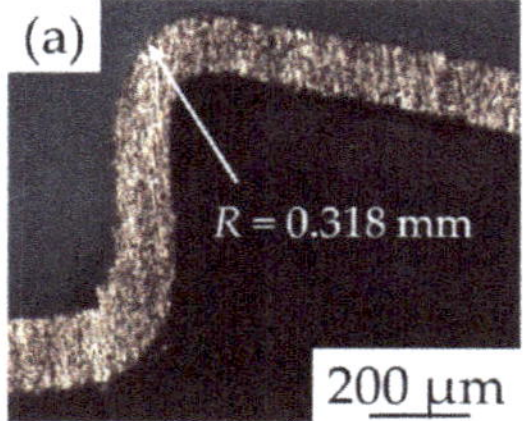

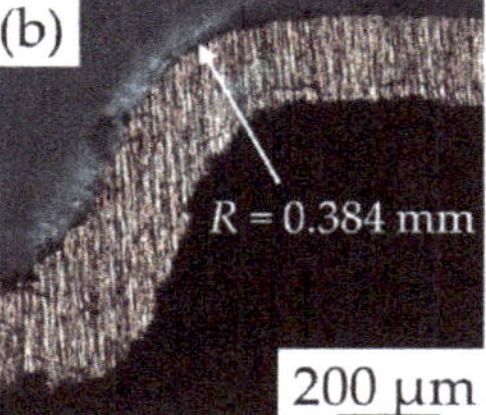

Figure 28. Radius shape of the corner of the deformed tubes: (**a**) taper-shaped tube; (**b**) cylinder-shaped tube.

4. Conclusions

In this study, warm THF at 250 °C for small-diameter ZM21 magnesium alloy tubes with an outer diameter of 2.0 mm and a wall thickness of 0.2 mm was examined by experimentation and FEM analysis. The influence of the loaded internal pressure and the axial compressive displacement on the deformation characteristics, the forming limit, and the forming defects of the tubes were examined and evaluated. The results obtained in this study are summarized below.

1. The fracture and buckling limit of the tube during warm THF were clarified. When the employed loading internal pressure condition α, which is a percentage of the internal pressure p_h to the bursting pressure p_b of the initial tube, was 90%, hydroforming of the ZM21 tube was impossible due to a fracture of the material during the axial feeding. On the other hand, warm THF, which was employed under the condition of α = 50%, caused buckling of the tube, and the tube was fractured by a severe folding deformation.
2. When the loaded internal pressure condition α was 70%, the tube with wrinkles was obtained. The warm THF under the condition of α = 80% was successful, without fracture and wrinkles.
3. The effect of the axial feeding direction on the deformation characteristics of the ZM21 tube was confirmed. The difference in the deformation behavior of the tube occurred around the axial feeding displacement ΔL = 1.0 mm. Additionally, the taper-shaped tube that was hydroformed successfully had grain with twin boundaries and refined grain in the bended and the expanded areas.
4. The observed forming defects of the ZM21 tube were fracturing of the tube during the axial feeding in warm THF using the cylinder-cavity die and pinching of the material in the section between the left and right die caused by die closing and an excessive expansion of the tube. The fracture of the tube could be caused by the severe thinning deformation of the wall with a maximum thickness strain of −31.5%. It is confirmed that a lower loaded internal pressure during the axial feeding stage is required to avoid the occurrence of these defects.
5. Warm THF for the tube using the cylinder-cavity die was successful, with the modification of the loading path. The thinning that appears during the axial feeding was suppressed by the loaded lower internal pressure. The hydroformed tube obtained by warm THF using the modified loading path had a maximum thickness strain of −2.5% after the axial feeding stage. The tube with the wrinkles could be expanded successfully into the taper and the cylinder shapes by increasing the pressure in the straightening stage.
6. The fracture limit of the tube during warm THF in Stage (3) was α = 200%. In addition, when the loaded pressure α was 200% or more, the tubes tended to fracture at the corner in the die cavity.

The improvement of the fracture limit of the tube in Stage (3), compared with that of the initial tube, is caused by the thickening of the wall, which occurs in Stage (2), and the high strength due to the finer grain.

7. The radius of the bended area in the taper- and cylinder-shaped tube were measured. It is considered that the forming limit of the tube with the die-corner shape was around $R = 0.3$ to 4.0 mm.

Author Contributions: Conceptualization, S.Y. and H.Y.; Methodology, H.Y.; Resources, T.F.; Material provision, T.F.; Experimental work, H.Y. and T.M.; Numerical analysis, H.Y. and T.M.; Data curation, H.Y.; Writing—original draft preparation, H.Y.; Supervision, S.Y., R.Y. and Y.I.; Project administration, S.Y.; Funding acquisition, S.Y. All authors have read and agreed to the published version of the manuscript.

Funding: This work was supported by The Die and Mold Technology Promotion Foundation.

Conflicts of Interest: The authors declare no conflict of interest.

References

1. Yamamoto, R. Biomedical application of magnesium alloys. *J. Jpn. Inst. Light Metals* **2008**, *58*, 570–576. [CrossRef]
2. Bommala, V.K.; Krishana, M.G.; Rao, C.T. Magnesium matrix composites for biomedical applications: A review. *J. Magnes. Alloy.* **2019**, *7*, 72–79. [CrossRef]
3. Yoshihara, S.; Furushima, T. Development and applications of magnesium alloy micro tubes: Efforts on development bio-absorbable material. *J. Jpn. Soc. Technol. Plast.* **2016**, *57*, 1028–1033. [CrossRef]
4. Choudhary, L.; Raman, R.K.S.; Hofstetter, J.; Uggowitzer, P.J. In-vitro characterization of stress corrosion cracking of aluminium-free magnesium alloys for temporary bio-implant applications. *Mater. Sci. Eng. C* **2014**, *42*, 629–636. [CrossRef] [PubMed]
5. Manabe, K.; Fuchizawa, S. Tube forming technology: Toward the highest card for weight savings. *J. Jpn. Soc. Technol. Plast.* **2011**, *52*, 33–41. [CrossRef]
6. MacDonald, B.J. Finite element simulation of hydroforming processes—A review and future directions. *J. Jpn. Soc. Technol. Plast.* **2012**, *53*, 176–182. [CrossRef]
7. Cahn, L.C. Finite-element damage analysis for failure prediction of warm hydroforming tubular magnesium alloy sheets. *J. Miner. Metals Mater. Soc.* **2015**, *67*, 450–458.
8. Manabe, K.; Fujita, K.; Tada, K. Experimental and numerical study on warm hydroforming for T-shape joint of AZ31 magnesium alloy. *J. Chin. Soc. Mech. Eng.* **2010**, *31*, 284–287.
9. Manabe, K.; Morishima, T.; Ogawa, Y.; Tada, K.; Murai, T.; Nakagawa, H. Warm hydroforming process with non-uniform heating for AZ31 magnesium alloy tube. *Mater. Sci. Forum* **2010**, *654*, 739–742. [CrossRef]
10. Liewald, M.; Pop, R. Magnesium tube hydroforming. *Mater. Sci. Eng. Technol.* **2008**, *35*, 343–348. [CrossRef]
11. Geiger, M.; Kleiner, M.; Eckstein, R.; Tiesler, N.; Engel, U. Microforming. *CIRP Ann.* **2001**, *50*, 445–462. [CrossRef]
12. Zhuang, W.; Wang, S.; Lin, J.; Hartl, C. Experimental and numerical investigation of localized thinning in hydroforming of micro-tubes. *Eur. J. Mech. A Solids* **2011**, *31*, 67–76. [CrossRef]
13. Manabe, K. Micro forming processing of metal tubes. *Jpn. Soc. Technol. Plast.* **2019**, *2*, 8–13.
14. Yasui, H.; Yoshihara, S.; Yamada, R.; Ito, Y. Deformation behavior on small-diameter ZM21 magnesium alloy tube in warm tube hydroforming. *J. Jpn. Soc. Technol. Plast.* **2019**, *60*, 346–351. [CrossRef]
15. Yasui, H.; Yoshihara, S.; Mori, S.; Tada, K.; Manabe, K. Material deformation behavior in T-shape hydroforming of metal microtubes. *Metals* **2020**, *10*, 199. [CrossRef]
16. Furushima, T.; Manabe, K. Fabrication of AZ31 magnesium alloy fine tubes by dieless drawing process. *J. Jpn. Soc. Technol. Plast.* **2010**, *51*, 990–992. [CrossRef]
17. Miyagawa, T.; Yoshihara, S.; Yamada, R.; Ito, Y. Influence of process condition on forming for taper shape using small diameter A1100 aluminum tube in warm tube hydroforming. In Proceedings of the 137th Conference of Japan Institute of Light Metals, Koganei, Japan, 1–3 November 2019; pp. 93–94.
18. Somekawa, H.; Singh, A.; Schuh, A.C. Effect of twin boundaries on indentation behavior of magnesium alloys. *J. Alloy. Compd.* **2016**, *685*, 1016–1023. [CrossRef]

MDPI
St. Alban-Anlage 66
4052 Basel
Switzerland
Tel. +41 61 683 77 34
Fax +41 61 302 89 18
www.mdpi.com

Metals Editorial Office
E-mail: metals@mdpi.com
www.mdpi.com/journal/metals

www.ingramcontent.com/pod-product-compliance
Lightning Source LLC
LaVergne TN
LVHW070148170726
843515LV00007B/1876

* 9 7 8 3 0 3 6 5 2 3 5 4 5 *